THE ETERNAL DARKNESS

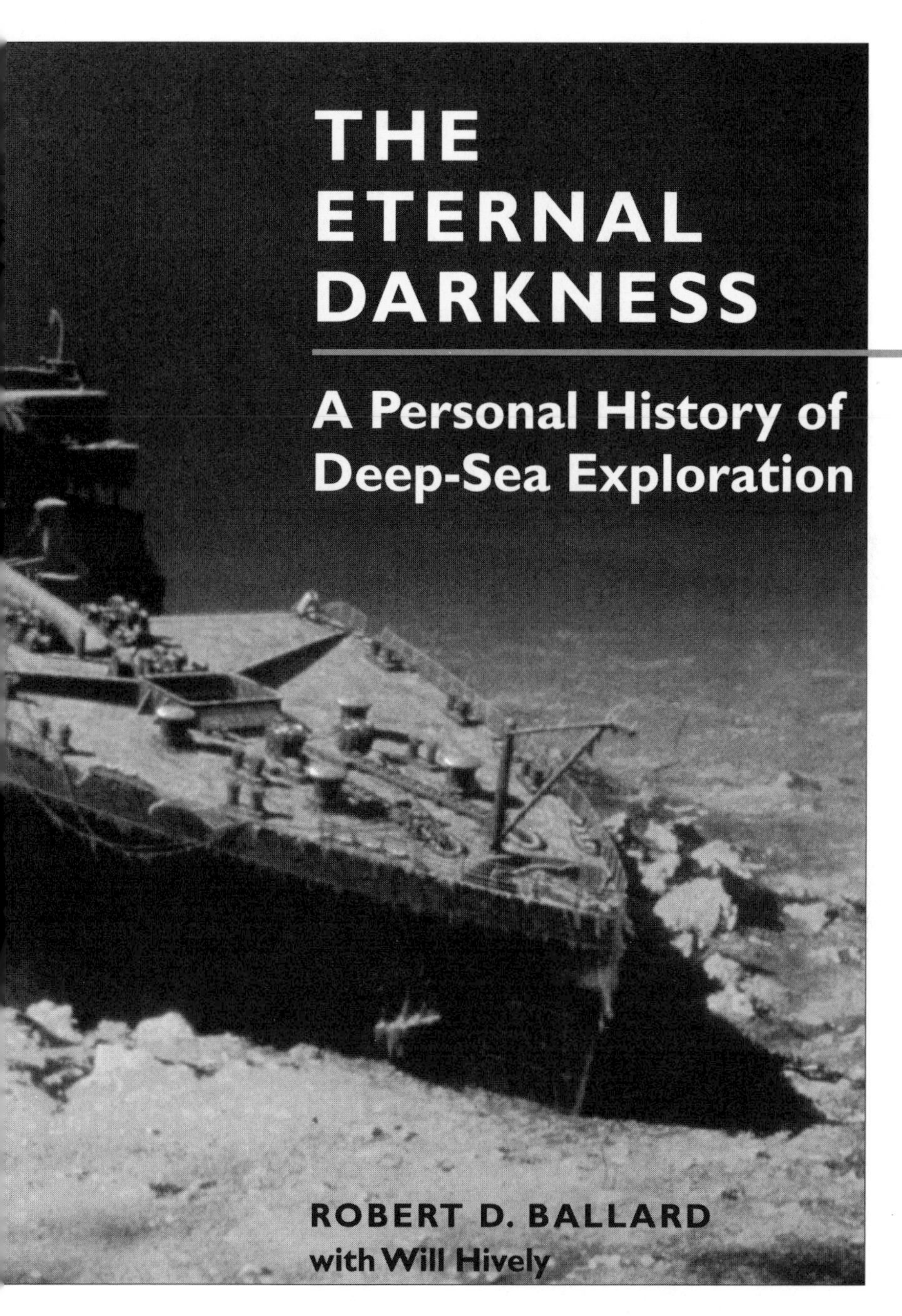

THE ETERNAL DARKNESS

A Personal History of Deep-Sea Exploration

ROBERT D. BALLARD
with Will Hively

PRINCETON UNIVERSITY PRESS
Princeton, New Jersey

Published by Princeton University Press, 41 William Street,
Princeton, New Jersey 08540
In the United Kingdom: Princeton University Press,
Chichester, West Sussex

Library of Congress Cataloging-in-Publication Data

Ballard, Robert D.
The eternal darkness : a personal history of deep-sea exploration / Robert D. Ballard with Will Hively.
p. cm.
Includes bibliographical references and index.
ISBN 0-691-02740-4 (cloth : alk. paper).
1. Underwater exploration—History. I. Hively, Will. II. Title.
GC65.B275 2000
551.46′07′09—dc21 99-43072

This book has been composed in Adobe Palatino and Gill Sans by Princeton Editorial Asociates, Inc., Roosevelt, New Jersey, and Scottsdale, Arizona

The paper used in this publication meets the requirements of ANSI/NISO Z39.48-1992 (R1997) (*Permanence of Paper*)

http://pup.princeton.edu

Printed in the United States of America

10 9 8 7 6 5 4 3 2 1

Frontispiece: Rendering by artist Ken Marschall of the *Titanic*'s bow section resting on the bottom of the ocean. (Illustration by Ken Marschall © 1992 from *Titanic: An Illustrated History,* a Viking Studio/Madison Press Book.)

CONTENTS

PREFACE

All of us are born explorers. From the very beginning, as infants and young children, we are a curious species of animal. We observe, reach for, and question everything around us. Many of us carry throughout our lives a series of fundamental questions that we seek to answer. Who are we? Where did we come from? Where are we going?

When we consider the intricate patterns of life on earth—how everything connects with everything else—our questioning becomes all-inclusive. Where did the earth, and the life forms on it, come from? Where are they headed? For answers to such cosmic questions, people have always looked toward distant worlds beyond their immediate reach. Traditionally, they have searched the heavens for clues.

I grew up at a remarkable time, when the heavens first became accessible. Unfortunately, they also became a potential battleground. When the Soviet *Sputnik* satellite flew over the United States in 1957, it caused Americans everywhere to turn and gaze toward the sky anew, in both fear and wonder. What might the Soviets send next? Weapons of mass destruction? The satellite galvanized the country to accelerate its race into space. Before long, the nation was building missiles capable of launching weapons as well as instrumented probes and humans. That quest inspired many explorers of my generation.

For some reason, my curiosity and wonder took me in the opposite direction. I dreamed of being an undersea explorer. Perhaps it was because the largest oceanographic institute in the world looked out on the sea a few miles away from my home in San Diego. And perhaps my vague ambition crystallized when the deep-diving bathyscaph *Trieste* left San Diego on a secretive mission in 1959—a daring attempt (as we later learned) to transport two men to the deepest part of the Pacific Ocean.

Whatever my initial motivations, I never gave up my dream to explore the deep ocean. It led me to earn both a commission in the U.S. Navy and a Ph.D. in marine geology and geophysics. It led me to spend thirty years at the Woods Hole Oceanographic Institution and to lead or participate in more than 110 deep-sea expeditions. Since 1997, it has continued to inspire me on expeditions and also in new educational programs for children at the Institute for Exploration in Mystic, Connecticut.

Because I have long participated in the history I am writing about, I have drawn heavily from my own experience to shape the contents of this book. And so it naturally becomes, in many places, a first-person narrative, especially in the latter chapters. This is not because I view myself as the most important deep-sea explorer. Far from it. Instead, I view myself as fortunate. I was lucky to become involved in the efforts of many persons—more than I can name here—who together designed, built, and operated deep-sea craft and made astounding discoveries. Many equally worthy persons and events receive little or no mention, simply because I knew them less well, and a book's length must be finite.

I am keenly aware of my status as an insider—a privileged witness to a fascinating burst of exploration. I feel it my duty now to tell that story as best I can. By right it belongs to us all, born as we are to become explorers. The ocean, however, must become more accessible if we intend to explore it in greater numbers, and this book is one part of my effort to make it so.

Although most of us have a chance at some time in our lives at least to gaze at the sea, and although many venture out across its surface, very few have been able to dive beneath the upper, sunlit layer. And yet most of our planet's solid surface lies beneath that watery blanket. It is a hidden world that in fact dominates, by overwhelming proportions, our living planet. The purpose of this book is to take you deep beneath the familiar surface of the sea to grasp the true character and true dimensions of this vast, eternal darkness—and to take you there in a way that retraces the steps of the pioneers who first explored that world.

Along the way, you will see how the nature of this journey has changed. It began as an elite quest that only a few could join. It is still a quest, but now a more common one, on the verge of being open to all. Ultimately, I hope that this book will help you to determine for yourself what role the deep ocean will or will not play in your future. But rest assured it will play a large role in the human future, as well as the future of every living creature on earth.

ACKNOWLEDGMENTS

How does one thank all the people involved in making this book possible? Not easily. To begin with there are those who helped make the field of deep submergence possible, as well as those who helped me become a part of that amazing history.

Clearly William Beebe, Otis Barton, Auguste and Jacques Piccard, and Jacques Cousteau were important pioneers during the early years of deep submergence. But so were those who followed in their footsteps—all of the pilots and technicians who mastered this unique art form and then went in harm's way to practice it.

Were it not for Andy Rechnitzer, Dick Terry, Bill Rainnie, and Earl Hays, I would not have ever had the opportunity to learn about deep-diving submersibles or been able to go beneath the sea to explore its many wonders.

But the people I am most thankful for are all the individuals who went to sea with me over the past four decades: the Cathys, Toms, Andys, Martins, Skips, Danas, Als, and all the rest. Whose sleep was constantly interrupted in the dark of night to roll out of bed and onto a rolling, wet deck to wrestle back aboard our priceless equipment from an angry sea. Who stood their watches day in and day out, night in and night out. Whose names are seldom heard or read in the press accounts of an important expedition in which they played a critical role. I thank you.

Thanks also to all the people and organizations that have stood by us through the years, supporting our expeditions: To the National Geographic Society, the U.S. Navy, and the Office of Naval Research. And to the members of the oceanographic community who made it possible for me to be a part of some of the most important moments in deep-sea exploration, from Project FAMOUS to the discovery of hydrothermal vents and black smokers—discoveries far more important than that of the *Titanic*.

I also thank Jack Repcheck and the staff at Princeton University Press, for hanging in there as this book progressed from an idea to a reality, and Steve Ramberg at the Office of Naval Research, who helped support this effort and had the patience to wait for its successful completion.

And finally and most importantly I owe my deepest gratitude to my family: to Barbara, Douglas, Benjamin, and Emily Rose, who kept the home fires burning while their husband and father took to sea to live his dream.

THE ETERNAL DARKNESS

INTRODUCTION

> They that go down to the sea in ships, that do business in great waters; these see the works of the Lord, and his wonders in the deep.
>
> —*Psalm 107:23–24*

At a time when most think of outer space as the final frontier, we must remember that a great deal of unfinished business remains here on earth. As robots crawl on the surface of Mars, as spacecraft exit our solar system, and as the orbiting Hubble Space Telescope pushes back the edge of the visible universe, we must remember that most of our own planet has still never been seen by human eyes.

It seems ironic that we know more about impact craters on the far side of the moon than about the longest and largest mountain range on earth. It is amazing that astronauts walked on the surface of the moon before any person saw those earthbound peaks. But it remains a fact that human beings crossed a quarter million miles of space to visit our nearest celestial neighbor before penetrating just two miles deep into the earth's own waters to explore the Midocean Ridge. And it would be hard to imagine a more significant part of our planet to investigate—a chain of volcanic mountains 42,000 miles long where most of the earth's solid surface was born, and where vast volcanic landscapes continue to emerge.

After the historic moon landings, humankind sent probes to the far reaches of our solar system and witnessed amazing, otherworldly scenes and events. Yet perhaps the most affecting image of all those ever returned from space was captured in 1968, when astronauts aboard Apollo 8, on their way to orbit the moon, turned their cameras back on their home planet and revealed it for what it is—a majestic blue-green marble suspended in a cold black void. From that day forward, our world seemed more fragile and finite than ever before. Despite the huge dimensions of the universe, our little planet suddenly became more impressive—a precious jewel.

That picture from Apollo 8 also revealed the vastness of the earth's oceans. We had all been taught that water covers 71 percent of the earth's surface, but now we had a stark visual representation of that figure. Great swaths of solid blue showed more powerfully than ever just how much more of the planet remained to be explored. This new view of our home did not mean that we should cease to wonder about the heavens. Someday, perhaps, astronauts will walk on Mars. But clearly it is the earth where, for at least the foreseeable future, the vast majority of the human race will live out their lives. And for this reason the oceans of the world must receive greater attention than they do at present.

Actually, the figure we so often see quoted—71 percent of the earth's surface—understates the oceans' importance. If you consider instead three-dimensional *volumes*, our landlubbers' share of the planet shrinks even more toward insignificance: less than 1 percent of the total. We and all other creatures that walk, slither, burrow, or fly occupy only a thin layer of soil and air, whereas marine life roams through the oceans' full volume. Most of that enormous volume, roughly 330 million cubic miles, lies deep below the familiar surface. That upper sunlit layer, by one estimate, contains only 2 or 3 percent of the total space available to life. The other 97 percent of the earth's biosphere—the volume of space in which life exists—lies deep beneath the water's surface, where sunlight never penetrates. It is a world that humans rarely glimpse, a realm of eternal darkness.

This hidden deep-sea environment dwarfs all other earthly habitats combined. It is the ultimate reservoir from which life everywhere draws sustenance—a fact we should keep in mind in this age of growing populations and pressure on resources. The planet's entire water supply cycles through the oceans. Some of it evaporates into the atmosphere and returns in rain and rivers; the rest sinks beneath the seafloor and returns through deep-ocean hot springs. The deep sea, in fact, seems to be the planet's central clearinghouse for nutrients and minerals essential to life. It is also a haven for tens of millions of species by some estimates, most of which have never been seen—a greater diversity of animals than in any other ecosystem. The first life on earth may well have started on the deep seafloor. Certainly dry land was colonized by a few life forms that came out of the ocean—one branch of which eventually evolved into our own species. We need to keep in better touch.

Until recently, it was impossible to study the deep ocean directly. To view what lies in the depths of the sea requires us to enter a world in many ways more alien than Mars. Try to imagine, if you will, what ancient mariners must have felt as they left the safety of their settlements and ventured out on the constantly shifting watery surface. What must have gone through their minds as the land's silhouette sank slowly toward the horizon and the color of the sea changed from greenish brown to the transparent blue of the open ocean? With no land in sight, they must have gazed down as far as they could see and tried to picture how deep the ocean was and what lay on its floor—if it had one. At times it must have seemed like a bottomless pit, at other times a ghoulish graveyard of lost souls where great monsters lived. The surface of the sea was to be crossed and recrossed as quickly as possible. To linger was to tempt violent forces controlled by the gods.

Yet even from the earliest historical records, we know of rare individuals who dove repeatedly beneath the surface. Often they went looking for valuable objects: ornamental shells, pearls, or sponges. The first efforts took the bravest divers only a short distance, probably just over 100 feet while holding their breath. Other schemes, such as breathing through reeds or from air-filled bags, would not have extended this depth. By the sixteenth century, diving bells allowed people to stay underwater longer: they could swim to the bell to breathe air trapped underneath it rather than return all the way to the surface. Later, other devices—including pressurized or armored suits, heavy metal helmets, and compressed air supplied through hoses from the surface—allowed at least one diver to reach 500 feet or so by the 1930s. For most, however, the limit remained between 200 and 300 feet, the maximum that compressed air would usually permit. Beyond that depth, so much oxygen accumulates in the blood that it quickly becomes poisonous. Divers who repeatedly went deeper than 200 feet sometimes breathed heliox, a mixture of oxygen and helium.

To protect us from the unforgiving pressures of the deep, engineers began building submarines. Inside their sturdy shells, occupants breathed air at a safe pressure of 1 atmosphere, the same as we breathe on the surface. But even a submarine will collapse under extreme deep-ocean pressures. Today's nuclear submarines can dive only 1,500 feet below the surface, slightly more than one-tenth the average depth of the world's oceans. Ninety percent of the total ocean volume remains beyond their limit.

It was 1930 when a biologist named William Beebe and his engineering colleague Otis Barton sealed themselves into a new kind of diving craft, an invention that finally allowed humans to penetrate beyond the shallow sunlit layer of the sea. They took their first deep plunge off the island of Bermuda, and it is here that my history of deep-sea exploration begins.

From those mile-deep waters, full of many surprises, the narrative traces a moving frontier, continually pushed back by new technology. It proceeds in three parts. In Part I it recalls the early, heroic days of magnificent men and their diving machines. Those dreamers pursued a difficult quest for greater depth and better mobility. Science then was largely incidental—something that happened along the way. In terms of technical ingenuity and human bravery, this part of the story is every bit as amazing as the history of early aviation. Yet many of these individuals, and the deep-diving vehicles that they built and tested, are not well known.

It was not until the 1970s that deep-diving manned submersibles were able to reach the Midocean Ridge and begin making major contributions to a wide range of scientific questions. A burst of discoveries followed in short order. Several of those discoveries profoundly changed whole fields of science, and their implications are still not fully understood. For example, biologists may now be seeing—in the strange communities of microbes and animals that live around deep volcanic vents—clues to the origin of life on earth. No one even knew that these communities existed before explorers began diving to the bottom in submersibles.

Part II follows the teams of scientists who learned how to use those machines to their best advantage, often in tense rivalry with other groups. In terms of scientific achievement, this era was every bit as important for the earth sciences as the race into space was for astronomy. In fact, some of these deep-sea missions began to complement space exploration.

Meanwhile new kinds of deep-diving craft continued to drive back frontiers. Technology again plays a leading role in Part III, most notably in the race to find the *Titanic.* This part of the book introduces a new paradigm for deep-sea exploration. The advances described in the final chapters may allow either you or your children to join a mission someday—not merely as distant observers but as active participants.

It would have been impossible to write a comprehensive history and still keep the book manageable. The contents are of necessity highly selective—more and more so as the history progresses. However, I spent almost as much time compiling the suggestions for further reading as I did writing the chapters themselves, and in that listing I have indeed tried to be comprehensive. A great deal of the knowledge we have gained from the use of deep-diving vehicles appears in those references. Although technology drives the narrative, scientific results dominate the Further Reading section. I believe that this is the first time such a list has been attempted.

Although the actors change as the story evolves, one presence haunts this book from beginning to end: the ocean itself. Entering the deep, black abyss presents unique challenges for which humans must carefully prepare if they wish to survive—and this, sadly, has not always been possible. It is an unforgiving environment, both harsh and strangely beautiful, that few who have not experienced it firsthand can fully appreciate.

First of all, the abyss is dark. And that darkness seems much more oppressive than the blackest chamber inside any cavern on land. Even the most powerful searchlights don't penetrate very far in the deep abyss—typically only tens of feet. One reason is that suspended particles scatter the light. Another is that water itself is far less transparent than air; it absorbs and scatters light. The ocean also swallows other types of electromagnetic radiation, including radio signals. That is why many vehicles described in this story dangle from tethers. Inside those tethers, copper wires or fiber optic strands transmit signals that would dissipate and die if broadcast into open water. (Another strategy for deep-sea communication relies on sound waves, which travel through water much farther than light, but also much more slowly.)

A second challenge we must merely endure: the abyss is cold. The temperature near the bottom in very deep water typically hovers just four degrees above freezing, and submersibles rarely have much insulation. Since water absorbs heat more quickly than air, the cold down below seems to penetrate a diving capsule far more quickly than it would penetrate, say, a control van up above, on the deck of the mother ship.

And finally, the abyss clamps down with crushing pressure on anything that enters it. This force is like air pressure on land, except that water is much heavier than air. At sea level on land, we don't even notice 1 atmosphere of pressure, about 15 pounds per square inch—the weight of the earth's blanket of air. Yet any vessel that dives below the ocean's surface adds, in effect, 1 more atmosphere for every 33 feet it descends—the weight of the water above the craft. As a submersible goes down, the pressure on its hull keeps increasing: 30 atmospheres at 1,000 feet, 300 atmospheres at 10,000 feet, and so on. In the deepest part of the ocean, nearly seven miles down, it's about 1,200 atmospheres—18,000 pounds per square inch. A square-inch column of lead would crush down on your body with equal force if it were 3,600 feet tall.

Fish that live in the deep don't feel the pressure, because they are filled with water from their own environment. It has already been compressed by abyssal pressure as much as water can be (which is not much). A diving craft, however, is a hollow chamber, rudely displacing the water around it. That chamber must withstand the full brunt of deep-sea pressure—thousands of pounds per square inch. If seawater with that much pressure behind it ever finds a way to break inside, it explodes through the hole with laserlike intensity. A human body would be sliced in two by a sheet of invading water, or drilled clean through by a narrow (even a pinhole) stream, or crushed to a shapeless blob by a total implosion. The thought of such a leak actually comforted William Beebe, when it first occurred to him at a depth no living human had ever reached. "There was no possible chance of being drowned," he wrote, "for the first few drops would have shot through flesh and bone like steel bullets."

It was into such a terrifying environment—dark, cold, and mercilessly crushing—that the first twentieth-century explorers ventured. The vehicle for their daring plunge? A simple tethered sphere.

PART I

DEPTH

CHAPTER 1

A SIMPLE TETHERED SPHERE

When once it has been seen, [the deep ocean] will remain forever the most vivid memory in life.

—*William Beebe*

On June 11, 1930, the first humans entered the world of eternal darkness and returned alive. Off the island of Bermuda, two men, Charles William Beebe and Otis Barton, descended 1,426 feet, nearly three times deeper than any previous diver. Their mode of transport resembled the stuff of legend: in the fourth century B.C., Alexander the Great had supposedly reached deep waters in a glass barrel lowered by chains. Beebe and Barton's remarkably similar craft, made of steel rather than glass, dangled just as precariously on a cable. A crew on a surface ship lowered the two men deep into the Atlantic, then hauled them back toward sunlight and air. The ancient Greeks, given access to modern industrial tools and steel, might have managed the feat in much the same way.

William Beebe was a fascinating mixture of scientist, poet, showman, and explorer. Born in Brooklyn, New York, in 1877, he seems never to have lost the curiosity of an excited child—a quality he was to combine later in life with the bravery of a confident professional. Instead of completing his final year at Columbia University, Beebe went to work full time in 1899 as assistant curator at the New York Zoological Park. He was soon tracking rare birds in Mexico, Trinidad, and Venezuela and

William Beebe and Otis Barton standing aboard ship next to the bathysphere, 1930. (Source: Corbis/Ralph White.)

studying wild pheasants in the Far East. On a trip to the Galápagos Islands, Beebe hiked up an erupting volcano. In the 1920s he began diving in the ocean, using a helmet, and he said that someday he hoped to go deeper in a new kind of diving vessel. His popular writings bubbled over with enthusiasm, encouraging readers to enter his newly discovered underwater paradise. "Don't die," he advised them, "without having borrowed, stolen, or made a helmet of sorts, to glimpse for yourself this new world."

Years before those words appeared in print, a teenage boy named Otis Barton had done precisely what Beebe suggested: he had made his own helmet, a crude wooden box with glass windows. In 1917, wearing his new headgear, he weighed himself down with rocks and sandbags and explored the harbor bottom at Cotuit, Massachusetts, breathing air forced down from the surface by means of a bicycle pump. Barton also dreamed of exploring the deep sea. However, unlike Beebe, he directed his passion toward the means of diving as well as the ends: he studied both engineering and natural science at Columbia. Yet Barton never held a job. Having inherited a large amount of money, he spent most of his life in pursuit of fantastic creatures—prehistoric fossils, rare wild animals, and "titans of the deep," such as sharks, octopuses, and giant squid. He was, altogether, a mysterious individual. During the years he worked with Beebe, we must rely for information chiefly on pictures of him toiling over his diving apparatus, or occasional comments that Beebe published indicating his confidence in Barton's engineering talents.

— The Bathysphere

In 1926 Barton read about Beebe's quest to explore the deep ocean. He wrote to the famous scientist/author, but Beebe never replied—having been inundated, it turned out, with all kinds of suggestions from crackpot designers of deep-sea vessels. Barton finally arranged a meeting through a mutual acquaintance in December 1928. Having done the necessary calculations and drawn up blueprints, he showed Beebe his idea for a diving craft. It was little more than a hollow sphere. Barton explained that a sphere would be the shape best suited to withstand the crushing deep-ocean pressure, which bears down equally from all directions. Beebe liked the design and agreed to test such a craft. He christened their diving chamber a *bathysphere,* joining the Greek word for *deep* to *sphere.*

Barton volunteered to pay for the craft's construction. Later, the New York Zoological Society and the National Geographic Society would fund the explorers' diving expeditions.

Barton's first steel chamber, cast in one piece at a foundry in New Jersey, weighed five tons—making it too heavy for the

Face of William Beebe peering out of the bathysphere. (Source: William Beebe/National Geographic Society Image Collection.)

barge, *Ready,* that would lower and raise it. The next version weighed half as much, but its small interior—only four feet nine inches in diameter—would barely accommodate two men. Had Barton made the hollow space larger, he would have needed a thicker, heavier hull. As it was, the steel wall measured one and a half inches thick—sufficient, Barton figured, to survive a descent to a depth of 4,500 feet.

A small, fifteen-inch circular opening in one side of the sphere allowed squirming passengers to enter head first. Around this entryway—a potential weak spot—ten large steel bolts protruded from the sphere. A 400-pound door fit over the bolts. After crew members had hoisted the door up and wrestled it into position, they screwed ten huge nuts onto the bolts. Then one last large bolt, held in place by a giant wing nut, plugged a small central opening in the door. (This allowed quick access, in case of emergency, to let out water or foul air.) A crewman then pounded the wing nut tight. Beebe referred to this moment as the most painful part of a dive. Ear-shattering reverberations bounced off the steel hull. Perhaps, at such moments, Barton's earlier name for the vessel seemed more apt: he had called it, simply but unpoetically, "the tank."

On the side opposite the door, three circular windows made of fused quartz, each eight inches in diameter and three inches thick, fit into "cannon-like projections"—short tubes that stuck out from the sphere like stubby telescopes to accommodate the extra thickness of the quartz viewing panes. Pieces from any of those windows would indeed shoot through the tubes like cannonballs—straight into the sphere toward the occupants—if some unseen flaw were to cause a window to crack and yield under pressure. Of the five quartz pieces initially made, only two passed all fitting and pressure tests, so Barton had to insert a steel plug into one of the three window openings for their first dives.

The interior of the bathysphere contained only the barest of necessities. Two small oxygen tanks, placed on either side of the windows, would keep the air sweet for eight hours, Barton estimated. One wire mesh tray above the windows contained soda lime to absorb exhaled carbon dioxide; another tray contained calcium chloride to absorb moisture. Believe it or not, on their first dives Beebe and Barton used small palm-leaf fans to circulate the air.

When underwater, the bathysphere hung from a 3,500-foot-long, seven-eighths-inch steel cable, raised and lowered by a steam-powered winch. Extra strands of cable, woven around the central core, were supposed to ensure that the sphere would not rotate. As the cable descended into the ocean, a solid rubber hose snaked down alongside it. Crewmen played out this hose by hand; it contained two wire conductors for a telephone and two for an electric light. Since the advent of underwater lighting technology was still many years away, Beebe and Barton simply mounted a 250-watt spotlight inside their sphere and aimed it out through a window.

After submerging, Beebe or Barton, wearing headphones, communicated via the telephone link with Gloria Hollister, an assistant. During many of those conversations, Beebe made observations and Hollister transcribed his words. Every sec-

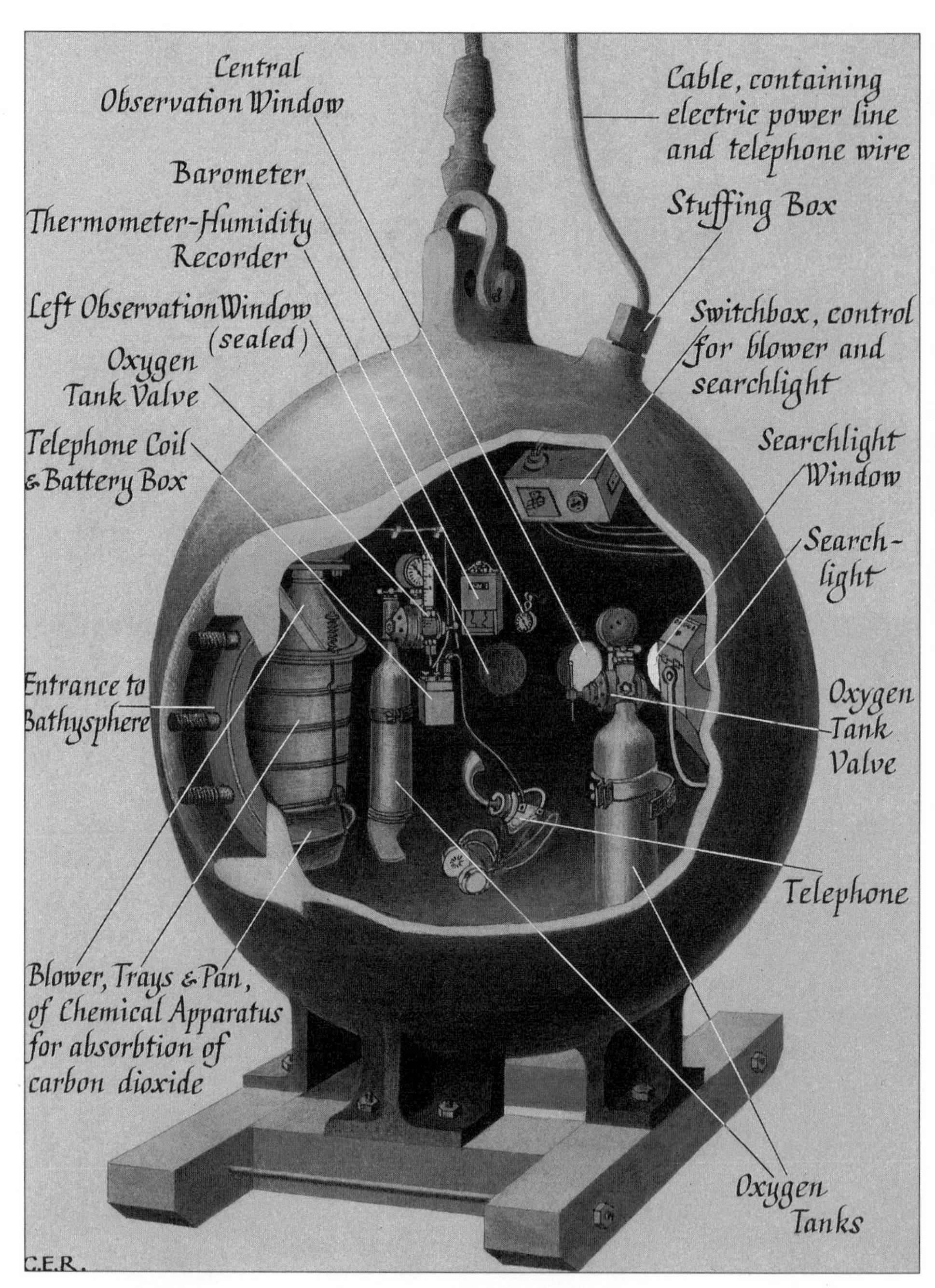

Detailed illustration of the bathysphere and its interior. (Source: Charles Riddiford/National Geographic Society Image Collection.)

ond counted, since the air supply was limited and observation time was short. On the positive side, the cramped occupants of the bathysphere had little to do but observe. They wasted no time fiddling with controls because there were none. They had no ability to maneuver their craft and no depth gauge. When the sphere approached the bottom, crew members on the support barge made depth measurements every few minutes, using a line attached to a small lead anchor.

— Light First Shines in the Eternal Night

The first series of deep-water dives took place in June 1930 near Bermuda, where Beebe had been studying marine animals. He had cast nets into the ocean and examined what they brought up, but he could not be sure how many creatures had eluded his nets. Opinions differed on this matter. Folktales suggested that strange, unknown kinds of life inhabited the deep sea. Alexander the Great, during his legendary dive, had supposedly seen a fish so large that it took three days to swim past his glass barrel, and tales of other deep-dwelling monsters had been passed down through the ages. Many contemporary researchers, on the other hand, believed that the ocean depths were too cold and dark to support any significant amount of life. Beebe had no idea what might await him below.

The first deep test of the bathysphere, conducted on June 3, nearly ended in disaster. After lowering the empty sphere to 2,000 feet, the crew could not haul it back; the steel cable and rubber hose had become so tangled that the cable would not rewind on its reel. Foot by foot, with great effort, crew members pushed the snarled hose down the length of the cable, which they gradually winched up and rewound. When they finally got the sphere back on deck, they found that the "non-twisting" steel cable had turned some forty-five times, twisting the rubber hose tightly around itself as it turned.

The rubber looked chafed and stretched—would the wires inside it be broken? Fortunately they were not, and the twisting forces seemed to have played themselves out with this lowering.

Several other tests followed. Finally, on June 6, with all the problems apparently fixed, the bathysphere was ready to take human passengers down into the eternal darkness.

Beebe later described the first attempt: "I . . . crawled painfully over the steel bolts, fell inside and curled up on the cold, hard bottom of the sphere. . . . Otis Barton climbed in after me, and we disentangled our legs and got set." Crew members lifted the heavy door and carefully slid it over the projecting bolts; then they screwed on the huge nuts. Although his vision was limited, Beebe could see the launch crew scurrying about the deck. Feeling trapped inside the small capsule and wanting to calm his nerves, he followed Houdini's technique of regulating his breathing and conversing in low tones. It helped.

At precisely 1:00 in the afternoon, *Ready*'s captain signaled with his hand to the crew chief. The cable attached to the top of the bathysphere suddenly went taut; the capsule rose quickly into the air and then began to drop. "I sensed the weight and sturdy resistance of the bathysphere more at this moment than at any other time," Beebe wrote. "We were lowered gently but we struck the surface with a splash which would have crushed a rowboat like an eggshell. Yet within we hardly noticed the impact, until a froth of foam and bubbles surged up over the glass."

As the bathysphere began to sink deeper, Beebe saw the familiar outline of the support ship's keel. He watched it grow smaller and fainter until it blended into the greenish glow of the surface. With its disappearance passed the last visible link to human beings in the upper world. From then on, their only reassurance that those on the surface had not lost control at their end of the tether would be the soft voice of Gloria Hollister relaying depth, speed, or information on surface weather conditions.

At 200 feet the bathysphere stopped. As it hesitated in midwater, crewmen clamped the steel cable and rubber hose together, giving the rubber more support so that it would not break under its own weight. (They would repeat this procedure every 200 feet.) Beebe and Barton's journey into the abyss then resumed. For long stretches, their only sense of motion was the constant movement of small organic detritus, "marine snow," falling not downward but upward as the diving craft dropped through it.

Before they reached their next scheduled pause at 400 feet, Barton startled Beebe with a sharp exclamation. Turning quickly, Beebe saw a trickle of water meandering down the wall beneath the door. For a moment they watched this little stream in horror, but then Beebe began to relax. "I knew the door was solid enough," he wrote, and he realized that higher pressure outside the sphere would only seal it more tightly. Rather than cancel the dive, he asked for a quicker descent.

At 400 feet the bathysphere plunged through the lower limit that submarines could reach in those days. Beebe and Barton soon passed 525 feet, the greatest depth any person had reached alive in an armored suit. At 600 feet the water took on a shade of blue no conscious human had ever seen. Beebe described the luminous color as pouring into the sphere through the viewports—a confusing otherworldly glow that seemed constantly bright as his eyes adjusted. "The blueness of the blue" became almost mystical: it seemed to pass "materially through the eye," Beebe remarked, "into our very beings." As the twilight deepened, "we still spoke of its brilliance," he continued. "It seemed to me that it must be like the last terrific upflare of a flame before it is quenched."

On a hunch, Beebe called for a halt of several minutes at 800 feet. The leak seemed no worse, the oxygen was flowing, and yet he decided that the dive should end. "Some mental warning," he wrote—"which I have had at half a dozen critical times in my life—spelled *bottom* for this trip." The bathysphere returned to the deck of *Ready* only an hour after it had plunged into the water. Its two occupants untangled themselves, crawled stiffly out, and noticed right away the unbelievable yellow hue that suffuses our sunlit world. Later, Barton packed soft lead into a groove around the door's edges, hoping that it would form a watertight seal on future dives.

The expedition put to sea once again on June 10. After being lowered empty to 2,000 feet, the bathysphere returned with no evident leaks. Beebe and Barton then climbed in, ready to make another attempt to reach total darkness. At 150 feet, however, the telephone began to crackle, and at 250 feet Barton cried out that the line was dead. Beebe also felt alarmed. "The leak on our other trip," he wrote, had been "annoying but not terrifying." Suddenly he understood why. All along, he knew he could quickly relay an urgent request to anyone in the twenty-six-person surface crew—especially to the man controlling the seven-ton winch. Down below, nothing seemed more reassuring than the sound of a human voice.

Beebe blinked the spotlight for all he was worth. This caused a bulb on *Ready*'s deck, wired into the same circuit, to flicker, which, by prearranged signal, meant trouble down below—bring us back! By the time the sphere resurfaced, the worried crew was showing considerable strain. Yet what they found inside was Beebe's smiling face, pressed against clear quartz. Later, they scrapped fifty fathoms of hose containing a broken wire.

The next day, June 11, seemed perfect. A fire on board *Ready* the night before had charred part of the deck, but no matter. It was only wood. This time Beebe took a chance and skipped the preliminary empty trial. He and Barton climbed into the sphere at 9:50 that morning, the bolts and nuts were screwed down by 10:00, and four minutes later they hit the water.

After the sphere splashed down, Beebe gazed up. The air-sea interface looked like a slowly waving pale-green canopy. Dancing on the puckered watery ceiling, clusters of golden sargassum reminded him of mistletoe hanging from a chandelier at Christmas. Through this translucent quilt, rays of sunlight penetrated the depths from above like shafts of light in a vast cathedral.

In his writings afterward, Beebe mixed poetic accounts of the historic dive with awed descriptions of the life he observed. "Long strings of salpa drifted past," he wrote, "lovely as the finest lace, and schools of jellyfish throbbed on their directionless but energetic road through life." Small "vibrating motes" passed by in clouds, "wholly mysterious." Looking more carefully, Beebe recognized them as pteropods, or flying snails, "delicate, shield-shaped shells driven along by a pair of flapping, fleshy wings."

They made the usual pauses at 200 and 400 feet and were now beginning to see their first true creatures of the deep: a lantern fish and bronze eels. The descent continued. At 800 feet Beebe caught his first glimpse of hatchet-fish in the distance, "heliographing their silver sides." A little later, "shrimps and jellies drifted past."

With the gloom growing ever darker, the pale blue glow up above seemed bright, almost skylike by comparison. The shift in color confused the senses. "Again," Beebe wrote, "the word 'brilliant' slipped wholly free of its usual meaning." Although the light seemed bright enough to read by, neither man could see print on a page. "We looked up from our effort to see a real deep-sea eel undulating close to the glass." Beebe recognized it from his studies—"a slender-jawed *Serrivomer,* bronzy-red as I knew in the dimly-remembered upper world, but here black and white."

At 1,000 feet the two explorers had a momentary sea-monster scare when a length of rubber hose above the sphere worked loose. It formed a black, sea-serpenty loop that swung past their viewport, backlit by a growing starfield of bioluminescent creatures too small to discern. Then a line of luminous dots suddenly appeared. Beebe turned on his spotlight, which revealed another school of hatchet-fish, from a half to two inches in length, swimming in formation. Had these denizens of the deep, with their sabertooth-like jaws, been the size of sharks, Beebe would have died of fright.

The scene outside his viewport now resembled a kaleidoscope of constellations in the night sky. These were living, glowing creatures, their patterns constantly changing as the bathysphere fell. Every so often the spell was broken as the voice of Gloria Hollister briefly interrupted, by now as faint as if it were coming from hundreds of miles away.

At 1,100 feet Beebe and Barton carefully inspected their steel chamber. The temperature inside the sphere had plummeted, causing moisture to condense on its cold wall. Rivulets of water streaked down, forming a growing puddle beneath the floor planks. To prevent his viewport from constantly fogging, Beebe tied a handkerchief around his face just below the eyes, bandit-style, directing his breath downward. From 1,250 to 1,300 feet the bathysphere passed through an empty "transparency" of water, devoid of life, which only heightened Beebe's discomfort as he sat, chilled to the bone, in the clammy sphere.

At 1,400 feet they paused as usual, dipped a few feet lower, and then jerked to a final stop. This time Beebe gazed down rather than upward. Although the water below him looked like "the black pit-mouth of hell itself," he felt a longing to go deeper. Barton meanwhile, without a tinge of emotion, casually informed Beebe that the pressure on every square inch of the viewport he was looking through exceeded 650 pounds. He added that each window was now holding back a total pressure of nine tons. From then on, Beebe wiped moisture off the quartz with a softer touch.

With the bottom of the ocean still thousands of feet below them, Beebe and Barton had reached their goal. They had dropped 1,426 feet, making them the first humans to see the world of eternal darkness. Almost immediately, the surface crew began reeling them back to the world of sunlight. After what had seemed a lifetime of descending, the return trip to the surface took only forty-three minutes. Twice during their ascent, large, shadowlike forms darted around in the distant gloom—stalkers of the deep, Beebe imagined, waiting for an opportunity that never came.

— The Living Ocean Gains a Third Dimension

The bathysphere's primary mission was to observe marine life in the open ocean, while suspended between surface and bottom. This was the midwater, or pelagic, world. But Beebe also used the sphere actually to explore the bottom, particularly in shallow water during an operation he called *contour diving*. It was an activity the two explorers pursued when the weather kicked up and they couldn't venture out to the open ocean.

When contouring, the tug *Gladisfen* towed the support barge *Ready* slowly out from shore, with the bathysphere being lowered as the bottom became deeper. This procedure was not as simple as it sounds; there was always a chance that the sphere might hang up on an uncharted coral head. Should that happen, the deck crew was instructed to let out cable and drop the sphere to the bottom while the barge and tug moved back into shallower water. The crew might then be able to lift the sphere free—provided, of course, that the cable had not snapped. Fortunately, the crew never needed to try this maneuver, although on one dive Beebe did see a coral barrier looming ahead. He had just enough time to phone the surface with urgent instructions to raise the cable.

Obviously, Beebe felt that the risks were worth taking. Many previous attempts to study life near the bottom had failed: nets and dredges had torn or tangled, cables had broken, and gear had sunk. Taking the same chances with his new bathysphere—its human occupants helpless inside—Beebe succeeded where others had not. During his relatively shallow dives, he observed fish behavior not previously known, such as the systematic cleaning of parrot fish by wrasse. On one contour dive he discovered an ancient beach at a depth of sixty-five feet, formed thousands of years earlier when the giant ice sheets that covered the continents locked up enough water to lower the sea level that much and more worldwide.

Geology, however, was not Beebe's main interest. "These shore dives," he wrote, "opened up an entirely new field of possibilities, the opportunity of tracing the change [in life forms] from the shallow-water corals and fish to those of the mid-water, with ultimately the . . . change of fish into the deep-sea forms. . . . Such things as these could never be seen except from a sphere such as ours."

The deeper dives proved equally rewarding. For two years, Beebe had been studying the ocean off the island of Nonsuch, his laboratory base in Bermuda. Being particularly interested in the vertical distribution of life, he had chosen a circular area in the open Atlantic and probed it deeper and deeper by dragging a net back and forth from the surface. The area he was studying thus gained a third dimension: it extended downward as a cylinder. Those earlier results had made Beebe pessimistic about the amount of life he might see on his dives with the bathysphere.

Clearly his pessimism had been unfounded. Now that Beebe had immersed himself within the deep world, the thought of dragging a net through this three-dimensional space reminded him of a biologist stuck on the ground with his eyes closed, trying to catch butterflies that soared up and down through a vast field of flowers. Now the biologist had wings, enabling him to mingle with the objects of his study in their own realm. He felt like an astronomer who had rocketed to the planets and back, appreciating for the first time the great distances separating those points of light. And he knew the joy of the paleontologist who had watched time run backward and his fossils come alive.

Before diving in the bathysphere, Beebe had seen deep-dwelling fish only as lifeless forms pulled from his nets. Now he could picture them darting about his viewport, displaying modes of swimming, behaviors, and colors that no one had ever witnessed. The bizarre shapes, the globular and angular forms, had to him suggested slowness. Yet their level of activity and agility seemed far greater than he had previously imagined. He concluded that this agility might help explain why many of the organisms he saw down below had escaped his net tows.

Prior to these early dives, some researchers had questioned whether the apparent light organs on the sides of pelagic fish actually emitted light in deep water. Beebe's experience proved beyond doubt that the organs do in fact emit light. Eventually—though not right away—marine scientists came to accept that truth.

Beebe also realized that life in the midwater depths was not uniformly distributed. He had seen layers of richness and paucity, shifting and changing with time. Furthermore, he was also surprised to discover deep-dwelling fish roaming in shallower water than the depths he had probed to capture them in nets, convincing him that they were able to see the nets' entrance in the sunlit upper layers and swim out of their way. Farther down, in total darkness, they would swim blindly across the path of the sweep nets. And finally, Beebe explained the presence of even larger creatures, giants he had never caught at any depth. He felt that they avoided capture by swimming faster than the heavy nets could be towed.

—— Going Deeper

Beebe and Barton continued to press the limits of their bathysphere. After the dives in 1930 they prepared for a similar series in 1932, making a number of changes to the sphere. The fact that they were now veterans of the deep also helped. In 1932, they entered the deep abyss with a new perspective, one common to the seasoned veterans of today. "The unproductive 'Oh's and Ah's' of my first few dives," Beebe confessed, "were all too vivid in my mind."

Though the men were less giddy, the thrill of exploration persisted. Before a dive on September 22, which would take him to a new record depth of 2,200 feet, Beebe described "a feeling of utter loneliness and isolation" that gripped him when dangling in midocean, "akin to a first venture upon the moon or Venus." Yet outwardly he was still the showman. Beebe and Barton not only established a new depth record that

September day, they also conducted the first live radio broadcast from the deep sea. Produced by NBC, it proved a fascinating experience for the millions who listened. At one point, Beebe announced to the spellbound world that he was looking through the viewport at a six-foot-long "sea dragon"—an unlikely beast that would in time return to haunt him.

In 1934, Beebe carried out his final series of dives with Otis Barton in their bathysphere. They achieved a depth of 3,028 feet—more than half a mile down. The crew captain allowed them to stay for all of three minutes before giving the order to haul up, fearing that the very end of the cable, which was now almost completely unwound, might not be fastened securely to the reel. Indeed, shortly after the winch started pulling, the crew heard a "terrific snap," which traveled all the way through the telephone line to Beebe as a "metallic twang." That ghastly sound came not from the cable, however, but from a snapping rope that helped guide it onto the reel. The cable itself held firm, and it continued winding back.

After each new dive series, Beebe continued to compare his observations from the bathysphere with the results of the more than 1,500 net tows he eventually made in the same area. He consistently saw major discrepancies, particularly the absence of larger and more active animals from the nets. Clearly, the fauna at midocean depths were far more abundant than he would ever have determined by previous methods. (In fact, some researchers now believe the ocean depths to be the dominant ecosystem on the planet, making up 97 percent of the biosphere by volume and containing more diversity of life in some areas than even the fabulous rain forests.)

Beebe himself identified many new life forms. On one dive, he observed a large fish at 1,500 feet that he had never seen before. It was at least two feet long, without lights or luminosity, and had a small eye, a large mouth, and a long, wide, filamentous pectoral fin. He called it the pallid sailfin, with the scientific name *Bathyembryx istiophasma,* which is Greek for a fish that comes from the abyss and swims with ghostly sails.

Soon after his first series of dives, and for the next several years, Beebe published numerous scientific and popular articles, followed in 1934 by a wonderful book titled *Half Mile Down*. In the book he described several "outstanding moments" that bathysphere diving had etched in his mind. Bioluminescence, that "first flash of animal light," topped his list. Next came the perception of utter blackness, "the level of eternal darkness" that only he and Barton had ever seen. A third outstanding moment came with "the discovery and description of a new species of fish"—many times.

No other earthly environment, Beebe insisted, could match the strangeness of the dark abyss. "The only other place comparable," he wrote, "must surely be naked space itself, out far beyond [the] atmosphere, between the stars, where sunlight has no grip upon the dust and rubbish of planetary air, where the blackness of space, the shining planets, comets, suns, and stars must really be closely akin to the world of life as it appears to the eyes of an awed human being, in the open ocean, one half mile down."

Not everyone was awed, however. One scientist who reviewed the book, Carl Hubbs, scoffed at Beebe's claim to have seen a six-foot sea dragon. He suspected, instead, that two fish had been swimming close together. Another animal, which Beebe had named a "Constellationfish," struck Hubbs as even more absurd. Beebe described it as having rows of glowing yellow dots surrounded by "very small, but intensely purple lights." That fish, Beebe continued, "will live throughout the rest of my life as one of the loveliest things I have ever seen." It seemed more likely to Hubbs that Beebe had seen a "constellation" of less remarkable creatures—invertebrates such as jellyfish—"whose lights were beautified . . . in passing through a misty film breathed onto the quartz window by Mr. Beebe's eagerly appressed face." Hubbs indignantly charged that Beebe had no right "to describe and assign generic and species names 'for animals faintly seen through the bathysphere windows.'" Another scientist, John Nichols, a curator at the American Museum of Natural History, hinted that *Half Mile Down* should be classified as fiction rather than fact, because Beebe wrote in "dramatic fashion rather than meticulous." To this day suspicion lingers that Beebe may have willfully invented a few species—"impossible" fish that he claimed to have seen but could not collect, and that no one has ever seen since.

In other respects, Beebe seems more assured in his status as a pioneer of ocean ecology—largely because of his systematic sampling, his many indisputable specimens, and his ability to inspire others to follow him into the abyss. John McCosker, for example, chair of Aquatic Biology at the California Academy of Sciences, dispatched this "Letter from the Field" in the fall of 1995:

> At 2,000 feet, it's 39 degrees F outside and getting cold and foggy inside. It is the darkest darkness . . . and the bursts of living light outside only accentuate the blackness. As I stare out the window, my childhood fascination with a well-read 1934 copy of *National Geographic* comes crowding back. In his article, "A Half Mile Down," William Beebe described . . . in vivid detail . . . the fascinating creatures that swam or hovered within his searchlight beam. The descriptions, complete with creative monikers like abyssal rainbow gars, sabre-toothed viperfishes, scimitar-mouths, great gulper eels, long-finned ghostfishes, gleaming-tailed serpent dragonfishes, and exploding flammenwerfer shrimps, were carried to a surface audience which included artist Else Bostelmann, who painstakingly took notes in order to illustrate the explorer's observations. Those paintings, which appeared in the *National Geographic* and in a book, also entitled *Half Mile Down,* whetted the appetite of a 10-year-old explorer-to-be. . . . Many years later, as a graduate student, I was told by my professors that much of what Beebe saw was colored by his vivid imagination. . . . Nonetheless, it was now my turn to experience firsthand what life is like in a realm so foreign to our light-dependent lives.

—— A Retreat from the Deep, and a Legacy

Beebe would have welcomed more dives like McCosker's. In the heyday of his career, he was convinced that scores of bathyspheres would soon be diving throughout the world, producing a burst of knowledge about the global distribution of marine life, in three dimensions. A true picture of the ocean's habitats would require many more dives in many more areas than the Atlantic Ocean near Bermuda. But Beebe's dream never came true. Most of the deep sea has yet to be explored—which means, if Beebe was correct, that some of its largest and most agile creatures may yet await discovery.

Several decades after Beebe and Barton's historic dives, both Japan and the Soviet Union made limited attempts to learn more: they developed various tethered diving capsules to carry researchers into deep waters. In the early 1950s, reports began coming out of the Soviet Union that researchers were using bathyspheres called hydrostats to observe fish behavior, in hopes of improving commercial fishing techniques. However, the greatest depth reached by a Soviet scientist using such a hydrostat fell short of Beebe and Barton's achievement: it happened in the Atlantic Ocean in 1960, when V. P. Kitaev descended to 1,970 feet. Later in the decade, the Russians developed new tethered vehicles, capable of being towed, for the commercial fishing industry.

Japanese scientists also carried out extensive observations from their bathysphere *Kuroshio* in the mid-1950s. Like the Soviets, they studied the behavior of commercial fishes, but they also conducted basic research. In particular, they showed great interest in the nature of suspended organic matter in waters adjacent to Japan, and they published some early scientific papers on this marine snow, an important component of the deep-sea food chain.

Despite these later Russian and Japanese expeditions, no one really continued in Beebe's footsteps as he had hoped. No one explored the pelagic depths with any degree of thoroughness. In fact, a long inactive period, from 1934 to 1948, followed

Beebe and Barton's dives in their bathysphere. The reason for this fallow period is unclear. Certainly World War II tied up foundries and shipbuilders, among other huge disruptions, but that alone cannot explain the loss of momentum. Perhaps most countries concluded that deep-sea exploration would never yield commercial benefits. The task was left to dreamers like Beebe and Barton, who would risk their lives to reach their goals—dreamers who are always few in number.

CHAPTER

2

BATHYSCAPHS RACE TO THE BOTTOM

> To climb the highest peaks, to travel through . . . celestial space, to turn our searchlights upon domains of eternal darkness, that is what makes life worth living.
>
> —*Auguste Piccard*

With the end of World War II, dreamers could turn their talents loose on the next frontier: the deep abyss. William Beebe and Otis Barton had never approached it. For all their cleverness and bravery, they had barely dimpled the upper bounds of a vast new world. Inconceivable challenges remained. In order to reach even the average depth of the world's oceans, someone would have to dive four times deeper than Beebe and Barton's record of 3,028 feet. That would already be a significant achievement—arriving at 12,000 feet, the true midocean. At that depth, the pressure bearing down on a hollow sphere, three tons per square inch, would be far greater than any vessel with humans aboard had ever withstood. And yet half the abyss would still lie below.

In the late 1940s Otis Barton resumed the quest. "Since coming home from the war," he confessed, "I had been living in a world bereft of meaning." Barton engaged the same foundry that had cast his bathysphere to make a sturdier version he called a benthoscope, and in August 1949 he began diving off the coast of California. His surface flotilla consisted of a yacht with comfortable living quarters and a hefty barge with a powerful winch. A third boat trailed those two, carrying a flock of reporters. Barton, always superstitious, found their vulture-like presence worrisome, but by the middle of the month he had given them a good story: he had taken his improved steel tank to a new record depth of 4,500 feet.

Ironically, this impressive dive would signal the beginning of the end for tethered spheres. Although similar capsules would continue to be used for many years, the problems to be overcome in order to drop them very much deeper seemed insurmountable. The hull would have to be thicker and heavier, the winch to lift it more powerful, and the cable far longer, which would add still more weight. The whole enterprise, in short, would grow more expensive, more difficult, and more dangerous for every foot added to the record.

If the great depths were ever to be explored, a different kind of diving craft would be needed. It would have to be stronger and heavier to withstand greater pressure, yet easier to bring back to the surface—two goals that seemed contradictory. Yet one such vessel made its debut in 1948, almost a year before Barton's new benthoscope would reach its deep-water zenith.

On November 3, the Belgian cargo ship *Scaldis* drifted slowly in a current off Dakar, the westernmost point of Africa. Resting on deck, a blimplike diving craft twenty-two feet long sat ready for launching. The small crew cabin attached to its underside looked in many ways like a bathysphere. It was, in fact, a pressure-resistant steel sphere, stronger and heavier than the one Barton was building and therefore capable of diving deeper. The large, hollow, football-shaped structure attached to its roof would also resist pressure, but not because of thick walls. For one thing, it would be filled with gasoline, which, like all fluids, is difficult to compress. For another, it was vented underneath to let seawater force its way in or out. Since the pressure inside would always remain the same as the pressure outside, the upper part of this craft could not be crushed. And since gasoline is lighter than water, this gas-filled "blimp" would be buoyant, keeping itself as well as the crew cabin afloat. To descend, its pilot would dump gasoline or flood empty tanks with seawater, increasing the vessel's weight. To rise back up to the surface, he would drop ballast. The crew could, in theory, soar elegantly down and up in this way, like underwater balloonists—free and untethered.

The advantages of a balloon-style descent seemed obvious when compared with a tethered plunge. Barton's heavier-than-water benthoscope, once submerged, would stay submerged. It had to be dragged back up to the surface with considerable effort, and a broken cable would mean death for its occupants. A buoyant craft, by contrast, would have to fight to stay down. It seemed inherently safer. Yet no one had tried this idea before, except in shallow water. The first deep-water test of the new technique off Dakar would, understandably, be an unmanned one.

By 1:00 in the afternoon, with diving preparations well under way, a bespectacled man with longish hair checked the status of a timing device inside the steel pressure sphere. When it went off, at 4:40 P.M., weights would drop automatically from the sphere, ensuring the craft's return to the surface after plenty of time for a record-setting test dive. Or so everyone thought.

No sooner had the hatch been sealed than all sorts of delays began. As time slipped away, the floating vessel drifted with a surface current into shallower and shallower water. Contrary to its name, this so-called *bathyscaph,* or "deep boat," seemed unwilling to go down. It finally had to be towed back out to sea, wasting more valuable time—especially when the towing cable broke. Then someone noticed that the gasoline-filled hull was riding too high in the water. More ballast had to be added—more time lost. The clock was ticking; the ungainly craft remained above water. Finally, just forty minutes before its weights were scheduled to drop, it settled beneath the waves and disappeared.

Would this bathyscaph ever descend a respectable distance into the abyss? More important, would it return to the surface? Only time would tell. As crew members stood by on *Scaldis,* and as the world press waited to report the results, one person especially must have been watching the clock in agony—the bathyscaph's inventor, Swiss physicist Auguste Piccard. If all went well, he was planning to make the next descent inside it himself.

Diving Balloon or Ascending Submarine?

A daredevil already by reputation, Piccard did not look the part. Behind his large, bald forehead, long strands of hair tumbled, Ben Franklin–style, to the base of his neck. Horn-rim glasses and often a tie, sometimes askew, gave him the appearance of a professor engrossed in his work. He had, in fact, once collaborated with Albert Einstein, a master of relatively safe "thought experiments." Yet Piccard seemed determined to escape his earthbound limits, not only with his mind but also with his body—and not only downward into the sea but upward as well.

Born in Basel, Switzerland, in 1884, Piccard earned a Ph.D. from the Swiss Institute of Technology, and by 1922 he was a professor of physics at the University of Brussels in Belgium. His collaboration with Einstein involved the study of cosmic rays, which, it was thought, could help explain certain puzzling electrical discharges seen in the upper atmosphere. Piccard developed instruments for measuring radioactivity and later set off to observe what actually happens when cosmic rays strike the upper atmosphere. He intended to go there himself with his instruments—up into the stratosphere, where no human had ever ventured.

On May 27, 1931, as Beebe and Barton were preparing for their second series of deep-sea dives, Piccard and his assistant, Paul Kipfer, climbed aboard a hydrogen-filled balloon in Augsburg, Germany, that rose more than 50,000 feet into the atmosphere. It was nearly their last experiment. Released too early, without warning, the balloon shot up nine miles in twenty-eight minutes. All was apparently well, though, because Piccard had designed a spherical aluminum gondola to hold pressurized oxygen. With 90 percent of the atmosphere below him, Piccard noted that the sky looked a very dark blue, almost black. He also noticed a leak through which his life-sustaining supply of oxygen was escaping. That wasn't the only problem. After plugging the hole with a paste they had brought along, Piccard and Kipfer discovered that the rope controlling the balloon's gas-release valve had fouled. Relying on a combination of skill and luck, they were able to land the balloon that night on a glacier high in the Austrian Alps. With the help of twenty soldiers and twenty local peasants, they carried it off the mountain the next day.

This spectacular ascent shattered the previous record for altitude, but Piccard was not happy; he had not had a chance to observe cosmic rays. With his twin brother, an aeronautical engineer, Piccard ascended again the following summer, to 55,800 feet. On that calmer voyage he completed his cosmic ray studies while pioneering technology that would eventually allow humans to travel into space inside pressurized capsules. Years later Piccard's son, Jacques, would write that these daring trips to the stratosphere also influenced his father's ideas about descending into the ocean. The bathyscaph was simply a balloon in reverse.

In his father's mind, however, the sequence ran the other way. His idea for building a blimplike diving craft had actually come first, and it eventually inspired his trips into the stratosphere.

As a young college student in Switzerland, Auguste Piccard had read about a German oceanographic expedition aboard a ship named *Valdivia.* Biologists on that cruise had brought up strange, glowing fishes, only to watch them die in the trawling nets. In order to view them alive in their natural habitat, scientists would need a new kind of submarine. Piccard thought about the problem and decided that a massively thick cabin, buoyed by a gigantic float, would do the job.

This dream churned within his brilliant mind long after he left college. Piccard later explained his chain of ideas in a book titled *Earth, Sky, and Sea:* "Far from having come to the idea of a submarine device by transforming the . . . stratospheric balloon, as everyone thinks, it was, on the contrary, my original conception of a bathyscaph which gave me the method of exploring the high altitudes." He realized that his hypothetical submarine cabin, meant to withstand hundreds of atmospheres of external pressure, could be made much lighter to withstand less than one atmosphere of internal pressure and suspended from a balloon. "In short, it was a submarine which led me to the stratosphere."

The Bathyscaph *FNRS-2*

In 1933, soon after his remarkable balloon flights, Piccard traveled to the World's Fair in Chicago, where his stratospheric gondola was put on display next to William Beebe's bathysphere. While he was there he met Beebe. It must have been quite a meeting between the record-holders for altitude and depth. Piccard would undoubtedly have absorbed Beebe's enthusiasm for deep-sea exploration, and he would have studied Barton's sphere very closely. One problem that has hampered explorers of all eras, however, is lack of funding, and in this respect Auguste Piccard was no different. His deep-sea balloon would remain a thought experiment for the moment.

Finally, after years of planning and waiting, the celebrated ballooning physicist forced his own hand in a chance encounter. At a reception Piccard was attending one day in 1937, King Leopold III of Belgium asked him about his work. The king and everyone in his entourage expected a discussion about stratospheric research. Instead, Piccard told them he was planning to build a bathyscaph, "for diving to the very bottom of the sea." Spellbound, the king pumped the scientist for more and more information. The next day Piccard assembled his assistants in his laboratory. "I told the king yesterday," he said, "that we are going to build a bathyscaph. We have no choice now but to do it."

Piccard next approached the Fonds National de la Recherche Scientifique (FNRS; the Belgian National Fund for Scientific Research) with a proposal to build a diving craft, to be named *FNRS-2*. (His stratospheric balloon had been *FNRS-1*.) Knowing that experts in a group sometimes impress one another by shooting down unusual ideas, Piccard relied on a divide-and-conquer strategy. He believed that the supporters he needed, when approached as individuals, would stick up for their own areas of expertise. First he asked scientists if they would be interested in a chance to observe "abyssal animals." Certainly they would. Then he asked engineers if someone could build a successful diving vessel along the lines he had suggested. Yes, they believed it might be possible. Finally he approached the FNRS directors. Were they capable of raising money, Piccard asked, for a project that Belgian scientists and engineers overwhelmingly supported? Yes, of course they could fund it.

At last, a new kind of diving machine was about to begin its journey from dream to reality. Piccard designed *FNRS-2* and tested the underlying concepts as well as he could in his laboratory. But no sooner had he ordered materials and begun construction than the outbreak of World War II, in 1939, put the project into suspended animation. No further progress could be made until the war ended.

Six years passed, the most horrible of the century for Europe. By 1945 the war was over. By 1948 the nations of Western Europe were beginning to get back on their feet—and the bathyscaph project was back on track.

Finally, on November 3, 1948, at 4:00 in the afternoon, a boatswain on board the support ship *Scaldis* chopped a line with an ax, and Piccard's bathyscaph plunged into the deep Atlantic. After far too many maddening, last-minute delays, the timer sealed inside *FNRS-2* had ticked down to just forty minutes. At 4:40 P.M., weights would drop and the buoyant craft would return to the surface. Was there still enough time for the bathyscaph to reach its target depth of at least 4,000 feet? As everyone waited and watched, the blimpish shape suddenly reappeared on the surface, a reassuring sight indeed—but much too soon. The time was only 4:29. The bathyscaph should still have been underwater, still going down. What had happened? How deep had it actually gone?

The answer was sealed in the pressure sphere, where an instrument had been set to record the depth of descent. Yet problems continued. A hose designed to remove the bathyscaph's huge load of gasoline could not be attached, and there was no way the crew of *Scaldis* could lift the fifteen-ton vessel back aboard when it was fully loaded. Its 6,600 gallons of gasoline—buoyant underwater but heavy above—first had to be pumped into holding tanks. But the sea was too rough and the hose too heavy; it refused to attach.

Piccard consulted the captain of *Scaldis;* they decided to try towing *FNRS-2* back to port. For several hours, Piccard watched his precious bathyscaph struggle through pounding waves. It seemed about to break up at any minute. When he could no longer stand the suspense, he ordered its gasoline dumped into the ocean and the bathyscaph hoisted aboard.

At long last, with *FNRS-2* safely on deck, Piccard climbed into the pressure sphere. Moments later he emerged with a large smile on his face. His bathyscaph had reached a depth of 4,500 feet. Since *FNRS-2* was capable of reaching far greater depths, it was clear to Piccard, to his son Jacques, and to others who had come along to observe that a new era in deep-sea exploration had begun.

Piccard's sense of triumph, however, proved fleeting. With no more gasoline to fill the flotation tank, no more dives could take place. Moreover, Piccard had exhausted his meager funds; when *Scaldis* returned to port he could not make the extensive repairs his bathyscaph now needed, tank up with gasoline, and go back out to sea. The first manned descent would have to wait. As a result, the dazzling series of dives that everyone had expected, culminating in a record-shattering human descent, ended instead with a single unmanned test, followed by substantial damage as *FNRS-2* tossed about in heavy seas.

The media treated this historic demonstration harshly. Unlike Beebe after his first dives off Bermuda, Piccard returned to shore not to headlines trumpeting success but to stories of failure. Almost immediately the Fonds National began to distance itself from his project. Soon the Belgian government would pull out of funding undersea exploration altogether. The following year, Otis Barton's solo plunge to the very same depth in his benthoscope would seem, by comparison, a rousing success.

Auguste and Jacques Piccard had just learned a valuable lesson, one they would remember for years to come. Bad publicity can be far more lethal to fulfilling a dream than any technical obstacle. Physics is predictable; the media are not.

—— The Birth of the Bathyscaph *Trieste*

Yet Piccard's accomplishment did not go unnoticed. Oceanographic experts knew that the dive had been a true success. In particular, French naval officers who had observed the test stood up for Piccard and began to argue the bathyscaph's merits. One of them, a young captain named Jacques Cousteau, had himself won respect for co-inventing a new diving technique during World War II. His scuba technology (later commercialized as the Aqua-Lung) allowed navy frogmen to breathe pressurized air from tanks strapped onto their backs. Cousteau had liberated divers from tethers and hoses; he grasped right away the advantage of Piccard's untethered bathyscaph.

Nevertheless, certain problems remained. Although the idea of a balloon-style diving craft now seemed feasible, Piccard needed to make *FNRS-2* more seaworthy. The pressure capsule had performed fairly well underwater, but the blimp-like flotation tank—which made up most of its bulk—proved to be vulnerable on the surface. His bathyscaph would need a much stronger flotation tank to withstand prolonged towing in rough seas. Such an overhaul would clearly be expensive.

The French seemed ready, at first, to take the place of the Belgians. They strongly supported rebuilding, but they balked at having the effort supervised by a foreign physicist. They wanted the project transferred to the port of Toulon and placed under French naval control. Piccard viewed that prospect as tantamount to piracy; he knew he could never function as an underling to French bureaucrats. After much frustration and countless, ultimately fruitless, discussions, Piccard shipped the bathyscaph to his home in Switzerland. The French then arranged to buy out the Belgians. They would attach a new flotation tank to the pressure sphere from *FNRS-2,* and the new bathyscaph, christened *FNRS-3,* would be transferred to the French navy after its first three dives. Piccard was heartbroken, but he continued to advise the French as they started their own program.

Meanwhile, Jacques Piccard was finishing school in Trieste, Italy. There, during the winter of 1951–52, he met a Professor de Henriquez, who directed the city's War and History Museum. Proud of his native Trieste and impressed by the bathyscaph idea, the professor offered to help the Piccards raise funds to build a new vessel, as long as it bore the name *Trieste.* Before long the Piccards had patched together a thin budget from scattered sources, including Italian and Swiss research foundations, manufacturers, the city of Trieste and several Swiss cantons, a few schools, and private citizens.

By 1953 *Trieste*'s pressure-resistant sphere was taking shape at the vast Terni mills north of Rome. Made of a hard, forged steel alloy—stronger than the cast steel used in *FNRS-3*—it measured seven feet two inches in diameter, weighed ten tons in air, and had walls three and a half inches thick. This magnificent ball seemed a "real jewel" to Jacques Piccard, but its

The sphere of the bathyscaph *Trieste* is inspected following a dive. (Source: U.S. Navy/National Geographic Society Image Collection.)

windows, he wrote in a book about *Trieste,* were "perhaps its finest feature." Submarine windows made of fused quartz, which is brittle, had always been worrisome; a barely perceptible scratch could cause them to shatter. Piccard instead chose six-inch-thick Plexiglas, a type of plastic that would deform under pressure. Each Plexiglas window plugged its viewport like a transparent cork—sixteen inches in diameter on the outside end, tapering down to four inches inside. As pressure increased, Piccard correctly reasoned, these pliant, cone-shaped windows would wedge more tightly into place.

Although more spacious than Beebe's cramped quarters, *Trieste*'s crew capsule would not be especially comfortable. It could accommodate two persons, but Piccard jammed half the interior full of equipment, including an underwater (acoustic) telephone, canisters of soda lime for absorbing carbon dioxide, two oxygen bottles, silica gel for removing moisture, and gas masks in case of a fire. He also placed silver-zinc batteries inside to power the bathyscaph's electrical systems. In one small compartment, Piccard's chronometer brought back memories of the earlier ballooning years.

The rest of *Trieste* looked simple but elegant. Its giant steel flotation tank, fifty feet long, could survive a towing in thirty-foot seas. It held 22,000 gallons of gasoline. An entrance tunnel passing through it from top to bottom allowed access to the cabin attached below. After the divers entered, this tunnel could be flooded with seawater to prevent its collapsing under pressure—the idea being never to leave a hollow, crushable space outside the crew's well-fortified chamber. Above the entrance tunnel, a small cubicle that acted like a conning tower kept waves from breaking over the opening while the craft was floating on the surface.

At both ends of *Trieste*'s large flotation unit, air-filled tanks added buoyancy. To descend, the pilot flooded those tanks. Then, to fine-tune the vessel's buoyancy, he could release up to 1.4 tons of gasoline from a central storage compartment. (As the gasoline flowed out, seawater entered from below.) To float back up to the surface, the pilot released iron pellets from two large ballast hoppers on either side of the cabin. Electromagnets held nine tons of the BB-like shot inside the hoppers, fusing the pellets temporarily into one large, immovable mass. When the pilot turned off the current, the pellets would regain their character as individual BBs and begin to trickle from their two containers, like sand through an hourglass.

The need to conserve battery power placed severe limits on all the bathyscaph's operating systems. Small mercury-vapor and incandescent lamps cast barely adequate beams of light through the dark abyss. Electric motors bathed in oil drove propellers for maneuvering at depth. Each motor delivered a miserly two horsepower, giving the bulky *Trieste* a maximum horizontal speed of 1 knot. The bathyscaph, often praised for

Trieste in Apua Harbor, Guam, following Piccard and Walsh's record dive. (Source: Thomas Abercrombie/National Geographic Society Image Collection.)

its untethered freedom to maneuver, was actually more an elevator than a submarine.

The Race Begins

The simultaneous creation of two bathyscaphs, one French and the other Swiss/Italian, couldn't help but touch off a race to the bottom. Although neither group said so publicly, each had the same ultimate goal: to go as deep as possible, and to do it before the other group. Soon both teams began testing their new vessels near their home ports of Toulon, in southern France, and Castellammare, on the Bay of Naples.

The contest heated up late in the summer of 1953, when both bathyscaphs were ready to dive deep. First to break Barton's benthoscope record of 4,500 feet were two French naval officers, Lieutenant-Commander Georges Houot and marine Engineer-Lieutenant Pierre Henri Willm, who in mid-August descended in *FNRS-3* off Toulon. All went well at first, as Houot wrote later in his book *2000 Fathoms Down*. They turned on the searchlight whenever the urge struck them. In its beam, Willm spotted "strange little vertical fish," sinking at the same speed as the bathyscaph, perhaps attracted to its light. "Beebe has drawn them in his book," Willm remarked. They chatted and joked their way down to 6,930 feet—some 300 feet above the floor of the Mediterranean—at which point the echo sounder failed. "Try to stop the machine while I fix the sounder," Houot told Willm. He then proceeded to blow a fuse with the tip of his screwdriver, and no amount of swearing after that would help with the sounder. Thinking that their craft might be damaged if it struck bottom too hard, Willm and Houot cut the dive short. They returned to the surface without having reached, or even sighted, the seafloor.

Next came Auguste and Jacques Piccard, who had launched their gleaming new *Trieste* on August 1. After testing it in shallow water, they scheduled a deep plunge for late September. Auguste Piccard, nearing seventy, had planned this event as a showpiece: the last of *Trieste*'s inaugural test runs, as well as the final dive of his career. On September 30, *Trieste* arrived at the designated site, off the island of Ponza, and promptly reached the bottom of the Tyrrhenian basin. Its sixty-three-minute descent ended with a slight jar as the bathyscaph plopped down on soft ooze. As Jacques Piccard noted in his book, they landed a bit too fast, becoming "slightly buried." In contrast to the French divers, the Piccards could hardly fail to see the bottom. "We desperately needed an echo sounder," Jacques wrote, "as soon as we could afford one." Nevertheless, they set a new record: 10,390 feet.

The ability of both bathyscaphs to go deeper would soon reach its first limit—which was simply the depth of the Mediterranean. Eager to press beyond that limit, the French began eyeing the crystal-clear waters of the Atlantic. In December they shipped their bathyscaph to the same berth in Dakar where Piccard had prepared *FNRS-2* for deep-water testing. Early in January 1954 they began towing their bathyscaph to the same area, near the Cape Verde Islands, where the earlier craft had met its demise.

The French group, like Piccard in 1948, decided to start with an unmanned test dive. In many respects it proved as challenging as Piccard's earlier test, demonstrating that the French navy had not solved all the problems of handling a bathyscaph in angry seas. After three days of towing, waves were running at a height of twelve feet when the tow cable parted, casting *FNRS-3* adrift in the night. What followed was a nightmare, as the crew on board the support ship *Elie-Monnier* struggled with a giant metallic balloon bobbing helplessly on the waves. Fortunately, they managed to reattach the tow cable. The hammering waves had caused minimal damage—as far as anyone could tell—and the bathyscaph descended several days later to a depth of 13,450 feet. Another unmanned test the next day proved equally successful, clearing the way for the first human attempt to reach the average depth of the oceans.

Early on the morning of February 15 the French pilot and engineer, Houot and Willm, crawled into a tiny dinghy for their trip to the waiting bathyscaph. After several hours of delay, which included a tricky repair to a ballast tank that was leaking iron pellets, the bathyscaph slipped quickly beneath the waves.

By 11:00 in the morning, *FNRS-3* had reached a depth of 3,280 feet when oil began dripping from a pressure gauge. Houot and Willm calmly took turns tightening a nut until the dripping slowed to a tolerable "oil shampoo" misting down on whoever had to kneel at the spot underneath it. The dive continued.

At 12:45 P.M. the depth gauge read 11,500 feet. Houot turned on the echo sounder, and ten minutes later it revealed the bottom coming up to meet them, quickly. It was going to be a vast flat plain, according to soundings made earlier from the surface. Slowly, Houot began to release ballast, arresting the bathyscaph's free fall. At less than fifty feet above bottom, a white sandy seafloor came into view. It looked rippled, and indeed a slight current was running, evidenced by the gentle swaying of sea anemones, which Houot referred to as "tulips of crystal." A moment later Willm reported seeing a shark about six and a half feet long swimming in and out of their artificial circle of light. They had reached a depth of 13,287 feet.

Less than forty minutes after the bathyscaph reached bottom, it was shaken by a tremor, followed by another. The exterior lights went out, and the eerily lit landscape disappeared into utter blackness. Willm sensed that a fuse had blown, and he quickly figured out what had caused the tremors: the electromagnets had cut out, and two heavy batteries had crashed to the seafloor. They were rigged to drop off in case of electrical failure, as a safety measure to lighten the bathyscaph and hasten its ascent. ("The ship is *too* safe," Houot had once joked.)

The bathyscaph now demonstrated, beyond all doubt, the chief advantage of positive buoyancy. Having accidentally shed some 2,500 pounds, it instantly shot back toward the surface, arriving there just seventy-five minutes later. During their rapid, gratifying ascent, the two passengers ate a tasty lunch and polished off a bottle of wine. All in all, little had been accomplished, but the mere fact that two men had come back alive from the bottom of the Atlantic Ocean marked the French dive as a success.

Trieste, meanwhile, remained in the Mediterranean. Its dives were not "newsworthy," Jacques Piccard admitted, but he insisted they were no less important. "Establishing records for records' sake was never our aim," he asserted. As the French pursued depth and glory, the Piccards, short of funds as always, conducted shallower research dives for scientific institutions. The following year, again short of money, they dry-docked the bathyscaph. In 1956 they resumed operations, taking scientists as far down as 12,110 feet to study the geology of the seafloor and the creatures that lived there.

The new era in deep-sea exploration, which had started so awkwardly in 1948, now seemed clearly under way. Soon the seas would begin to reveal their secrets. Going down to the depths first made accessible by bathyscaphs, scientists would find the ocean floor much more rugged, alive, and surprisingly strange than their colleagues confined to the surface had thought. But in the late 1950s, the remotest parts of the deep abyss still remained out of reach. Although the French bathyscaph crew and the Piccards had both descended more than two miles, a good deal of the seafloor sank lower than that, with canyons and trenches plunging far, far deeper. The deepest crevasse on earth, in fact, descended nearly seven miles.

The U.S. Navy Joins the Race

In 1956 Jacques Piccard spent a hundred days in the United States, traveling to the major centers of oceanographic research and trying to sell them on the virtues of the bathyscaph. To his pleasant surprise, he found a small contingent of would-be explorers eager to enter the world that only he and a few others had seen. His visit came to an optimistic end at a meeting of the National Academy of Sciences in Washington, D.C., when Piccard and Robert Dietz, a highly respected marine geologist, presented papers on the bathyscaph's potential value for deep-sea research. Their papers would be prophetic. Later in his career, Dietz became one of the founders of a revolution in the earth sciences known as plate tectonics—a geological theory that deep-sea exploration would help to verify.

Piccard also won the support of Willard Bascom, a respected ocean engineer from the Scripps Institution of Oceanography in La Jolla, California. Bascom spearheaded an effort to unite ocean researchers behind a written document, presented at the same meeting. Allyn Vine, an equally respected geophysicist at the Woods Hole Oceanographic Institution in Massachusetts, helped to draft the resolution, which stated:

> The careful design and repeated testing of the bathyscaph have clearly demonstrated the technical feasibility of operating manned vehicles safely at great depths in the ocean.
>
> The scientific implications . . . are far reaching.
>
> We, as individuals interested in the scientific exploration of the deep sea, wish to go on record as favoring the immediate initiation of a national program, aimed at obtaining for the United States undersea vehicles capable of transporting men and their instruments to the great depths of the oceans.

A scientist spoke up from the back of the room: "Is it intended that this be a number of vehicles?" Another voice: "If we vote for this, does this commit us in any way to go down in it?" In the end, all voices voted in favor, none opposed.

The Office of Naval Research (ONR) responded quickly. It drew up a contract, in February 1957, directing the Piccards to conduct a series of dives with their bathyscaph in the Tyrrhenian Sea off Naples, "so that American scientists could carefully evaluate its potential."

From July to October of 1957—the same year the Soviet Union launched *Sputnik*, the world's first orbiting satellite—the bathyscaph *Trieste* made twenty-six dives carrying biologists, geologists, physicists, underwater acoustics experts, assorted VIPs, and other navy personnel down to varying depths in the Mediterranean. These dives proved a sparkling success.

They also occurred at an opportune time. Two nations had emerged from the ruins of World War II as rival superpowers: the United States and the Soviet Union. In the 1950s, as they began their race to dominate the "high frontier" of space, a similar contest seemed about to pit them against one another in the oceans. The ONR believed that the bathyscaph held promise for opening up this deep frontier. It seemed capable of advancing not only basic scientific research but also studies that could be applied to military objectives, such as improved underwater surveillance. Wanting more than ever to use the bathyscaph, the navy dangled an offer that the Piccards found attractive. There was a small wrinkle, however—one the Piccards had heard before. The navy didn't want to lease *Trieste* and operate it out of Italy; they wanted to buy it and move it to the United States.

After much negotiating, the Piccards agreed, with one important stipulation: Jacques Piccard would have the option to go down in *Trieste* on "dives presenting special problems."

The question now was where *Trieste* would be based. At the time, the Naval Electronics Laboratory in San Diego seemed a logical choice. It was a hub of activity for naval research, with the largest U.S. oceanographic institute, Scripps, just a short distance away. But before any meaningful program could get under way, a great deal of work had to be done to prepare the navy facilities in San Diego for *Trieste*'s arrival. Finally, the bathyscaph arrived amid what seemed to Jacques Piccard a "flood of publicity." From December 1958 to September 1959 several test dives were conducted, first in the harbor and then in deeper waters off San Diego.

As *Trieste* went through its tests, one thought constantly lurked in everyone's mind. Although it was never openly acknowledged, a provocative lure waited far out in the Pacific Ocean, two hundred miles off the island of Guam. It was Challenger Deep, the deepest spot in the world, some 35,800 feet below the ocean surface. Since its discovery in 1949 by the British research vessel HMS *Challenger II*, a descent into Challenger Deep had become the ultimate goal within the small and elite fraternity of deep-sea explorers. This was the Mount Everest of the ocean, a hole to be "scaled" in reverse—from its broad opening in the dark depths of the Mariana Trench to its narrower base, thrusting down from the deepest part of that trench.

If *Trieste* did not capture the ultimate prize, it seemed likely that some other bathyscaph would. Beginning as early as 1957, oceanographers had seen news reports about French plans for a "super bathyscaph" that would be able to descend to at least 35,000 feet. The Soviets, too, might become contenders, if they accelerated the development of their deep-ocean capabilities, as was being rumored. "We all knew what we were shooting for," Piccard wrote. "That late summer in 1959, one could sense the increasing tension in the air."

Quietly, so as not to arouse the press too soon, Jacques Piccard went about supervising the construction of a new pressure sphere for *Trieste.* The Italians at Terni had declined to bid on his proposal, so Piccard approached the Krupp steelworks in Essen, Germany—a renowned manufacturer of military armaments. Krupp had sent its heaviest steel-making equipment to Yugoslavia as war reparations, but its engineers claimed they could nevertheless fashion a sphere in three pieces and join them together with no loss of strength. This new pressure capsule, if it performed as planned, would make *Trieste* capable of reaching the bottom of Challenger Deep. Weighing thirteen tons in air and eight tons in water, its hull would be five inches thick, swelling to seven inches at the viewports. All this steel—and the Plexiglas in its smaller-than-usual windows—would need to hold up under nine tons of pressure per square inch, three times more than the French bathyscaph had endured at the bottom of the Atlantic Ocean.

After much preparation, including construction of a larger flotation tank holding an extra 6,000 gallons of gasoline, *Trieste* departed San Diego on October 5, 1959, loaded aboard the USS *Santa Mariana.* From November 1959 to early January 1960, several preparatory dives off Guam tested the capabilities of craft and crew. The most ambitious of these dives reached the bottom of a hole called Nero Deep, 23,000 feet down. As congratulations arrived on the setting of a new record, the navy teased reporters that an even bigger story would soon be breaking. Everyone now knew the ultimate goal.

The "Big Dive" was set for January 23, 1960. When news of the schedule reached the diving group, tensions mounted still higher—not because of the date but because of the singular noun: *one* dive. Expedition plans had originally called for three dives into Challenger Deep. This would have given each of the five principal persons associated with *Trieste*'s diving program an opportunity to make history: Lieutenant Don Walsh, naval officer in charge; Andreas Rechnitzer, head of the Deep Submergence Program at the Naval Electronics Laboratory and scientist in charge; the geologist Robert Dietz; Kenneth Mackenzie, a physicist; and Jacques Piccard, *Trieste*'s most experienced pilot. Five men had been primed to go. Now only two would get the chance.

When Piccard found out, a few days before the dive, that Walsh and Rechnitzer had been selected, he was stunned. But he was not so overwhelmed that he forgot his contract with the navy. Most especially, he recalled the clause giving him the option to dive on any mission presenting "special problems." After a great deal of communication back and forth between expedition members and navy brass in Washington, a final decision came back. Walsh and Piccard would take the plunge.

The Big Dive

Blue water crashed over the bow of the USS *Wandank* as it pulled *Trieste,* riding six hundred feet astern. The two-hundred-mile tow from Guam was a rough one. When the towboat and bathyscaph arrived at the dive site on the evening of January 22, a destroyer escort was busy dropping more than eight hundred TNT charges in an attempt to find the deepest hole. Finally, early the next morning, echoes rebounding from the distant bottom revealed the basement of the world, at an estimated depth of 35,700 feet. Flares and bright-colored dye marked the spot as twenty-five-foot seas hammered a small rubber boat struggling to reach *Trieste.*

Piccard and Walsh both managed somehow to board the bathyscaph. What they saw on deck was not reassuring. Pounding waves had washed away part of a surface telephone during the tow, making it impossible to talk to the support crew outside *Trieste* once the two men were sealed inside. A tachometer intended to measure the speed of descent had been damaged beyond use, and another instrument for measuring vertical currents hung by its wires.

It was 8:00 in the morning. To make the fourteen-mile round trip into Challenger Deep and return before dark, Walsh and Piccard had to begin their dive within the hour. If they surfaced at night, their support crew might not find them. Postponing the Big Dive or starting right away, without making repairs, were the only choices. Piccard and Walsh began their descent at 8:23.

The bathyscaph now demonstrated, beyond all doubt, one of the chief disadvantages of positive buoyancy. The trouble starts at the ocean's surface, constantly disturbed by wind and waves. This wave action has little effect on a submerged vessel, but it thoroughly mixes the upper layer of the sea, giving the water a fairly uniform temperature. Below that layer, one encounters colder water. The boundary between the layers, called a *thermocline,* separates the warm and less dense upper layer from colder, denser water below. On this particular day the thermocline was well developed, and Lieutenant Walsh was about to learn some physics. At a depth of 340 feet, *Trieste* literally bounced off the cold thermocline, bringing its descent to a halt. Unable to wait until the gasoline cooled, Piccard jettisoned some of it, which meant that his vessel now held less buoyancy in reserve—the margin of safety was slimmer. The bathyscaph continued to fall. It had to fight its way down through similar thermal barriers at 370, 420, and 515 feet, losing valuable time and precious gasoline.

Finally *Trieste* reached its terminal velocity of three feet per second, about as fast as an elevator falls. Piccard now turned his mind toward other worries. He began to think about the narrow slot they were aiming for, less than a mile wide. What about lateral currents? And what about the unknown terrain inside Challenger Deep—would they smash against steep canyon walls? He knew that such fears, common to all explorers as they push beyond the limits of the known, would have to be held in check.

By 11:30 the divers had reached 27,000 feet. More than 99 percent of the water in the world's oceans was above them. They continued to fall. At 32,400 feet a strong explosion rocked the bathyscaph. Walsh and Piccard exchanged wide-eyed glances, waiting for something to happen. Nothing did.

They continued to fall, keenly aware that something outside their cabin had broken. A new mystery soon overshadowed that disturbing thought. The depth gauge now read 36,000 feet, putting them some 200 to 300 feet beyond the lowest point on earth, according to previous soundings. Yet they still saw nothing on their depth sounder. Was the bottom so soft that the sounder could not pick it up—a nebulous ooze that would engulf and trap them? With more than two days' worth of oxygen, suffocating was not an immediate concern, but freezing to death was. The abyssal cold had penetrated the bathyscaph's heavy Krupp steel, chilling everything. Tension rose inside the tiny capsule, which now seemed more like a spherical coffin dripping with water.

Finally, at 12:56 P.M., a black echo appeared on the graph paper scrolling through the sounder: forty-two fathoms (252 feet) to go. Piccard began releasing steel shot to slow the descent. Suddenly, at three fathoms, a light-colored ooze came into view. The depth gauge now read 37,800 feet. (Later the mystery was solved. Piccard and Walsh learned that the gauge had been calibrated in Switzerland, in fresh water. Their true depth was 35,800 feet.)

Piccard peered out the forward viewport as *Trieste* fell its final fathom. There, lying on the bottom, he saw a solelike flatfish with two round eyes. What would those eyes be good for, he wondered, in such utter darkness? As the bathyscaph gently touched down, the fish rose up slowly and began to swim away. Piccard felt surprisingly happy. Life exists, he realized, even at the deepest spot on earth. He looked over at Walsh, and the two divers shook hands. They had made it!

Lieutenant Walsh, looking out the other viewport, which angled upward, then discovered the cause of the explosion on the way down. The large plastic window in their antechamber, just outside the pressure capsule, had cracked.

More fears to control! Piccard analyzed them as calmly as he could. The antechamber, being flooded, would not collapse. It separated their pressure capsule from the entrance tunnel above it, which was also flooded. No immediate danger there. But if the cracked window could no longer hold its seal, Piccard and Walsh, once they returned to the surface, would find it impossible to pump seawater from the space above their closed hatch. They would be trapped inside their steel fortress, several feet below the ocean's surface. Would support crews be able to get them out? Probably not. Piccard considered the next possibility—a harrowing tow back to Guam while sealed inside—and then caught himself. Those events, if they did become real, were still hours away. His thoughts returned to the chores at hand.

The fish still intrigued Piccard. The more he saw of their landing site, the more he wondered about the fish. Why had it descended so incredibly far, to such a desolate spot? Unlike the Mediterranean seafloor, no mounds or burrows adorned this ooze. There were a few marks—perhaps animal tracks. But to survive here, Piccard reasoned, animals would need a current capable of bringing them oxygen and nutrients. He checked the meter showing horizontal current: it read zero. If there was a vertical current, he would never know; that instrument had been smashed by the waves.

It was time to go. After only twenty minutes on the bottom, Piccard dropped more ballast to begin the slow ascent, still feeling anxious about surfacing before dark. Gradually the bathyscaph's speed increased, rising to five feet per second, but it then slowed as *Trieste* began to break through the thermoclines. After many hours at near-freezing depths, the gasoline in *Trieste*'s float was now colder than the water it was trying to enter—the reverse of the situation going down. But

Piccard and Walsh wave from *Trieste* after surfacing from their historic deep dive, January 23, 1960. (Source: Thomas Abercrombie/National Geographic Society Image Collection.)

the direction of travel had also reversed, so the thermoclines remained a barrier.

Walsh and Piccard finally reached the surface at 4:56 P.M. Their luck had held, and so had the plastic window. Would it hold up now to a pressure differential as they pumped out water? Slowly, they cleared their escape route. Carefully, they opened the hatch. As they clambered up a ladder to fresh air and sunlight, two navy jets zoomed overhead, dipping their wings in salute.

The race to the bottom was over. True scientific exploration could finally begin.

CHAPTER

3

THE TRAGIC DAWN OF THE MODERN DEEP SUBMERSIBLE

One should not attack the sea. You need to make love to it.

—*Jacques Cousteau*

As a boy growing up in Pacific Beach, a suburb of San Diego, I was glad when the bathyscaph *Trieste* came to town. My friends admired other kinds of ships, especially the huge aircraft carriers that headed out toward the broad horizon from navy bases nearby. I enjoyed watching them too, but what appealed to me more was the ocean itself, a mysterious depth beyond the grasp of their keels. I hoped someday to dive beneath the surface and explore that undersea world. Submarines fascinated me, and *Trieste* was no ordinary sub. From the day it arrived in 1958, the local newspaper treated every outing as a major event. Even a test dive in the harbor, seventy feet down, rated headlines.

I was a senior in high school when *Trieste* fulfilled its superb potential, making the ultimate plunge into Challenger Deep. With no part of the abyss too remote to be reached, the kind of career I had dreamed of when younger now actually seemed feasible. I could imagine a new phase of history beginning on the deep frontier: scientists and other explorers pushing on through eternal darkness, establishing human outposts, perhaps even working and living underwater. I wanted to be part of it.

Yet history would not follow my imagined script. For one thing, the idea of working and living undersea in any great numbers turned out to be impractical, and in the long run unnecessary. For another, the very ship that inspired my dreams turned out to be a technological dead end, like the bathysphere before it.

In retrospect, Lieutenant Walsh and Jacques Piccard seem lucky to have survived their trip into Challenger Deep. They certainly returned in better shape than the bathyscaph. On the ocean's surface, heavy seas had ripped instruments apart; down below, a window had cracked. The toll could have been far worse. After the Big Dive, experts reanalyzed the mighty Krupp sphere and decided that it wasn't safe. The navy retired it in 1961 and never attempted to build a better one. The deepest spot on earth then receded from human contact. Although a Japanese robot visited Challenger Deep many years later, no living person has ever been back.

Shallower destinations, meanwhile, remained within reach, and bathyscaphs continued diving. There were never many of them. Only two, at most, were ever in service worldwide at any one time: one American and one French. With the race to the bottom finished, both nations used their bathyscaphs to embark on long and illustrious research programs. Although a few deep trenches remained off limits, bathyscaphs could still descend at least 20,000 feet after 1960—far enough to explore 98 percent of the world's seafloor. The scientists who had originally clamored for access still wanted to go there, and some of them did. Yet as bathyscaphs made more and more dives, two great limitations became obvious: their huge size and their lack of horizontal mobility.

The lack of mobility occasionally became dangerous. Submarine canyons, which turned out to be popular research sites, proved especially challenging for bathyscaphs to navigate. The absence of good topographical data made such dives all

Trieste II being taken aboard its mother ship after a dive by the author. (Source: Claude Petrone/National Geographic Society Image Collection.)

the more risky. Both *FNRS-3* and *Trieste* frequently collided with steep canyon walls, touching off landslides that went roaring downslope, kicking up sediments and reducing visibility to zero. I witnessed one such collision firsthand on board a second-generation bathyscaph, *Trieste II,* during a 1977 descent into the Cayman Trench in the Caribbean Sea.

As we neared bottom, our bathyscaph drifted too close to a steep volcanic slope. The pilot quickly dropped ballast, but the huge ship had too much momentum; its front end crashed into the slope and continued to scrape its way down the rocks. I watched the scene outside my viewport as metal slowly crumpled; then I saw shimmering bubbles. "Gasoline," I cried out, "in the water!" That was our buoyancy bleeding away—the only force that could bring us back up. Without hesitation, the pilot and copilot dropped every remaining ounce of ballast, including emergency weights. As the bathyscaph began to rise, we kept our eyes glued to the ascent meter. If more gasoline had escaped, we might never have resurfaced.

For many years, researchers accepted such risks. As clumsy as the bathyscaphs were, they at least took scientists deep under water. But on the surface, out of their element, the very features that allowed bathyscaphs to plunge deep worked against them. Their heavy pressure spheres and huge flotation tanks made them too massive to carry aboard support ships. Instead they had to be towed, slowly, to a diving site. That sometimes meant towing in rough waters, where the wallowing, blimplike vessels became vulnerable to pounding waves. They were also difficult to maintain while partly submerged in the open sea, many miles from a base. In effect, bathyscaphs also limited the mobility of their support ships.

If deep-diving submarines were ever to have true global coverage, they would need to be small enough to be carried on board surface ships. The support ship could then travel more quickly to the diving site, and between dives its crew could haul the sub back aboard for maintenance and repairs. Yet bathyscaphs were large for good reason: their heavy crew capsules required huge flotation tanks for buoyancy. A lighter capsule on a smaller, more mobile craft would not be able to descend very far before being crushed.

Technology had reached an impasse: deep-sea vessels, it seemed, could opt for depth or mobility, but not both. Fortunately, several ideas for resolving this conflict were already taking shape even before *Trieste* reached the bottom of Challenger Deep.

— A Diving Saucer

Jacques Cousteau acted first, using his own ideas to design a radically new kind of diving machine. Cousteau had received his introduction to deep-sea diving in 1948, when, as a young French naval officer, he participated in the much-maligned test of *FNRS-2* off Dakar. He could see from the deep descent and rapid return of Auguste Piccard's bizarre-looking craft that bathyscaphs held great promise, even though *FNRS-2* never completed another dive.

Jacques Piccard remembered, years later, how flamboyantly Cousteau spoke of his father's bathyscaph. "Professor," he told Auguste Piccard in 1948, "your invention is the most wonderful of this century." Evidently his admiration was genuine. Cousteau had been hoping to dive in *FNRS-2* himself, and he did dive later in *FNRS-3*, the French bathyscaph. Yet by the early 1950s Cousteau was expressing different opinions about the ideal characteristics of a deep-diving craft. He instructed his engineers accordingly. "Throw the classic idea of submarines out the window," he said, "and start with what *we* need."

What Cousteau needed became apparent one day while he was cruising the Red Sea on board *Calypso*, a World War II minesweeper he had converted into a research vessel. As he hunted for ancient shipwrecks, he was growing frustrated with scuba techniques, which confined his bottom-searching to the shallow coastal margins. In his ship's mess, as the story goes,

Jacques Cousteau with *Calypso*. (Source: Corbis/Bettman.)

Cousteau picked up two soup plates and held them together rim to rim. He wanted a sturdy diving craft like that, he said—a *soucoupe* (saucer). It should be easily maneuverable, with room for one or two persons, and light enough to carry on board *Calypso.*

After returning to his home port of Toulon, Cousteau found backers to fund a company that would build such a diving saucer. The investors hoped eventually to manufacture and sell the saucers in quantity. Cousteau's ideas for this ship seemed eminently practical, like the Aqua-Lung he had developed earlier. And his knowledge of diving seemed impeccable. Indeed, Cousteau's instructions to the young engineer who would lead the effort, Jean Mollard, reflected his frustrating experiences aboard *FNRS-3* as well as his years of scuba diving. A diving vessel should maneuver not like a huge balloon but gracefully, he felt, the way scuba divers moved—"like angels."

Cousteau thought long and hard about how to achieve this. "Put all you can of the power plant and auxiliary systems outside the hull," he told Mollard. "That's the main lesson we have from the bathyscaph." The hull would still have to accommodate the crew, of course, because humans must breathe air. But why stuff equipment, too, inside such a heavy, armored shell, forcing the crew compartment to be larger, if that equipment could be made pressure- and waterproof so as to function just as well outside?

Making the diving craft lighter did not mean making it faster. "Pay no attention to speed," Cousteau advised. "It isn't needed in an exploring submarine. We want agility, perfect trim, tight turns, and hovering ability. Let the men look out with their eyes and make them more comfortable than the awkward kneeling attitude in the bathyscaph. Put them on their bellies on a mattress. Give them a new kind of ship's log: still and movie cameras with lighting systems, a voice recorder, and a claw to pick up things outside."

The result, as it began to take shape on drawing boards, was the prototype of all modern submersibles. Research and development took several years. Engineers had to reinvent motors, pumps, and instruments to make everything smaller. Then, part by part, they had to build and test each item to make sure that it would work not only in their labs but also in the unforgiving deep environment.

Cousteau, meanwhile, focused more of his efforts on underwater photography. In 1953 he began collaborating with Harold Edgerton, a renowned inventor and electrical engineer at the Massachusetts Institute of Technology. "Doc" Edgerton had pioneered the development of deep-sea cameras and strobe lights, earning himself the nickname Papa Flash. Together he and Cousteau mounted a camera on an underwater sled, added a sonar device to trigger a flash when the camera got close to the bottom, and learned how to tow this sled back and forth, up and down, above fractured seafloor. In the summer of 1956, far out in the Atlantic Ocean, they lowered their sled four and a half miles into the Romanche Trench, snapping the first pictures that showed life at such great depth. They also took the first pictures of the volcanic terrain that makes up the Midocean Ridge.

In 1957, sensing that his career had permanently stalled, Cousteau resigned his commission in the French navy and accepted an offer to direct the Oceanographic Institute of Monaco, the oldest marine museum in the world. A new chapter in his diving life also began in that year, when his crew finally hoisted a yellow, saucer-shaped diving hull aboard *Calypso.*

At the heart of the hull was a slightly squashed (wider than tall) pressure capsule, made of three-quarter-inch steel. A fiberglass fairing surrounded it, extending the saucerlike shape. This hull sacrificed any pretense of achieving bathyscaph-like depths in favor of small size, light weight, and exquisite handling. Just nine feet long, the first *Soucoupe* would weigh, when finished, only four tons. Air inside the crew capsule gave it all the buoyancy it needed.

Early tests in shallow water near Toulon went well. With the little saucer secure on its fantail, *Calypso* then headed farther out into the Mediterranean. Cousteau's destination was the deep water off the island of Riou, a familiar diving area.

As usual with new diving vessels, the first deep descent would be unmanned. Once on station, crew members weighted the craft with heavy ballast hung on a short cable, attached a long tether to the saucer, and lowered both the saucer and its ballast over the side of *Calypso*. Slowly, the long tether spooled off its drum as the odd-looking hull sank deeper and deeper. Finally it reached 2,000 feet, well above bottom but twice the hull's designed operating depth. There the diving capsule "soaked" for a very long fifteen minutes as Cousteau and his engineers prayed it would resist the powerful pressure squeezing it like the jaws of a vise.

At last the tether reversed itself on the drum, and before long a welcome yellow shape could be seen coming up toward the surface. The rest of the operation would be routine. When the hull had arrived within fifteen feet of the surface, Albert Falco, Cousteau's constant diving companion, descended a ladder to release the ballast weights. Just at that moment a violent swell lifted *Calypso*'s stern—and the saucer with it. As *Calypso* fell downward into the trough beyond the swell, the saucer's lighter hull began to kite, gliding instead of dropping. The tether went slack. As *Calypso* rose suddenly on the crest of the next swell, the tether snapped, "like a violin string," and the saucer disappeared underwater again, this time for good.

Although the research group in Toulon had nearly completed a second hull, it seemed foolish to proceed without knowing the fate of the first, which had been outfitted with expensive strain gauges. They might have recorded precisely how the hull had failed as it plunged far below its intended test depth, and this information might help engineers improve the design. As Cousteau and his crew returned to port, a recovery operation began to take shape in their minds. They would use two minesweepers to dredge the site. They hoped they could drag a very long cable, connected to both ships on the surface, that would loop beneath whatever was left of the saucer.

The next day Cousteau and company returned to the site and began carefully surveying the area, using a precision echo sounder aboard *Calypso*. To their surprise, the outline of the whole saucer appeared on the echogram, at a depth of 3,247 feet—which was 33 feet above bottom, the exact length of its ballast cable. The pressure hull was still intact, buoyed off the bottom like a weather balloon. The hull and its viewports had survived a pressure more than three times greater than they would need to withstand at the proposed diving depth of 1,000 feet. Only a bathyscaph could go deeper than this jaunty little saucer now sitting on the bottom; no conventional submarine would have remained intact.

With this reassuring knowledge in hand, Cousteau called off the recovery in order to save funds, now needed to complete what would become the first fully functional *Soucoupe*, using the newly completed diving saucer number two. Hull number one would be left forever on the bottom of the sea.

The final version of the *Soucoupe* resembled no other vehicle on earth. It had no propellers, rudder, or planes to maneuver the hull through water; those were too clunky for Cousteau. Instead, hydrojets mounted in the saucer's bow allowed it to spin on its axis and so make extremely tight turns. For basic propulsion, a jet pump mounted on the stern ejaculated water streams through flexible plastic pipes. The pilot could clamp either pipe, diminishing the flow to make gradual turns. He could also tilt the submarine up or down by pumping mercury ballast between fore-and-aft trimming tanks. With all these features, Cousteau's strange little saucer had the ability to climb and dive at near-vertical angles as well as bank, unlike most deep submersibles to follow.

Cousteau's engineers placed power and hydraulic systems outside the pressure sphere, as he had instructed, in a flooded box covered by a fiberglass fairing. This made the hull both hydrodynamic and snag-proof. The submersible carried two 55-pound pig irons as ordinary ballast and a 450-pound emergency weight that hung beneath the hull. The two occupants fine-tuned the saucer's buoyancy using a remarkably small twelve-gallon tank situated between them; they could add seawater to the tank when the sub was light or remove it when the sub was heavy.

As opposed to the one or two viewports typical of bathyscaphs, Cousteau's submersible had ten. Three hemispheric ports in the dome allowed the occupants to see overhead; there were two more forward ports for the passenger and pilot and two for various camera systems. The other three had sonar transducers mounted in them that could transmit up, down, and forward. Cousteau also improved the interior view, making all tasks easier: the pilot lay within an encircling instrument panel. Buttons for the cameras, lights, and tape recorder also lay within easy reach.

One of those buttons triggered an Edgerton camera on the starboard bow, synchronized with a flash tube on the port side. Engineers less knowledgeable about underwater photography might have mounted camera and flash at the same place, but the floating particles in seawater would have scattered much of the light right back at the camera, creating the same effect as headlights in fog. Moving the flash to one side reduces this backscattering, and it also lengthens the shadows in a scene, bringing out the three-dimensional character of a shipwreck or seascape. Cousteau's early recognition of the need to separate the camera from its light source led to excellent underwater photographs, a hallmark of his work. A separate movie light, mounted on a retractable hydraulic boom, added versatility for filming. On the other side of the vehicle, a second hydraulic arm with an elbow and grasper enabled the pilot to pick up specimens and tuck them into a collecting basket through a spring-hinged lid.

The *Soucoupe* that boasted these features, constructed with hull number two, began diving in 1959. It performed brilliantly. During its first tests, conducted in shallow Caribbean waters off the islands of Puerto Rico and Guadeloupe, Falco "flew" the saucer like an airplane. He glided, tilted, and hovered; climbed over reefs; dove to the bottom; and sat there. The *Soucoupe* performed, Cousteau gloated, like "a loiterer, a deliberator, a taster of small scenes." Soon, however, the experimental nickel-cadmium batteries began to exhibit an annoying tendency to explode and start fires, forcing Cousteau to terminate the tests. Secured somewhat less than triumphantly aboard *Calypso,* the diving saucer returned to Toulon on Christmas Day. In the following year, Cousteau's engineering team installed standard lead-acid batteries. The *Soucoupe* was finally finished.

Over a long and productive career spanning three decades, Cousteau's yellow saucer visited continental shelves around the globe and explored some of the most impressive submarine canyons. It also tended an experimental colony of French aquanauts living 100 feet below the surface. In all, the *Soucoupe* made more than a thousand dives. But with its operating depth limited to 1,000 feet, it never quite reached the world of eternal darkness—the world into which its predecessor, the empty hull number one, had accidentally plunged during its initial deep-water test.

The task of carrying humans much deeper than 1,000 feet in mobile diving craft was left to later saucerlike submersibles such as *Deepstar 4000* and *Cyana.* Both would leave lasting marks on deep-sea exploration. Cousteau himself, the pioneer of the modern submersible, never attempted to extend the depth his *Soucoupe* could reach. Instead he turned back to the sunlit realms, which still harbored countless unexplored shipwrecks, intriguing scientific puzzles, and ample opportunities for stunning photography.

Incredible Boat-Shrinking Foam

For the same reasons that Cousteau became disenchanted with the French bathyscaph, Americans familiar with *Trieste* began dreaming of their own small submersible. Unlike Cousteau, however, none of them had the authority, individually, to act on those dreams. Instead, the early U.S. effort to build a small submersible proceeded like a fitful attempt to invent gunpowder. An unstable mixture of people, ideas, and opportunities had to combine in just the right way—and still there were misfires.

No sooner had *Trieste* completed its Big Dive than several members of the San Diego group responsible for its operation began to argue the merits of something smaller. These included Andreas Rechnitzer, a marine biologist who headed the Deep Submergence Program at the Naval Electronics Laboratory, and Don Walsh, the navy officer who had dived aboard *Trieste* into Challenger Deep. Harold "Bud" Froehlich, a General Mills engineer who had built *Trieste*'s mechanical arm, listened with interest to these discussions. Soon he was circulating designs for a small submersible he called *Seapup* to anyone who was interested.

While the *Seapup* design began to attract notice, other sparks were flying on the East Coast. Around 1960 Charles B. "Swede" Momsen, Jr., chief of undersea warfare in the Office of Naval Research (ONR), received a proposal from the Reynolds Aluminum and Metals Company to build an aluminum submarine. Although this diving vessel, to be called *Aluminaut,* would be made of a lightweight metal, Reynolds engineers had not envisioned a light submersible. Instead they proposed using aluminum by the ton to enclose a crew capsule much larger than any attached to a bathyscaph. The plans for *Aluminaut* specified a hull fifty-one feet long that would carry six persons. It would be buoyant without a separate flotation tank, but it would have to be towed out to sea, like a bathyscaph, and could not be brought aboard its mother ship for maintenance and repairs.

As a veteran submarine commander, decorated during World War II, Momsen felt comfortable with *Aluminaut*'s large size. He also believed that the navy needed something more agile than a bathyscaph to carry humans into the deep—where they might, for example, continue the studies of acoustics begun with *Trieste* or tend new equipment like underwater listening arrays. As far as Momsen was concerned, there was only one problem with the *Aluminaut* proposal: ONR was not in the business of building and owning submarines. It could, however, rent one, if he could find scientists outside the navy interested in maintaining and using it.

Momsen made his first sales call to the Scripps Institution of Oceanography in La Jolla, just north of San Diego. Surprisingly, his idea received a cold reception. Scientists at Scripps had stood by while the researchers who operated *Trieste* became well known for diving exploits that accomplished very little, in the opinion of many at Scripps. They were not enthusiastic about finding new ways to put humans underwater. Instead, oceanographers at Scripps expressed far more interest in unmanned vehicles: platforms that could tow cameras and instruments deep underwater. They wanted something that could survey large areas of the ocean floor rapidly; they saw little scientific value in manned submersibles, large or small, that poked around the bottom with their observers' faces in the mud.

The resistance Momsen encountered at Scripps was typical among physical scientists at that time, aside from a small core group of deep-sea dreamers. Even the navy, which had been so eager to acquire *Trieste*, now seemed reluctant to broaden scientists' access to the deep with smaller, more mobile submersibles. Having proved they were capable of touching the bottom, most navy brass saw no compelling military reason to stay there.

Rebuffed at Scripps, yet unshaken in his conviction that the navy needed to extend its mastery of the oceans far below the surface, Momsen approached researchers at the Woods Hole Oceanographic Institution on Cape Cod, Massachusetts. There Allyn Vine welcomed the idea of a new submersible. Vine, a geophysicist, was a special breed of genius. In 1956 he had been one of the scientists who had prodded funding agencies to build or acquire a bathyscaph. Earlier, he had pioneered the study of thermoclines—sharp boundaries between layers of warm and cool water—and he had gone on to apply this knowledge to practical submarine problems during World War II. He had shown, for instance, that underwater listening devices could detect sounds reflected from thermoclines, vastly extending their sensitivity and range. Vine now believed that the best way to understand any part of the ocean was for researchers to go there themselves with their instruments.

Paul Fye, the director of Woods Hole, agreed, and he offered his institution as *Aluminaut*'s base. Unlike many of his peers at other research institutions, Fye believed in the scientific value of sending humans to the bottom of the sea. He also realized that such missions, more glamorous than unmanned surveys, would appeal to the public and thus win recognition for Woods Hole.

Yet after Momsen provoked this initial burst of enthusiasm, discussions between Reynolds, ONR, and Woods Hole bogged down. The sticking point was the ultimate ownership of *Aluminaut*. Reynolds wanted to retain title, while ONR wanted ownership eventually transferred to Woods Hole. As time ticked on, the idle but eager engineers that Fye had hired to run the *Aluminaut* program began to question the sub's fundamental design. They had developed the same concerns as Cousteau about building and operating large-hull submersibles.

More than three years passed in this way, with everyone negotiating more or less nonstop yet still disagreeing about ownership and, increasingly, basic design. Paul Fye, having already formed his Deep Submergence Group at Woods Hole, felt that something new had to be tried, and soon. Meanwhile, in Washington, rival groups within ONR were trying to pry loose the monies Swede Momsen had been squirreling away for the project. Momsen acted quickly: he authorized Fye to cut loose from Reynolds and request bids for building its own submersible. Now that Woods Hole engineers were calling the shots, the specifications that went out to bidders were not for a sub like *Aluminaut* but rather for a much smaller vehicle, similar to Froehlich's *Seapup.*

A completely new substance, a sort of lightweight packing material, inspired confidence in the push for a smaller, less conventional design. Using this new wonder material, engineers could put thick steel pressure hulls into small submersibles, allowing them to dive deep, yet dispense with the large, gasoline-filled floats that bathyscaphs needed for buoyancy. The miracle of modern chemistry making all this possible—the key technical breakthrough—was known as *syntactic foam.*

Most people think of foam as light, soft, and crushable, full of air bubbles that easily collapse. Syntactic foam is light, but not soft and not crushable. Instead it consists of rigid bubbles—in the form of hollow glass microspheres—embedded in a hard epoxy resin. A hollow glass sphere, when microscopic in size, is capable of resisting enormous pressure. Millions upon millions of such tiny spheres, all of them significantly lighter than water, permeate the syntactic foam that fills every available nook and cranny in the hulls of modern deep-diving submersibles. Thanks to its strength and lightness, the amount of this foam necessary to compensate for a heavy crew capsule and other support systems is much smaller and lighter than a bathyscaph's gasoline float. A cubic foot of syntactic foam in the 1960s weighed only thirty-two pounds, half as much as seawater. A block the size of a large refrigerator could buoy a ton of submerged weight.

For the first time, a diving craft could be made small enough to be lifted out of the water by support crews working in the open sea, yet sturdy enough to be truly deep-diving. With such new capabilities stirring strong excitement, Woods Hole sent its request for bids to seven companies. Only two of those companies—General Mills, where Bud Froehlich worked, and North American Aviation—submitted bids to construct what would become the most productive deep submersible ever built.

During the summer of 1962, with the bidding process well under way, Andy Rechnitzer hired me to work with his team of design engineers at North American Aviation. I was a sophomore at the University of California, Santa Barbara, studying for a degree in the physical sciences. My father, Chet Ballard, was chief engineer of the Minuteman Missile program at North American. He knew Rechnitzer, hence my summer job.

In my eyes, Rechnitzer was a dazzling deep-sea explorer. He had been diving on *Trieste* since 1957, and he had almost ridden it down into Challenger Deep; Jacques Piccard had taken his place just days before the Big Dive. More recently, though, Rechnitzer had been trying to persuade the navy to build a small submersible for his Deep Submergence group at the Naval Electronics Laboratory in San Diego. Eventually he gave up and resigned from his job there. When I arrived at the site of his new job in Anaheim, California, his team had almost finished its proposal for building the Woods Hole submersible—the kind of diving craft he had long been dreaming of. I played a minor role in North American's final bidding effort, which unfortunately it lost to General Mills.

Aside from my personal stake in the matter, I could see that a remarkable new diving machine was going to be built, one way or another. Working with Rechnitzer's team at that pivotal moment gave me my first introduction to the world of deep submergence. From that summer forward, I would never leave it.

— "Test Depth"

General Mills began building the submersible, as yet unnamed, later in 1962, but then promptly sold the contract to Litton Industries. As work on this trailblazing new sub progressed, however, profound changes transformed the entire deep submergence community. The pace of events in the larger world quickly outstripped the plodding pace of the little sub's design and construction.

The USS *Thresher* at sea. (Source: U.S. Naval Historical Center.)

In 1963, a stunning cold war tragedy changed the course of deep-sea exploration: the USS *Thresher,* a sleek and powerful nuclear submarine, went down off the coast of New England. More accidents and sinkings followed, some of them involving fearsome nuclear weapons. Ironically it was these disasters, and not the science-laced arguments of deep-sea advocates, that would confirm in many minds the importance of deep submersibles—and the wisdom of building many more of them. And it was the terrible doom of *Thresher*'s crew that set those wheels in motion.

The story of *Thresher*'s demise began at the Portsmouth Naval Shipyard in New Hampshire, where the sub was being overhauled early in 1963 after more than a year of shakedown cruises, tests, and service at sea. During this overhaul the shipyard took apart and examined all the sub's operating systems, making repairs or replacements as needed. Anxious to have *Thresher* back on patrol, the navy personnel in charge were rushing to finish the lengthy procedures. They persisted in this effort despite troubling results from a recent testing of silver-brazed pipe joints. The pipes carried seawater throughout the submarine—into the ballast tanks, for instance—and they cooled the nuclear reactor. Of the 145 joints inspected by a new ultrasonic procedure, 14 percent failed, although they had passed a hydrostatic pressure test earlier. The older tests were deemed sufficient, however, and 2,855 more joints underwent no further testing.

Unfortunately, the cold war years were a time when risk taking seemed unavoidable. Lieutenant Commander Wes Harvey, *Thresher's* captain, reviewed reports on its status and decided to take it back out to sea. A broken pipe joint would not necessarily mean catastrophe. Flooded compartments could be sealed if necessary, valves could shut off the flow of water, and under most circumstances crewmen could make repairs or at least contain the damage while the sub resurfaced and returned to port.

Thresher cruised out of Portsmouth Harbor in the early dawn of April 9, 1963, with 129 personnel on board. So crowded was the submarine with extra shipyard technicians that the torpedo room had been converted into a dormitory, its racks holding bunks instead of torpedoes. Waiting offshore was the submarine rescue ship *Skylark,* which would stay with *Thresher* as it conducted sea trials over the next several days. But calling *Skylark* a rescue ship implied that should *Thresher* get into trouble, *Skylark* could come to its rescue. Nothing could have been further from the truth.

Skylark was equipped with a McCann rescue chamber dating back about twenty-five years. This chamber could operate at depths up to 850 feet. It could be lowered to a submarine resting on the bottom of the ocean and mated to one of its hatches. The submarine hatch could then be opened and its crew transferred to the chamber for transport back to the surface. But in the 1960s a modern nuclear submarine spent most of its time in waters where the bottom lay below 850 feet, too deep for the McCann chamber to reach.

Thresher's test depth—the official estimate of its maximum safe operating depth—remained secret. It was probably around 1,000 feet. The sub would be required to approach this depth before being cleared to return to duty. If *Thresher* experienced a problem during that final test dive, it was on its own. No existing rescue technology could save it. The captain and crew would have to rely on its powerful nuclear reactor to thrust it forward and upward, away from the deadly pressures below.

Following shallow-water tests in the Gulf of Maine, *Thresher* informed *Skylark* that it was heading into deep water off the continental shelf. Early the next morning, April 10, *Thresher's* crew heard their skipper call out, "Rig for deep submergence."

As *Thresher* slipped down into deeper water, the daylight outside faded to blackness, and a growing pressure tightened its grip on the hull. Inside the sub, crew members must have sensed the rising pressure—metallic creaks and groans would remind them. Nevertheless, they worked routinely through their checklists. At 0900 hours, Quartermaster Jackie Gunter called *Skylark* to convey a terse announcement: "Proceeding to test depth."

For the next few minutes, all went well. But then something happened. Experts who later investigated the disaster believed that a silver-brazed pipe broke inside the sub's auxiliary machinery space. Most likely a noise rang out, a sharp bang. An ominous hissing, similar to that of a powerful steam locomotive, would have followed. Seawater makes that sound as it sprays into a submarine under high outside pressure. It also forms a chilly, dense fog.

The leak would have been the least of the captain's worries. The fog apparently condensed on walls and equipment and shorted a nearby electrical panel. An alarm would have warned officers that electrical circuits on the port side had "tripped out." Their natural reaction would have been to shift the electrical load to the starboard bus; they would also have informed Commander Harvey. But such a shift in the power load would trigger automatic safety features, leading to a reactor scram, or shutdown.

As nuclear fission ceased, the sub's propeller blades would have slowed to a halt. The crew, having been drilled for such emergencies, probably rushed to close valves and compartments around the broken pipe. With the leak isolated, they would seek permission to restart the reactor. But it would take seven long minutes for power to return.

Ordering a full rise on both planes and as much up-angle as possible would have been a logical response, with everyone on board now praying that the sub's momentum would carry them out of danger. *Thresher* would have to glide, forward and upward in a swooping curve, until the crew had enough time to bring the reactor back on line. Then they could power themselves out of trouble.

It was not meant to be. *Thresher* apparently stalled as it began taking on water, and then drifted backward. Already at test depth when the pipe failed, it quickly approached crush depth. As a final desperate attempt to gain a few more seconds to start the reactors, Commander Harvey would have ordered the main ballast tanks blown. At 0913 hours, he called his rescue ship on the hydrophone: "*Skylark, Skylark,* this is *Thresher.* Have positive up-angle; am attempting to blow."

Harvey undoubtedly knew that conventional ballast tanks will not "blow" at *Thresher*'s test depth. Fighting against high outside pressure, compressed air would have emptied the tanks, at best, only halfway. But experiments performed months later suggest that even half-empty seems too optimistic. As air expands from compression, it cools. Probably, when the first blast of air slammed into the ballast tanks, ice formed on a new, fine-mesh strainer that had been added to protect the air valves from debris. The ice would have instantly cut off further air flow, sealing the fate of the sub.

Thresher continued to sink. Its crew continued to work, as fast as humanly possible, on the nuclear reactor. At 0917 hours, another message came up from the depths. Personnel on board *Skylark*—by now riveted to their listening posts—heard only garbled sounds and noises, among them the words "test depth."

Moments later the navy's secret underwater listening system, designed to track Soviet submarines, picked up a powerful implosion.

Exuberance Mixed with Sorrow

Thresher was dead. Every piece of it, and every person on board, had vanished without a trace, as if swallowed by the earth itself. For many long months no one knew why the submarine had sunk. Its loss shocked the nation.

The navy, caught up in the logic of the cold war, had additional worries. *Thresher* had been its newest attack submarine, designed to run quietly on nuclear power. It could stay underwater for months at a time stalking Soviet ships. The navy had been planning to build a new class of submarines based on the *Thresher* design. Navy brass needed to know what had happened to their prototype; they had to find the wreckage.

At first, the search for *Thresher*'s remains led only to deeper frustration and embarrassment. For several weeks, dozens of ships and thousands of men could not find it. A camera lowered from one ship seemed to have photographed a piece of *Thresher,* but that turned out to be the ship's own anchor. It might help, some thought, to have humans on the bottom looking. Only a bathyscaph could take them there. The navy summoned *Trieste.*

Starting in June, the sturdy behemoth dove to the bottom—a mile and a half down—five times. It lumbered over the large search area, about a hundred square miles, at its usual slow pace, with crewmen peering out from the restrictive viewport. The seafloor looked much the same all over. With the primitive methods available in those days for keeping track of position, it was hard to tell precisely where the craft was at any moment, or where it had been. Nevertheless, the navy had no alternative for searching the seafloor. In August *Trieste* returned to the site, and on the third dive it happened to drift over the remnants of *Thresher.*

On several subsequent dives, *Trieste* revisited the wreckage. Much of what the navy learned from those excursions remained secret, as did the exact location of *Thresher*'s debris. Clearly, though, the site was a mess. According to a navy commander who had gone down in *Trieste* to inspect the scene, *Thresher*'s scattered, twisted, shredded wreckage looked like an "automobile junkyard." Details remained sketchy, in part because the bathyscaph was not mobile enough to probe the entire site. Yet whatever the navy might eventually have learned about design flaws in its nuclear submarines, the attempt to find *Thresher*'s wreckage exposed an entirely different weakness: inadequate, outdated deep-ocean technology. In the future, navy investigators would surely want to find and sometimes recover other military hardware from the deep seafloor, if only to keep it from falling into enemy hands. But just as surely, they could not rely solely on the bathyscaph *Trieste.*

Stung by criticism of every aspect of the disaster, the navy mounted intense new efforts to improve its deep-ocean capabilities. Two weeks after *Thresher* went down, the secretary of the navy formed a panel of fifty-eight experts to recommend what steps the service should take if a similar disaster struck again. After nearly a year of analysis, the panel issued its top-secret report. One immediate result was the formation of the Deep Submergence Project Office, which issued proposals for new research, new technologies, and new submersibles. One long-term result was to galvanize, with cash and official sanction, the formerly small and obscure community of deep-sea advocates: scientists, inventors, and dreamers.

The navy's new deep-ocean initiative, although born of tragedy, blended well with other, more exuberant, crusades in the 1960s. These were the glory days for American engineers, who responded to challenges on every front with some of their greatest technological triumphs. With a strong economy and the full support of the government, the military-industrial complex was running at full throttle, building many kinds of sophisticated vehicles: fighter aircraft, nuclear submarines, ICBMs, spaceships. In this can-do atmosphere, the deep-sea initiative naturally expanded. The idea became—at least in some quarters—not merely to improve techniques for rescue and salvage but also to explore, claim, and conquer.

A notion that had existed all along among a hard core of deep-sea dreamers now gained wider acceptance. It seemed only logical that if the United States could launch spacecraft toward the outer reaches of the solar system and prepare to land astronauts on the moon, Americans could also explore the inner reaches of the oceans. Discussions of a "Wet NASA" swirled around the corridors of power in Washington, among academic research institutions, and through the media to the American public.

Given all that territory underwater, there must be something in the abyss worth pursuing, or so the thinking went. By the mid-1960s, an assorted group of scientists, analysts, and futurists were selling the deep sea as America's next manifest destiny, waiting to reward the first to arrive in the dark abyss with untold riches—new commercial fisheries, oil reserves, precious gems and metals. The geologist John Merrill, for instance, wrote an influential book in which he described manganese nodules just there for the taking. Other visionary types, such as Willard Bascom of Scripps, advanced schemes to harness the energy of the tides and waves. Popular magazines such as *Time* and *National Geographic* devoted glossy pages to the emerging field of oceanography, and funding for blue-water research was flowing out of ONR and the National Science Foundation.

Sensing a potential bonanza, companies all over the United States scrambled to get into the game. Aerospace firms, in particular, believed that their expertise at manufacturing complex systems-in-motion could be applied to the world's oceans. Building on pioneering work by the Piccards, Cousteau, Froehlich, and others, they began designing deep-diving submersibles, hoping to sell or lease them to researchers and entrepreneurs. A swarm of new vehicles soon appeared, ready to test the market. In 1965 Westinghouse launched its *Deepstar 4000,* which could carry a crew of three down to 4,000 feet. In 1966 General Dynamics followed with *Star II* and *Star III.* A year later Lockheed launched *Deepquest,* which could take four occupants to 8,000 feet. In 1968 North American Rockwell (so named after a merger in 1967 of North American Aviation and Rockwell) put its *Beaver Mark IV* into the water; Grumman launched *Ben Franklin,* a "mesoscaph," or middle-depth submersible, designed by Jacques Piccard; and General Dynamics finished two submersibles for the navy, *Sea Turtle* and *Sea Cliff.* Smaller companies also took the plunge into deep-sea technology, as did several groups in Japan, an island nation with an abiding interest in the oceans.

Bathyscaphs, too, came up for a burst of renewal, in both France and the United States. In 1961 *Archimède,* which in principle could dive to 36,000 feet but in practice never did, replaced *FNRS-3.* In 1964 *Trieste II,* designed for a crew of three and rated for 20,000 feet, replaced the older *Trieste.* The second version weighed eighty-eight tons and needed 66,000 gallons of gasoline for buoyancy.

Even the powerful father of the U.S. nuclear navy, Admiral Hyman Rickover, joined the deep-sea push. In the 1960s he took time off from his massive supership projects to commission the world's smallest nuclear reactor, to be placed inside the world's deepest-diving nuclear research submarine. Ready for service in 1969, *NR-1* was capable of reaching 3,000 feet. Unfortunately, Rickover classified as secret even its most basic features for many years. This decision severely limited its usefulness to science, depriving researchers of access to such astounding capabilities as a luxurious 130-foot hull for a crew of eleven and life-support systems enabling forty-five days of continuous submergence.

— *Alvin*

Yet before any of these new players could get their plans off the drawing boards, two other groups had nearly finished building their new submersibles. Reynolds and Woods Hole had long since bought into the deep-sea groundswell. Both launched their vehicles in 1964, ahead of everyone else. The hefty Reynolds *Aluminaut* weighed in at seventy-six tons. The Woods Hole submersible, by contrast, weighed only fourteen tons. Overcoming one last obstacle, bickering factions there finally reached a historic agreement: deciding on the name

Alvin at its commissioning, June 5, 1964. (Source: Woods Hole Oceanographic Institution.)

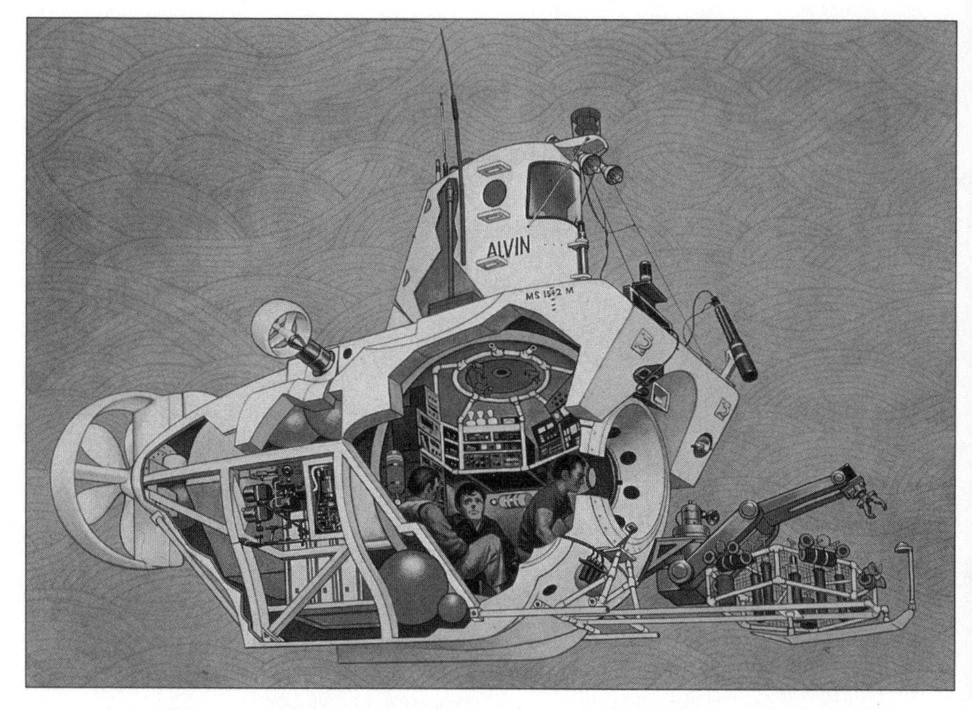

Cross-sectional diagram of *Alvin.* (Source: Davis Meltzer/National Geographic Society Image Collection.)

they would give it. They called it *Alvin,* as a contraction of Allyn Vine's name and *not,* most insisted, a reference to a popular cartoon character, an irrepressible chipmunk.

On June 5, 1964, with flags and bunting and navy brass attending in full uniform, little *Alvin* received its christening at Woods Hole, to fanfare worthy of a battleship.

Although small, *Alvin* was no toy. Its white fiberglass hull, twenty-two feet long, contained a steel pressure sphere six feet ten inches in diameter with walls almost two inches thick, designed for a test depth of 6,000 feet. Five big balls of syntactic foam filled much of the remaining space in the hull. The sub would carry a crew of three: a pilot, copilot, and scientific researcher. (Eventually the ratio would shift to two scientists and one pilot.) A large propeller at its stern and two smaller

Alvin sitting on *Lulu*'s cradle while under way to a dive site, circa 1968. (Source: Woods Hole Oceanographic Institution.)

props on either side gave *Alvin* its horizontal mobility. And, showing that it really meant business, a remote-controlled arm with a powerful crablike pincer jutted out from the bow.

It seemed only fitting that Allyn Vine and Bud Froehlich, along with pilot Bill Rainnie, make the first two dives in *Alvin.* They were largely ceremonial, however. Without a support ship to carry it into the ocean, the little submersible could barely get wet. But at least it stayed wet, dipping and dunking most of the summer. After a total of seventy-seven tethered descents at or near Woods Hole—to maximum depths of seventy feet—*Alvin* made its first free dive on August 4, to a depth of thirty-five feet. On that restrained and cautious note, the first diving season ended.

The following spring, in March 1965, *Alvin* headed out to sea on board *Lulu,* a strange-looking catamaran that a local Cape Cod builder had cobbled together from two surplus navy pontoons. *Lulu* had a slot of open water in the middle, between the two pontoons, into which a support crew could raise and lower the sub.

After stopping at Port Canaveral for more tests, *Lulu* and *Alvin* moved on to the Bahamas, where a tethered, unmanned descent to 7,500 feet proved that *Alvin*'s hull was sound. The sub now had two pilots, Rainnie and Marvin McCamis. On July 20 they made their first 6,000-foot dive for the navy, demonstrating that they could operate *Alvin* at test depth. During this series of dives they also inspected an underwater listening array off Andros Island, a task that met the original justification Momsen had used to obtain funding. The 1965 diving season ended with *Alvin* well tested and certified—just in time for the next disaster.

The Missing H-Bomb

On January 17, 1966, a squadron of B-52 bombers took off from Goldsboro, North Carolina, and headed east on a route known as Chrome Dome. Each B-52 carried four enormous hydrogen bombs. Their target: the Soviet Union. After two midair refuelings, the squadron neared the bootlike shape of the Italian peninsula. It was then and only then, as they were ordered to bank south and return to the United States, that the crew on each plane knew for certain that this had been just another practice run.

With the squadron homeward bound, the last refueling was scheduled to take place high in the air south of Spain. The KC-135 tanker sent up to satisfy the bombers' thirst for fuel, a military version of the Boeing 707, arrived on schedule. The fueling boom was lowered into place, and the first bomber, commanded by Captain Charles Wendorf, crept upward. "Watch your enclosure," the pilot of the KC-135 cautioned him. Wendorf throttled back, but he was too late. The boom missed the refueling hole on top of his plane, striking instead the plane's longeron, or backbone. In seconds the B-52 broke in half, as crew members struggled to parachute to safety.

The crew of the KC-135 never had a chance. Flames shot up the boom to the fuel storage tanks above. It was all over for them within a matter of seconds, as a brilliant fireball lit up the sky. Seven of the eleven men on both planes perished in that explosion.

The wreckage of the two planes began to fall 30,000 feet to the earth below, along with four fliers lucky enough to have escaped the fiery death of their comrades. Also plummeting to earth were the B-52's four hydrogen bombs, together powerful enough to wipe out the population of southern Spain. Fortunately, their detonating systems were not armed. Even so, a hydrogen bomb is not something to be left lying about the countryside for anyone to pick up.

In the frantic search that followed this accident, three H-bombs would quickly be found on land, near the Mediterranean coast. The fourth, however, was missing. Clearly it had fallen into the sea. The pressure to find it mounted as alarming stories appeared in the press and tourists began to flee Spain. Once again, a tragic accident reminded the U.S. military that it was sorely in need of a deep submergence capability to react quickly in times of emergency.

The air force asked for help from the navy to find the last bomb. As with *Thresher* two years earlier, the navy had a large area to search, more than a hundred square miles. Scuba divers probed the bottom in shallow waters but found nothing; the search would have to go deeper. And deeper it could go with the arrival of *Mizar*—a former cargo ship that had been fitted with deep-sea lights and cameras in the aftermath of the *Thresher* incident. The photographs *Mizar* was able to take showed that the bottom was rough: rolling plains crossed by canyons. Yet they did not reveal any trace of the bomb.

In February *Alvin* arrived. After nineteen dives into murky water and clinging muck, Bill Rainnie and crew located the bomb on March 15. Marvin McCamis was the pilot that day. As the sub followed a suspicious-looking furrow down a slope so steep that even *Alvin* bumped and scraped it like a bathyscaph, one of the diving crew saw the blockbuster—or rather, its huge parachute—lying directly beneath his viewport in 2,450 feet of water.

Bringing this monster back up turned out to be even harder than finding it. In fact the bomb at first went the wrong way, sliding a few hundred feet deeper into the canyon. *Alvin*'s exhausted crew then spent several more days finding it again. The recovery ultimately required the finesse of a fifteen-foot experimental robot called CURV (Cable-controlled Underwater Recovery Vehicle), which had to be upgraded to dive deeper than its original limit of 2,000 feet and rushed to the site from a navy lab in California. The vicious-looking 2,000-pound robot chugged through the deep with battery-powered propellers, barely faster than a speeding bathyscaph. It inspected a worksite through television cameras and captured whatever its controllers directed it to grab with a single mechanical claw. When it joined the armada that was trying to raise the H-bomb, a surface crew lowered and manipulated it from a support ship while *Alvin*'s submerged crew offered feedback.

After several attempts, CURV managed to attach lines to the bomb and parachute and then intentionally tangle itself in the mess; powerful winches then hauled the whole bundle back up to the surface. It was a giant step forward—and downward—for remotely operated vehicles.

The recovery of the missing H-bomb confirmed more than ever the value of deep submergence technology. Within two years after *Thresher* had gone down, a new kind of deep submersible, a state-of-the-art robot, and other new gear deployed both above and below the surface had changed the way humans could work underwater. That new technology also permitted the people of Spain to put a terrifying disaster—one whose outcome could have been far more tragic—behind them. Surely more good would come of it.

PART II
DISCOVERY

CHAPTER

4

SCIENTISTS BEGIN EXPLORING THE DEEP

A profound thing . . . happens to every single person who gets in that sphere [*Alvin*'s crew capsule]. Even if they don't get samples, they come back a changed scientist.

—*Tanya Atwater*

On July 17, 1966, *Alvin* made its first deep dive for science. The biologist Robert Hessler descended 5,850 feet that day, to the bottom of the Atlantic, diving in the same waters off Bermuda that William Beebe and Otis Barton had visited decades earlier.

Like Beebe, Hessler had tried to haul dredges over the bottom. A few came back with fine samples of microbe-infested muck, but many returned empty or full of rocks, and others simply disappeared. When he went down to find the reason, Hessler saw a rugged terrain bearing little resemblance to the ocean "floor" as commonly imagined. "I was awed by the tremendous vertical precipices," he said, "and I finally understood why we had so much difficulty ever taking any samples from that area. That dive really taught me something. From then on, whenever I lowered a dredge into the ocean, I could close my eyes and picture what the bottom of the deep sea looked like."

Some fish that Hessler never saw may also have been lurking there. During several dives in the 1930s, Beebe had seen shadowy shapes in the distance, "large and very real living creatures," he insisted. They reappeared from time to time as if stalking his bathysphere. No such beasts made themselves known to *Alvin*'s scientific observers in 1966. The following year, however, one did.

On July 6, 1967, *Alvin* was making a routine dive at a depth of 2,000 feet on the Blake Plateau, off the southeast coast of the United States, when pilot Marvin McCamis saw what he had thought to be a large rock suddenly come to life at the edge of his field of vision. No sooner had it disappeared from his view than scientist-observer Rudy Zarudski heard a loud rasping sound on the hull. Looking out the starboard viewport, copilot Val Wilson yelled out, "We've been hit by a fish!"

A cloud of blood encircled the viewport as a large swordfish fought frantically to free itself, its sword somehow stuck in the sub. In the confused moments that followed, the shrill alarm of a leak detector reverberated within the pressure hull. The occupants were not sure of the alarm's cause, but their training had taught them to assume the worst. The steel sphere was not quite solid. Wires passed into it through tubes called penetrators, which required constant maintenance to prevent leaks from developing around their washers, nuts, and sealing materials. Perhaps the sword had somehow cut one of the wires passing through a penetrator, and seawater was now leaking into a vital electrical circuit, if not the pressure hull itself.

McCamis quickly reported the sub's condition to the support ship *Lulu* overhead, and Bill Rainnie ordered *Alvin* back to the surface. When it broke through, support divers found the swordfish still very much alive. But just as they attached a line to its tail, the fish, in one final dying flurry, tore itself free, leaving its sword still embedded in the sub's exterior. Thanks to the quick actions of the divers, everyone aboard *Lulu* that night enjoyed fresh swordfish steak.

During the postdive inspection, damage to *Alvin* proved superficial. The steel pressure sphere remained intact. The fish's one-and-a-half-foot sword had wedged, instead, into a

An unfortunate swordfish stuck in *Alvin*, July 6, 1967. (Source: Woods Hole Oceanographic Institution.)

seam separating two segments of the outer fiberglass body. (The alarm that had gone off a little later was unrelated, caused by a minor leak into the sub's oil-filled battery boxes.) While slashing into the fiberglass, the sword had just missed an important bundle of electrical cables. That was not, however, the close call that most concerned the inspectors. Swordfish have been known to sink small boats; their bills have reportedly penetrated wood a foot thick. What if the sword had hit one of *Alvin*'s viewports? Might the Plexiglas have given way, instantly killing the sub's occupants? The answer to that question would have to await further research—or further field testing by enraged fish.

Submarine Canyons

Eleven days later, the geologist K. O. Emery and several members of a group he had formed at Woods Hole began using *Alvin* to explore submarine canyons. They made their first two dives off Cape Charles, Virginia, and proceeded to points farther north on the continental shelf. Swordfish was not on the menu, they hoped.

In theory, Emery's group was observing simple terrain, but recent work had caused geologists to question their understanding of how it had formed. The continental shelf, in most parts of the world, consists of soft sediments that have compressed into semi-consolidated rock. Its geological history seemed obvious: layers of sediment thousands of feet thick built up over eons on the seafloor, and under pressure those layers became rocks. In the conventional view, nothing much happened to the shelf rocks as long as they remained under the ocean.

The origin of canyons slicing through the layers of rock also seemed obvious: the canyons had formed during past ice ages, when glaciers on land locked up a great deal of water, causing the sea level to drop dramatically and exposing the relatively shallow seafloor around much of the continents. Rivers cut easily across this newly emerged surface, extending their length in some cases by hundreds of miles. When the most recent ice age ended, glaciers melted, the sea level rose, and this ancient surface became seafloor once again; the river valleys became submarine canyons.

This conventional interpretation of submarine canyons, though plausible, dated from a time when geologists had conducted their fieldwork on dry land. Would the new work of marine geologists—who had only recently gained access to the other 71 percent of the earth's solid surface—support the old ideas?

The first marine geologists were, understandably, specialists in sedimentary studies. Principal among them was Emery's mentor, Francis Shepard, considered by many to be the father of marine geology. The investigation of submarine canyons was a passion of his, and, working at the Scripps Institution of Oceanography in La Jolla, California, he had easy access to large canyons nearby. At the same time, scuba-diving geologists had come up with evidence contradicting the idea that submarine canyons were simply ancient drowned features—geological dead zones. Instead, they seemed to be active sites of sand transport and erosion. Struck by these discrepancies, Shepard realized that manned submersibles might allow observations beyond the range of scuba divers, and he began using Cousteau's *Soucoupe* to explore the Scripps and La Jolla Canyons off southern California.

Having visited the deep mouths of those canyons in the bathyscaph *Trieste* years before, Shepard was now more interested in the narrow, shallower portions of the canyons where a clumsy bathyscaph could not venture. It was at the heads of the canyons, not their mouths, where he thought undersea erosion might be most active. As he explored, Shepard found that the narrow ravines and overhanging walls observed by scuba divers continued the entire length of the canyons. In fact, in one portion of its run, a canyon nearly seven hundred feet deep was so narrow that *Soucoupe,* only nine feet wide, couldn't squeeze through to reach the bottom. To keep the sub from getting stuck, Shepard had to use its mechanical arm to push it off the canyon walls.

Such spectacular vertical mazes also exist above sea level—in the canyonlands of the American Southwest, for example. Shepard's research, which could only have been performed with a small submersible, clearly demonstrated that the submerged canyons were not just relics of the ice age but active conduits for the movement of sediments across the continental margin and into deep water. Just as biologists were finding more life in the abyss than they had expected, geologists were beginning to find the seafloor itself more active and interesting than they had thought.

While Shepard was exploring submarine canyons off the West Coast, his former graduate student, K. O. Emery, explored the continental shelf and canyons off the East Coast. As ocean currents cut their way through layers of rock, they expose in canyon walls the geological history of a region. Using *Alvin*'s ability to maneuver and hover near vertical walls, geologists in Emery's group were able to sample the exposed layers and add further detail to the recent stratigraphic history of the continental shelf.

Emery was also interested in the history of sea level fluctuations. Fifteen thousand years ago, the eastern shore of North America lay between 230 and 430 feet below its present elevation. As the sea level rose in fits and starts, a series of ancient shorelines were created and later flooded, leaving clearly recognizable features on what is now the continental shelf. During one *Alvin* dive in 1967, Emery and Robert Edwards found a submerged beach and oyster reef formed 8,000 to 10,000 years ago off Chesapeake Bay. On a ridgetop, which at one time had been dry land near the beach, they found heaps of oyster shells thought to be kitchen middens of the early humans who must have inhabited the area.

Like Beebe, who had found a submerged shoreline off Bermuda, Emery and Edwards simply noted their find and passed on. To this day, little work has been done at such prehistoric archeological sites. The continental shelves of the world may yet prove to be repositories of a significant part of the human story.

— An Epidemic of Sinkings

The early explorations of the 1950s and 1960s demonstrated the new capabilities of deep-diving vessels. After visionary pioneers had spent years perfecting submersible design, a few oceanographers were finally using the new technology to great advantage on the deep frontier. What they experienced in the vast eternal darkness was changing not only their knowledge but also the scientists themselves. As the Scripps-trained geophysicist Tanya Atwater put it years later, "some jobs cannot be done except with *Alvin*": first, to explore the undersea world itself in fine detail, and second, to "get your eyeball and your gut calibrated."

In the spring of 1967, I was preparing to join their ranks. After graduating from the University of California, I married and started working full time for Andy Rechnitzer's Ocean Systems Group at North American Aviation. They were developing the *Beaver Mark IV* submersible as part of an ambitious scheme to ferry aquanauts and supplies between the ocean's surface and underwater oil fields. As long as I remained on this project, the company would sponsor my graduate work toward a Ph.D. in marine geology at the University of Southern California.

My budding career as a deep-sea explorer seemed well under way until a navy officer knocked on the door of our apartment one evening. He was delivering an order to report for active duty. As it turned out, I spent the next three years in the U.S. Navy, far away from southern California—the cockpit, as I believed, of marine research and innovation.

While attending college at Santa Barbara, I had joined the Reserve Officers Training Corps, earning a lieutenant's commission along with my undergraduate degree. Although most ROTC trainees working on graduate degrees were allowed to defer their years of active service, the navy called me up for duty that spring of 1967. Ironically, my assignment took me across the continent to Massachusetts, where I would be a liaison between the Office of Naval Research and Woods Hole. I would be working closely, in other words, with individuals who had rejected the bid of my friends and colleagues at North American to build *Alvin*.

Nevertheless, the people I met and the projects I saw at Woods Hole fascinated me. Over the next two years I developed a deep respect for pilot Bill Rainnie, geologist K. O. Emery, and many others, as well as for *Alvin*'s capabilities. In fact I began to think about finishing my graduate work on the East Coast after leaving the navy, hoping that I might dive in *Alvin* someday myself.

Then, in 1968, just as the outlook for deep submergence was brightening after the *Thresher* sinking, tragedy struck again. One submarine disaster after another shocked navies all over the world. The losses began in March, when a Soviet Golf-II submarine sank in the Pacific. A few weeks later, in May, the USS *Scorpion*, an attack sub of an older design than *Thresher*, disappeared while on war patrol in the Atlantic, leaving few clues about what had happened. Ninety-nine crewmen aboard *Scorpion* perished, and a similar number probably died on the Soviet sub. Both ships were carrying nuclear weapons. Shortly after those two accidents, the Israeli submarine *Dakar* vanished on its inaugural voyage, prompting rumors about secret nuclear technology being brought into Israel. Then the French submarine *Minerve* went down in the Bay of Biscay.

The search for *Scorpion* got into full swing in June 1968. The usual flotilla of surface ships, planes, and submarines converged on an area some 400 miles southwest of the Azores, where underwater listening devices had detected a series of explosions that might have been caused by weapons. Soon *Mizar,* with its towed deep-sea cameras, began combing the bottom. All summer long—at a painfully slow pace, one visual frame after another—it examined a growing portion of the search area, more than a hundred square miles of seafloor. At last, late in October, several photographs revealed *Scorpion*'s hull, lying in water more than two miles deep. It looked surprisingly intact, broken cleanly in two as if the craft had been struck by a torpedo. That wreckage seemed to tell a story quite different from *Thresher*'s dreadful implosion, which had scattered debris all over the bottom.

Wanting to gather more information, and hoping to find two torpedoes tipped with nuclear warheads that the ill-fated sub had been carrying, the navy decided to bring in a deep-diving vessel. At that time, only one could do the job. It was not *Alvin,* which had never been certified for dives beyond 6,000 feet. *Sea Cliff* and *Sea Turtle,* two newer submersibles, were not yet ready for service. That left only the bathyscaph *Trieste II.* But the deep-diving blimp could not stand up to the heavy seas of approaching winter storms. It began to look as if any deep excursion to *Scorpion*'s wreckage would have to wait until the following summer.

By this time, however, a far more serious problem than *Alvin*'s limited test depth prevented Woods Hole from participating. I learned about it on October 16, two weeks before *Mizar*'s cameras located the remains of *Scorpion.* I was in Ottawa that evening, preparing to lecture before the Canadian Defense Force on the future of the U.S. Navy's manned deep submergence program. The future was looking good for the navy—which led the world in deep submergence activities—and increasingly better for me. Just before I stepped up to begin my talk, a secretary gave me a telephone message. Suddenly, half my speech evaporated, along with most of my dreams.

The message informed me that *Alvin* had joined *Scorpion, Thresher,* and the other large subs now lying on the bottom of the ocean. It had sunk 135 miles southeast of Woods Hole, in water a mile deep.

— Crush Depth of Dreams

Alvin's Achilles' heel proved to be not its viewports but rusting cables on its support ship, *Lulu.* From the beginning, almost everyone at Woods Hole had expressed doubts about *Lulu.* It looked more like a piece of floating wreckage than a sophisticated research ship. It was, in fact, a temporary launching platform meant to be replaced as soon as Woods Hole could come up with the necessary funding. When *Alvin* had finally been completed and tested in the mid-1960s, most of the funding for deep submergence was used up, leaving little for its support ship. And little is just what Woods Hole got. Constructed of two navy surplus floats designed to detonate mines, *Lulu* was registered in the state of Massachusetts as an outboard motorboat—the oddest-looking ship I had ever seen.

Launching procedures were equally awkward. A box-shaped cradle, suspended on four cables between *Lulu*'s twin catamaran hulls, lowered and raised *Alvin* into or out of the water. Before a dive, *Alvin* sat on this cradle. The copilot and scientific observer climbed inside and descended to the pressure sphere, carrying personal items and often a lunch. *Alvin*'s pilot remained standing in the submersible's sail (a small conning tower) above them, with the hatch to the pressure hull still open. One of the divers inside then passed up a small box to the pilot, containing remote driving controls attached to a long cable. When everyone was in position for a launch, *Lulu*'s chief engineer raised the cradle slightly so that crew members could remove safety blocks. The cradle and submersible now hung suspended on four steel cables, one attached to each corner of the cradle, high above the water's surface.

Normally, the chief engineer would lower the cradle slowly into the rectangular slot of water between the two hulls until the sub floated free. Then *Alvin,* steered by the pilot standing in its sail, would back out and into the open ocean, clear of *Lulu.* But that was not what happened on dive number 307.

It was supposed to be the last dive in the 1968 season, after which all four cradle cables were going to be replaced. Unfortunately, the project engineers had waited one dive series too long. Unlike chain, which rusts from the outside in, cables rust from the inside out because seawater penetrates to the innermost strands, which stay wet the longest. The corrosion is mainly internal; a visual inspection cannot tell how much strength has been lost.

As the chief engineer of *Lulu* raised the cradle to free the locking blocks, the forward port cable suddenly snapped, followed a fraction of a second later by the forward starboard cable. The rear cables held. When the first cable broke, the forward part of the cradle twisted left; when the second cable broke, it tipped violently downward. *Alvin* instantly slid out, tumbling toward the left while falling forward. Its fragile fiberglass sail, in which pilot Ed Bland was standing, crashed into *Lulu*'s two-inch-thick steel pontoon hull.

By sheer luck—and by instinctively crouching—Bland avoided being decapitated as the sub fell more than ten feet to the water's surface. As it continued plunging below the surface, water rushed into the open hatch, pinning its occupants against the interior walls. Once fully flooded, the sub would weigh more than ten thousand pounds in water, heavy enough to send it to the bottom. But the flooding would take several seconds, during which *Alvin* would retain enough buoyancy to rise back to the surface for a few brief moments—that is, if it didn't come back up beneath the port pontoon under which it had plunged.

Fortunately, *Lulu*'s captain was on the ball that day; otherwise all three of *Alvin*'s occupants would have died. He used *Lulu*'s powerful thrusters to move the ship sideways. Luck combined with the captain's skill: *Alvin* popped back to the surface precisely between the two hulls. Bland was dazed by a blow on the head but managed to jump onto one of *Lulu*'s pontoons. The alert crew chief, George Broderson, then pulled copilot Roger Weaver to safety by his arms, and science observer Paul Stimpson somehow followed him out of the hatch, possibly by hanging onto Weaver's legs.

In less than sixty seconds, *Alvin*'s brief reprieve above the surface ended. When it went down for the second time, it began a free fall to the ocean floor some 5,200 feet below.

Having spent years looking for objects in the deep, the crew aboard *Lulu* reacted quickly. They grabbed anything they could and threw it overboard. Lawn chairs, pieces of metal, a fifty-five-gallon drum—anything handy began raining down upon the ocean floor. Anything that would tell future searchers that *Alvin* was nearby.

The Impregnable Bologna Sandwiches

A hasty search effort failed to locate the downed sub. Two research ships went to the site, but their sonar detectors were not powerful enough to resolve an object as small as *Alvin* on the bottom. Woods Hole then rented a submersible called the Deep Ocean Work Boat from General Motors, but even when submerged it could not find *Alvin*—and during a second attempt, in November, its support ships nearly sank in hurricane-force winds with heavy seas. It wasn't until June 14 the following year that the navy's supersecret deep-towed camera system aboard *Mizar* snapped a photograph of *Alvin* sitting upright on the bottom, with its hatch still open.

Months before this, during the long winter, I and many others at Woods Hole had written off the little sub in our minds. That single image, however, convinced navy brass, after much heated debate, to go after it and attempt its salvage.

No untethered object as heavy as *Alvin* had ever been recovered from such a depth in the ocean. The plan was straightforward but by no means simple. It began with *Mizar* returning to the site in August to relocate the sub. That took three days. Then the submersible *Aluminaut,* which had been towed to the site, was to carry a 7,000-foot reel of heavy nylon line with a T-shaped toggle bar at the end down to *Alvin* and drop it inside the open hatch.

Aluminaut floated on the ocean's surface like a bathyscaph, too big and heavy to be lifted aboard a support ship. High seas—the constant bane of such large submersibles—then prevented anyone from attaching the reel of nylon line to *Aluminaut*'s bow. On the spot, a new plan was developed. The line would be lowered to the bottom from Mizar instead of being carried down by submersible. *Aluminaut* would then dive to the bottom, locate *Alvin,* locate the toggle bar at the end of the line, and drop the bar into the sub.

All went as planned except the last step. Dropping the line inside *Alvin*'s open hatch proved too difficult for *Aluminaut*'s mechanical arm; operators compared the task to pushing a limp wet noodle through a straw hole, with a current tugging it sideways. After twelve hours, the sub's power was running low, forcing it back to the surface. That gave the rough seas another opportunity to ruin everything: high waves made it impossible to recharge *Aluminaut*'s batteries. The tiny flotilla had to limp back to Woods Hole and wait for another chance.

That chance came when the weather improved late in August. This time the crew in *Aluminaut* successfully inserted the toggle bar through *Alvin*'s open hatch. With a securing tug, *Mizar* took in the slack and winched *Alvin* back to the surface.

Alvin immediately after its recovery, September 1969. (Source: Woods Hole Oceanographic Institution.)

But *Mizar* could not actually lift *Alvin* out of the ocean. In the water, fully flooded, the sub weighed ten thousand pounds; when lifted out, it would weigh much more. With *Alvin* still hanging underwater beneath it, *Mizar* proceeded slowly to the protected waters off Martha's Vineyard and placed the retrieved submersible on the shallow, sandy bottom. A barge and crane waiting in the island's lee then made the final lift. Safely resting aboard the barge, the tattered sub made the last leg of its voyage home up Vineyard Sound, to a waiting crowd on the dock at Woods Hole.

Incredibly, a significant bit of science came back with the submersible. When salvagers began clearing debris from the pressure capsule, they found a soggy lunch carried in by the crew on the day *Alvin* had sunk. The contents included a thermos full of broth, which had imploded, some apples, and three bologna sandwiches, which still tasted fresh to one intrepid engineer who took a bite. A skeptical biologist confirmed this and then took the recovered food to a lab. Eventually two Woods Hole microbiologists, Holger Jannasch and Carl Wirsen, published the results of this unintended deep submergence experiment:

> *Alvin* broke surface again in September 1969 after resting almost one year on the ocean floor. In the excitement over her successful recovery, the oceanographers almost overlooked the striking outcome of *Alvin*'s degradation experiment: the food in the box lunch was practically untouched by decay, although containing the usual amount of bacteria.
>
> The broth . . . was perfectly palatable. Four of us are living proof of this fact. The apples exhibited a pickled appearance. But . . . enzymes were still active, and the acidity of the fruit juice was not different from that of a fresh apple. The bread and meat appeared almost fresh except for being soaked with seawater.
>
> In conclusion, the food recovered from *Alvin* after ten months of exposure to deep-sea conditions exhibited a degree of preser-

Bologna sandwich recovered from *Alvin.* (Source: Woods Hole Oceanographic Institution.)

vation that, in the case of fruit, equaled that of careful storage, and in the case of starches and proteins appeared to surpass by far that of normal refrigeration.

Spurred by the still-edible lunch, microbiologists expanded their research. Over the next several years, Jannasch and others injected organic materials into the seafloor to see how quickly bacteria would grow on or decompose them. Jannasch later used these findings to challenge a notion that the deep abyss would be a suitable dumping ground for the world's solid organic wastes—the ultimate garbage pit. "If . . . the true removal of pollutants is intended," he wrote, "then the slow rates of microbial degradation argue clearly against deep ocean disposal."

Biologists Find Rot after All

In many respects, the lunch fared better on the bottom than *Alvin* itself. A year's soaking in brine had corroded metals and left filmy mineral deposits on other surfaces. Pressure had imploded instruments in the crew capsule, and *Aluminaut*'s mechanical arm, in the struggle to insert the toggle bar, had broken propellers and smashed holes in the fiberglass sail.

Recovered late in 1969, *Alvin* remained out of service all through the following year. It resumed diving on May 17, 1971, with biologists' experiments high on the agenda.

The episode of the well-preserved lunch had suggested an image of the deep ocean as an immense refrigerator. The near-freezing cold, along with low levels of oxygen, seemed capable of preserving organic matter indefinitely. Soon, however, shocking results from the experiments of another researcher sharply contradicted the image of a deep ocean in which biological processes shut down in suspended animation.

In 1972, Ruth Turner of Harvard's Museum of Comparative Zoology—the first woman to go on a science dive—used *Alvin* to sink a series of wooden panels into the bottom at a depth of 6,000 feet. After 104 days of exposure, several were recovered. They came back in very bad shape. The wood, riddled and weakened by wood-boring bivalve mollusks, "began to fall apart while being picked up by the mechanical arm of *Alvin*," Turner reported in the journal *Science*. The tiny openings of these animals' burrows peppered the wood's surface. Unlike the hapless surface bacteria that had gone down with the lunchbox and failed to consume its contents, these little mollusks had apparently evolved to cope very well with the deep environment. Years later, Turner guessed that symbiotic bacteria in the mollusks' gills were helping them digest the wood.

Turner's discovery of the voracious mollusks was only one of many confounding surprises presented to marine biologists by the ocean depths. Yet although marvelous in terms of biology, her findings seemed to be terrible news for marine archeologists. If these creatures could destroy fresh pieces of wood in a matter of months, what would be left of ancient ships that had sunk in very deep waters—ships made mostly of wood?

The site where Turner carried out her experiments was known as Deep Ocean Station number one (DOS No. 1), the first long-term research station established on the deep seafloor using manned submersibles. Researchers had selected the location a year earlier, in 1971, because it lay along a line between Woods Hole and Bermuda, two research bases that *Lulu* and *Alvin* often visited. While plying between those points in earlier years, researchers had conducted many net tows from the surface. Those tows had established a good database against which to compare the newer results from *Alvin*. Although other deep stations have since been established, scientists continue to return to DOS No. 1 even today.

Fred Grassle, a Woods Hole biologist, has been one of the most faithful users of DOS No. 1. In long-running experiments, he has placed trays with nutrients on the seafloor to study microbes and animals that naturally grow there—what kinds, how many, how fast, consuming which nutrients, and so on. Other biologists later put down small instrumented jars to measure respiration, and they found that deep-sea animals need ten to a hundred times less oxygen than their shallow-water counterparts.

Thus biologists quickly got over their "oh's and ah's," as Beebe once put it—the early, unproductive phase of sheer amazement. Within a few years after Hessler had made the first science dive in *Alvin*, the whole field of benthic biology was rapidly moving into a far more quantitative phase.

Illusionists Exposed

While some biologists expanded their work on the ocean floor, others used small submersibles to investigate a midwater mystery. Scientists had long known that certain layers of seawater above the bottom scatter sonar signals. These odd chunks of water appeared and then vanished, as if a layer of the sea had temporarily become a murky solid. Such deep-scattering layers, or DSLs, sometimes gave false bottom readings that confused submarines and surface ships making depth soundings. These layers of water had fascinated oceanographers since they first appeared on the crude echo sounders of the 1940s.

Fishermen who began using echo sounders quickly assumed that the DSLs were full of squid, which caused the sonar reflections, but they failed to capture many in their nets. Another suggestion came from scientists diving in the bathyscaph *Trieste* off California, who saw dense layers of prawns and other creatures. Could these be causing the deep scattering? Biologists concluded by the mid-1960s that the scattering layers did contain marine organisms, making nightly migrations toward the surface. They were probably small fish or siphonophores, using darkness to hide from larger predators while searching for food in the upper layers of the sea. So dense were the animals as they rose toward the surface that their small bodies collectively registered as false bottoms on echo sounders.

In 1967, Woods Hole biologists Richard Backus and Jim Craddock used a highly sensitive sonar device mounted on *Alvin* in an attempt to actually see the animals making up one of these layers. With lights turned off, they crept down about 2,000 feet in water southeast of Long Island and homed in on a good-size blob on their scope. When they turned on the outside lights, they saw thousands of lantern fish surrounding the sub. Backus described them as a "tremendous cloud. . . . Beautiful, beautiful fishes . . . swimming up, swimming down . . . getting thicker and thicker. You couldn't call it a school because they are moving in different directions, but it's a fantastic aggregation, fantastic numbers."

Mystery solved: the perpetrators were finally exposed, at least in one case. Rarely had any nets towed behind ships on the surface captured a single lantern fish, and yet *Alvin's* net brought up 744 on one dive alone.

Noisy Water and Inconstant Gravity

Since the navies of the world, particularly the U.S. Navy, were the primary sponsors of deep submergence technology, it is not surprising that much of the early research carried out by submersibles had a military agenda. In fact, the U.S. Navy's Deep Submergence Programs Office had a tremendous influence on the direction of deep-sea studies.

The field of marine research most interesting to the navy during the cold war was underwater acoustics. With the deployment of nuclear warheads aboard Soviet submarines in the late 1950s and early 1960s, antisubmarine warfare dominated navy thinking. One of the most ambitious goals was to know as precisely as possible, at all times, the location of every single Soviet submarine as it prowled beneath the waves. Achieving this goal involved the development of sophisticated listening systems, and interpreting the data required a detailed knowledge of underwater acoustics.

For that reason, the navy's Underwater Sound Laboratory used the *Soucoupe, Star III,* and *Deepstar 4000* in seeking to understand the effects of depth and surface waves on ocean sounds. The navy also sponsored research in bioacoustics—the study of cries or noises generated by living creatures, such as whales, dolphins, and shrimp. Since many submarine-hunting sonars operated at great distances and commonly bounced signals off the bottom, the navy was equally interested in the acoustical properties of seafloors. *Alvin* and *Deepstar 4000* measured the loss of acoustical energy under various bottom conditions and collected core samples for laboratory analysis. Geologists not associated with the military would have loved to get their hands on some of this data, but it was to remain classified for many years.

A new task considered to be of utmost urgency befell the navy with the development of intercontinental ballistic missiles in the 1960s. The military needed to have a precise knowledge of the earth's gravitational force all along the path a missile would follow from launch site to target. The pull of gravity on any object varies with the density of the materials underneath it, such as different kinds of rock or depths of water. Even the slightest variation will influence the final path of a descending warhead. The targets for most of the missiles in America's nuclear arsenal were not the large cities of the Soviet Union but the small silos in which Soviet missiles were hiding. To hit these "hardened" targets required great precision and a thorough mapping of the earth's gravitational field.

Measuring tiny variations in gravity while at sea, however, was a tricky business. The rolling and heaving of a surface ship had to be counterbalanced through the use of a sophisticated, stable platform on which the delicate gravimeter was mounted. Yet such motions were muted underwater, and therefore the military had been making gravity measurements for years in conventional submarines. But they dove to relatively shallow depths, where wave action could still affect measurements. In the 1960s, military groups in both France and the United States sponsored studies to determine whether manned submersibles diving to greater depths would make better platforms for gravity measurements. They soon discovered that the slightest movement by any occupant disturbed such a small craft enough to affect the measurements. To maintain accuracy, the submersible had to squat on the bottom and pump in as much seawater for ballast as possible. Then the crew had to remain as still as possible.

Such a procedure meant that only a few readings could be taken and a small distance traversed before the sub had to return to the surface. When orbiting satellites and large surface ships devoted to military research proved more efficient, the navy quickly lost interest in using small submersibles for this work.

A Mass Financial Extinction

Although the early dives in manned submersibles produced no earthshaking scientific breakthroughs, they established much of the technical know-how for major discoveries that were soon to follow. That was in itself a significant advance. In a way, the dispelling of false expectations for instant wealth was also a kind of progress. The early, sometimes faltering, and always expensive steps toward the deep abyss eventually discouraged all but the most determined fortune-hunters. The mineral deposits, the abyssal oil, the deep-sea fisheries that were supposedly just waiting for commercial harvest—none of those could be exploited. The costs would simply have been prohibitive.

As the optimism of the 1960s subsided, it became clear that too many deep submersibles were competing for too few users. The logic of supply and demand soon winnowed out the less competitive programs. By the early 1970s many if not most of the small submersibles built during the 1960s—more than a hundred worldwide—had been retired from service. At the same time, dreams of a Wet NASA were clearly doomed. Even the original NASA was facing cutbacks soon after humans had landed on the moon. There would be no space station or manned mission to Mars anytime soon. The war in Vietnam was devouring U.S. resources that might have gone toward such programs, but that was not the only explanation. Somehow, as often happens, society was losing its will to explore.

Just as plans for *Alvin* had surged forward on high expectations for deep-sea expansion, the little sub could not buck the ebbing financial tide when priorities changed. Once the navy took delivery of its own submersibles (*Sea Cliff, Sea Turtle,* and the nuclear-powered *NR-1*), it had little use for *Alvin,* the quirky pioneer. In the summer of 1969, the Office of Naval Research informed Woods Hole that funding for its submersible operations would be phased out over the next three years. I was one of the officers in the delegation that brought this heartbreaking news to Bill Rainnie, an *Alvin* pilot and also the project manager of Woods Hole's Deep Submergence Group. It was one of my last duties before I completed my service in the navy.

Meanwhile, I was making arrangements to finish my doctoral work at Woods Hole. Since my active service in the navy was nearing its end, I badly needed a job. Rainnie gave me one. Unofficially, I had already become what he called a "cheerleader" for *Alvin.* He now hired me to drum up paying customers who would keep the submersible busy—scientists, we presumed, who would spend their precious grant dollars on research trips using *Alvin.* It was I who asked Ruth Turner of Harvard, for example, if she would like the submersible to put wood on the bottom of the ocean to see if any mollusks would bite.

Day by day, week by week, I helped the *Alvin* group line up other users. Too often, none of us knew where the next money would come from. *Alvin* went on like this for years, at times on the verge of going under, financially, for good. More than once, Rainnie advised everyone in the group to start searching immediately for new employment. But before that long agony unfolded, one of the craft's earliest customers turned out to be me.

— Scientific Combat Training

By the summer of 1970, I had found sponsors—initially the Pentagon's Advanced Research Projects Agency (ARPA) and later the National Science Foundation—willing to support an ambitious mapping project in the Gulf of Maine. The location I wanted to study probably held no interest for ARPA; the military was interested instead in the navigation techniques we proposed to develop.

My mapping would unfold in two phases. First, I intended to use *Lulu* to make seismic soundings from the surface while *Alvin* was being overhauled after its yearlong dunking. The following summer, I would dive in *Alvin* to collect rock samples. This would help keep Rainnie's group in business at Woods Hole. It would also give me the data I needed for my doctoral dissertation.

Unlike most submerged continental margins, which consist of soft sediments, the Gulf of Maine contains hard bedrock—a seaward extension of the Appalachian mountain range. On land as well as undersea, these mountains consist of crystalline igneous and metamorphic rock dating back hundreds of millions of years. I believed that a thorough study of these submerged rock formations would support a new theory known as plate tectonics: the idea that the earth's continents ride on crustal plates that drift slowly over the surface. My findings might help clarify how continents now thousands of miles apart, on opposite sides of the Atlantic Ocean, had once been joined together in a single piece.

To investigate this area with the required precision, I would have to adapt traditional field-mapping techniques that geologists used on land. Such maps display geological information in three dimensions, including not only the surface exposure of rock formations but also their subsurface structure and composition.

Fortunately, as my starting point I had obtained detailed bathymetric maps showing the topology, or contours, of a region measuring some forty thousand square miles. These maps provided a picture of the seafloor's surface—the hills and valleys—but not its underlying structure.

Unfortunately, getting information about the deeper structure required the use of seismic profiling techniques. Basically, that meant cruising back and forth on the surface while shooting off a very loud air gun underwater every twenty seconds. These shock waves would bounce off the seafloor surface as well as its deeper layers, and instruments that analyzed the returning waves would reveal the underlying structure, in much the same way you can tell from a rap on a door whether it is made of wood or metal, and whether it is solid or hollow. But these very loud shock waves also reverberated on the hull of the research ship every twenty seconds, nonstop. Imagine trying to sleep through that.

By September 1970, *Lulu* had performed this part of the work from the surface. We survived the maddening noise without any murders—but just barely. One night, the sight of a seasoned crewman rushing toward me with flaming red eyes and a knife in his hand convinced me to call off the soundings for a while.

In July 1971, we returned to the Gulf of Maine with *Alvin* to sample the bedrock. This could not be done except by submersible, because the entire Gulf had undergone extensive glaciation during the ice age. As the glaciers retreated, they had dropped sand and gravel all over the area, and those deposits bore no relation to the underlying bedrock. As a result of this recent glacial activity, traditional dredging operations almost always collected rounded, glacier-borne rocks called "erratics" while missing the more difficult-to-sample bedrock sticking out in isolated areas. To find those outcrops with *Alvin*, we had to navigate precisely. Our seismic profiles helped us pinpoint the outcrops, and subsequent dives in *Alvin* brought back the samples.

Drilling cores from this rock proved challenging. Our situation while neutrally buoyant was similar to that of astronauts trying to use tools while weightless in space. Torque from the drill would often set *Alvin* rocking, but pilot Ed Bland became expert at working the controls, and sooner or later we always got the samples we needed.

The Gulf of Maine turned out to be a good training ground for dives we would make a few years later. Although I had no way of knowing it at the time, *Alvin* was about to escape the relatively shallow depths of the continental margins to dive in the open ocean. There the consequences of succeeding or failing to gather samples would become far more serious. Hypotheses, reputations, and careers would be at stake. Out in the deep Atlantic, in the largest yet least-known mountain range on earth, the long struggle between traditional geology and plate tectonics was reaching its climax.

CHAPTER

5

THE MIDOCEAN RIDGE
Womb of the Earth

Geologists have a new game of chess to play, using a spherical board and strange new rules.

—*Patrick Hurley*

A theory that most academics quickly branded eccentric, preposterous, and improbable took shape in the mind of a wounded German soldier during World War I. Alfred Wegener, the conscript who proposed it, had done wide-ranging research in the earth sciences. As he lay recovering in a military hospital, he mulled over a large number of observations that had intrigued him for years. Like others before him, he noted how the shapes of continents fit together. South America, in particular, fit almost perfectly into Africa once you removed the Atlantic Ocean. He also found matching geographical features. The Appalachian Mountains of North America, for example, seemed to continue in Scotland, and certain rock formations in Brazil lined up with similar formations across the ocean in Africa. Even the locations of fossils matched: Wegener pondered impressive lists of identical plant and animal species found on opposite coasts of the world's oceans—ancient forms of life that could not have crossed such vast expanses of water. What could explain all these interlocking patterns?

According to conventional views, little explanation was needed. Geologists tended to dismiss matching continental shapes as coincidence. The fittings were rarely exact, after all, and some looked highly unlikely. Matching geographical features could also be coincidence, or perhaps in a few cases real. Some parts of ancient, ocean-spanning rock formations might have sunk beneath sea level, while other parts lifted higher. There was nothing remarkable in that; basins have been sinking and mountains rising for eons. The same sort of vertical heaving and sagging could explain the presence of matching fossils on opposite coasts: ancient land bridges, now submerged, might have connected continents from time to time, allowing species to cross and mingle.

These reasonable explanations (or dismissals) of interlocking continental features never convinced Wegener. In 1915 he published *The Origin of Continents and Oceans,* a book that offered a strikingly new perspective. In his long treatise, Wegener proposed a simple explanation for all the apparent similarities across oceans. Long ago, he argued, there was only one continent, a gigantic supercontinent. Somehow, around 200 million years ago, it began to split into pieces. As the fragments slowly migrated to their current positions, new oceans such as the Atlantic filled the widening gaps between lands that had once been joined.

Wegener's theory became known as "continental drift." His book, which went through several editions in the 1920s, did seem to explain not only the coincidental features among the present-day continents but also absurd displacements over time, such as evidence of past glaciers near the equator and tropical fossils in frigid Antarctica. (These lands had drifted into new climate zones, Wegener suggested.) Yet most experts viewed the piles of evidence Wegener amassed as so many unrelated facts dumped together between two covers. That was because Wegener could not explain one key aspect of his theory: How could continent-size slabs of rock move sideways across the planet's surface, short of magic carpets or levitation?

Geophysicists rightly rejected every mechanism that Wegener suggested. In the first place, they argued, continents would never have survived the enormous stress of plowing through solid seafloor. If they had indeed moved so far, they would now be distorted and broken beyond recognition. More important, no conceivable force could possibly shift so much of the earth's crust. The gravitational pull of the sun and moon, for example, sloshes tides back and forth a few feet in the oceans, but tidal forces are much too weak to move solid earth any appreciable distance.

Wegener froze to death in 1930 while returning from a remote weather station on the ice cap of Greenland. His grand synthesis died with him. "Continental drift" entered geology textbooks not as a unifying theory but as a classic example of wild, overreaching supposition—a fantastic notion held up to scorn.

A Revolution in the Earth Sciences

As an undergraduate, I absorbed the conventional views about continental drift. On a winter evening in 1968, however, a respected geologist named Patrick Hurley shattered my sense of what's likely or absurd. I was listening to Professor Hurley's lecture at the Massachusetts Institute of Technology, trying to keep up my studies in geology while still serving in the navy. His research group had been calculating the ages of rocks using radiometric dating, a technique far beyond anything available during Wegener's time. Most crystalline rocks contain traces of radioactive elements, and their presence in varying amounts can be a telling signature, unique to the time they were formed.

Hurley's group had collected rock samples in the Appalachians and compared them with samples from similar formations in Europe. They had also sampled rocks in Brazil and compared them with rocks in Africa. Their results astonished me. These rocks from opposite sides of the Atlantic were not merely similar; they were the same. Displaying pictures of rocky terrain taken in the field, Hurley showed how a formation that ended on the east coast of Brazil reappeared in a bluff on the west coast of Africa. Those photographs showed powerful similarities. Now the radiometric comparisons showed identity. And this new evidence seemed irrefutable. These massive layers of rock had been formed at the same time and place, and yet now they were separated by thousands of miles. Somehow they had pulled apart, leaving the Atlantic basin between them.

In a few brief minutes I had lost my bearings. How could whole mountain ranges pack up and move—change their planetary address? Then my understanding of earth's long history began to rearrange itself, in accordance with new ideas.

What made the moment of insight even more exciting was my knowledge that Hurley was not alone in his challenge to conventional geological wisdom. In the 1950s, new evidence had begun to emerge that revived the debate about continental drift. By the late 1960s, more and more earth scientists were becoming convinced that the continents had indeed wandered. And most of the supporting evidence was coming from marine geology. New techniques for examining the seafloor were offering a second opportunity to test Wegener's bizarre ideas.

The first clues came from careful seafloor mapping. Before the nineteenth century, most people thought that the ocean floor was relatively flat. However, depth surveys made by dropping weighted lines in the Atlantic and Caribbean had revealed, by the nineteenth century, a highly irregular seafloor. Especially intriguing, a line of underwater mountains dotted the mid-Atlantic. The picture of these mountains sharpened after World War I, when echo-sounding measurements clearly revealed that they formed a long, continuous chain in the central Atlantic. In the 1950s, oceanic surveys conducted by many nations expanded the picture once again, showing that the Pacific, Indian, and Arctic Oceans contain similar ridges. In 1956, Bruce Heezen and Maurice Ewing of Columbia University noticed that earthquakes in the ocean floor occurred along a globe-encircling line that followed all these ridges. The ocean ridges were apparently a connected formation—a single mountain range, submerged on the deep ocean floor, that encircled the planet.

Called the Midocean Ridge, this immense global mountain chain—some 42,000 miles long and, at its widest, more than 500 miles across—zigzags between the continents, winding its way around the globe like the seam on a baseball. Though hidden beneath water, the Midocean Ridge is the largest geological feature on earth. Rising in places more than 15,000 feet above the seafloor, it covers about 23 percent of the planet's total surface. And the path that it follows often seems to mirror the shapes of the continents.

Dredging operations recovered basalt from the summits of these ridges. Basalt, a common volcanic rock, solidifies when magma cools. Mapping surveys in the Atlantic also revealed that a rift valley—a linear zone of volcanic eruption—runs along the length of the submerged mountains. By the late 1950s and early 1960s, a few scientists were suggesting that the Midocean Ridge marked the location of weak zones in the earth's crust—those areas where the surface stretched like taffy, became thin, and cracked. Magma then rose through the cracks and erupted into the rift valleys.

The geologists Harry Hess of Princeton University and Robert Dietz of Scripps (who had pioneered the scientific use of the bathyscaph *Trieste*) separately provided critical insights that explained these features. The earth's crust was thin at the Midocean Ridge, they proposed, because segments of seafloor on each side of it were pulling away from each other. Magma welled up in the gaps, forming new crust. This new crust in turn stretched thin as it welded itself to the older receding crust, allowing still more magma to continue welling up and creating ever new pieces of seafloor in assembly-line fashion. Meanwhile, the preceding batches of crust moved farther and farther away from the ridge. In 1962 Hess published these ideas in an article titled "History of Ocean Basins," and Dietz gave the theory a name that stuck: he called it seafloor spreading.

Hess and Dietz were among the few who understood the broad implications of seafloor spreading. If the earth's crust was emerging and expanding along the Midocean Ridge, Hess reasoned, it must be compressing and disappearing somewhere else. He suggested that while new crust spreads away from the ridges, older crust converges at the deep ocean trenches. When two approaching crustal slabs meet, one bends downward, squeezes under the other, and, in effect, recycles. The sum of all this spreading and converging describes a seafloor in continuous motion: it emerges in midocean, spreads away from either side of a ridge, and sinks into trenches somewhere else, moving across the earth's surface like a conveyor belt. Continents, which are made of lighter rocks than the seafloor, ride higher above it. Ocean basins open up between continents that are drifting apart and close between approaching continents. And the continents that once fit together, as Wegener suggested, are still drifting. They are not plowing through solid seafloor (geophysicists had been right about that) but instead riding over hot, pliable, semimolten rock tens of miles beneath the crust.

This entire mechanism—seafloor spreading, continental drift, the seafloor diving into trenches, and the volcanic recycling of crust—later became known as *plate tectonics.* According to this theory, the huge slabs of crust moving around on the earth's surface are rigid "plates" that include both continents and seafloor. Most of the boundaries between plates lie under the ocean, along its great ridges and trenches.

Hess based his ideas largely on intuitive geological reasoning, and at first he won few converts. Earth scientists could accept the fact that a great volcanic mountain range encircled the globe, yet reject the idea of crust moving sideways. The Midocean Ridge might be nothing more than a vertically rising welt on a static seafloor. It seemed that traditional views might still survive. But then an amazing flood of evidence all but clinched the case in the minds of many geologists.

The most stunning breakthrough came in 1963, when Frederick Vine and D. H. Matthews of Cambridge University in England published magnetic maps of an area in the Indian Ocean known as the Carlsberg Ridge. These maps revealed a new kind of interlocking pattern: a perfect match in magnetic properties between slabs of seafloor on opposite sides of the Midocean Ridge.

Apparently, the earth's magnetic field reverses polarity—the north and south magnetic poles "flip"—for reasons that are not completely understood. There have been 171 reversals in the past 76 million years. The history of these reversals can be seen in basalt because magnetized iron particles in the molten rock freeze in one orientation or the other as magma cools and solidifies. The particles behave, in effect, like fossilized compass needles. The Vine-Matthews maps revealed parallel strips of ocean floor with a zebralike pattern of magnetic orientation: a strip of one polarity alternated with the next strip of opposite polarity, in a pattern that became known as *magnetic striping.* Most astonishing to scientists, the stripes were symmetrical, or mirror images of each other, with the line of symmetry at the very center, or axis, of the Midocean Ridge.

Hess's concept of seafloor spreading explained this pattern. The moving deep-sea crust acted like a natural tape-recording of reversals in the earth's magnetic field. Each magnetic stripe on one side of the ridge matched a stripe on the opposite side because both had been formed at the same time and place, in the middle of the ridge, and then been ripped apart as the seafloor spread away in opposite directions. This evidence won many new converts to the theory of plate tectonics, but even more was to come.

In 1947 Maurice Ewing, leading an expedition on the U.S. research ship *Atlantis,* had found sediment layers on the Atlantic seafloor to be far thinner than expected. Many scientists believed that the oceans had existed without much change for at least 4 billion years, so the sediment layer should have been thousands of feet deep. Why then was there so little accumulation? Researchers looking for answers pulled up drilling cores from all over the seafloor, especially during the 1960s. They found that the ocean bottom was composed of surprisingly young rock, no more than 200 million years old—nowhere near the 4 billion years of traditional theory, and suggestively close to the length of time that Wegener had given for continental drift. It was the *Glomar Challenger* expeditions, starting in 1968, that eventually settled the issue. Samples of rock taken systematically from seafloor on both sides of the Midocean Ridge established a pattern similar to magnetic striping. The ocean floor nearest to the ridge had very little sediment. It was very young, and sediments had not had time to accumulate there. The seafloor got progressively older, with more and more sediment on top, as researchers took samples farther away from the ridge in either direction. Again, the concept of seafloor spreading could account for this new evidence.

Finally, as geologists got hold of more sensitive instruments to chart the locations of earthquakes and tremors in the 1960s, they found yet another astonishing pattern. As hundreds and thousands of dots marking the origins of earth-shaking tremors and volcanic rumblings accumulated on maps of the world, they clustered precisely along the boundaries proposed for tectonic plates: the great ocean ridges and trenches. Those lines marked the spots where the earth's crust was most actively slipping and sliding, without question. According to the new theory, those were precisely the places where the great conveyor belts started or stopped—the places where slabs of crust were ripping apart, slamming together, or grinding sideways against one another. Eventually, geologists realized that 90 percent of all volcanic and earthquake activity happens undersea—just as it should, if that's where magma emerges to form new crust, and where old crust descends back into the earth.

All these new lines of evidence were stirring passionate debates as I began my graduate studies. It felt as if a civil war had erupted between traditional geologists and those who accepted plate tectonics. The new theory of a young, dynamic seafloor and drifting plates of crust represented a revolution as comprehensive and far-reaching for the earth sciences as Newton's theory of gravity for astronomy and physics or Darwin's theory of evolution for the life sciences.

By the late 1960s, the balance seemed to be tipping in favor of the new theory. Certainly, the main battleground had shifted from dry land (the site of virtually all traditional fieldwork) to the oceans. In little more than a decade, the deep seafloor had completely reversed its image, from a rather boring geological dead zone to the hot spot of current research. According to plate tectonics, the continents (and all dry land) were mere passengers on the earth's crust—floating like so much solidified scum. The primary moving and shaking occurred undersea, in the abyss, and played itself out far more in the horizontal than in the vertical dimension. Suddenly, earth scientists had "a new game of chess to play," in Hurley's words. And play they did.

Project FAMOUS

The new theory had gained ground in the 1960s thanks to indirect observations, such as magnetic and seismic surveys. What was needed next was direct observation of the ocean floor, especially within the central rift valley of the Midocean Ridge, where many geologists now believed that new seafloor was being created. It would be hard to imagine more convincing evidence than the actual sighting of brand-new seafloor. Yet "sighting," for geologists, involves rigorous description with precise mapping and sampling—traditional, hard-rock fieldwork. No one had ever attempted such a thorough exploration in the deep ocean.

As the 1970s got under way, only two nations had the wherewithal to undertake such a mission. By then, the United States and France had emerged as the global leaders in the development of deep submergence technology. In France, a new organization named CNEXO, for Centre National pour l'Exploitation des Oceans, had consolidated the country's various marine programs. This was the French version of a Wet NASA, run by engineers. Its first project was to construct and launch, in 1970, a new deep submersible called *Cyana,* built around a high-strength steel pressure sphere with a 10,000-foot depth capability. It could dive nearly twice as deep as *Alvin,* far enough to reach the rift valley in the Mid-Atlantic Ridge.

Late in 1971, my academic mentor at Woods Hole, the geologist K. O. Emery, received a letter from Xavier Le Pichon, a former student of Maurice Ewing and a strong supporter of plate tectonics. Now in charge of a major new French marine laboratory, Le Pichon wanted to use the bathyscaph *Archimède* and the new submersible *Cyana* to investigate the Midocean Ridge. He was suggesting a joint program between France and the United States. He was keenly aware of Woods Hole's submersible *Alvin,* and he knew that we were using it for fieldwork on the continental shelf. But geologists all over the world, especially those interested in plate tectonics, were now eager to explore much deeper. In his letter, Le Pichon briefly explained what he had in mind—a detailed mapping effort in the rift valley of the Mid-Atlantic Ridge—and asked Emery if he thought submersibles were up to the challenge.

Since I was using *Alvin* in the Gulf of Maine, Emery asked me to draft a response. Our studies had clearly showed that a manned submersible could be used as a field mapping tool, in conjunction with surface ships. We had also demonstrated that scientists using *Alvin* could make precise, detailed observations and carefully collect rock samples from exposed outcrops. I drafted an enthusiastic letter, and Emery signed it.

The idea of a joint French-American expedition to the Mid-Atlantic Ridge had been brewing at the highest levels; government officials had been discussing it since the late 1960s. But before such a major program could receive funding, it needed a considerable amount of support within the earth sciences community. CNEXO, the new French organization to which Le Pichon's laboratory reported, could make decisions without significant outside review. In the United States, however, that was not how the system worked. The National Science Foundation was the obvious source of funding, and peer review played a large role in its decisions. The established titans in a field often determined which upstarts, if any, would receive support.

It took a minor revolution to get the titans' grudging endorsement for a major expedition using manned submersibles. Many earth scientists still considered these vehicles to be outrageously expensive, limited in range, and unproved in fieldwork—a risky gamble for important new research. The skirmishing grew especially heated over issues of money and control. As the fight to pry sufficient funding away from more conventional programs unfolded, with respected geologists and geophysicists arguing for both sides, I remember all too well the substantial finger of Maurice Ewing wagging in my face. Ewing was a strong champion of traditional oceanography: he had used surface ships, dredges, instruments, and probes in many pioneering studies that supported plate tectonics. As he waved his finger in my face, he was threatening to have *Alvin*'s pressure sphere melted down into paper clips if it failed in this mission.

The international project that eventually won funding became known as the French-American Mid-Ocean Undersea Study (Project FAMOUS). Its chief scientists were, to no one's surprise, Jim Heirtzler, chairman of Woods Hole's Department of Geology and Geophysics, and Xavier Le Pichon. Both had been strong supporters of plate tectonics, and both were graduates of the Lamont Laboratory at Columbia—one of the most respected institutions in oceanography, and one that Ewing himself had founded.

The area that the FAMOUS researchers selected for investigation lay between 36 and 37 degrees north latitude. They picked this segment of the Mid-Atlantic Ridge for several reasons. The ocean there often had calm, favorable weather in

Physiographic map of the Atlantic Ocean compiled by Bruce Heezen and Maria Tharp, showing the study area on the Mid-Atlantic Ridge for Project FAMOUS. (Source: National Geographic Society Image Collection.)

midsummer, and it was close to a good support base some 400 miles to the northeast, in the Azores. Most important, scientists determined that this site represented a typical segment of the Midocean Ridge. They wanted to avoid exceptional volcanic hot spots, like those in the Azores, but they still wanted to explore an active crust-forming zone.

All evidence confirmed that this was indeed an average part of the ridge, still playing tectonic chess. Aerial magnetic surveys, conducted both then and later, quantified that notion: they suggested that seafloor in this region was spreading at the rate of 2.3 centimeters, or nearly one inch, per year. Although that might seem slow by human standards—roughly the speed at which fingernails grow—one inch per year accumulates to huge distances over the immense time spans of geology. At that rate, the undersea conveyor belt would easily have traveled the whole width of the Atlantic Ocean in 200 million years—just right, according to theory.

A Long International Countdown

Never in the history of deep submergence had so much effort gone into prepping a dive site and divers. The preliminary work resembled the kind of planning, detailed study, simulation, and training that goes on before a major space mission. Teams of scientist-divers, French and American, traveled the world to learn as much as they could about volcanic terrains in Hawaii and Iceland, and along the Assal Rift in East Africa (a dry-land region below sea level where one can actually observe "seafloor" spreading in air). Since many of these scientists had little or no experience with manned submersibles, they also received extensive dive training.

Although Project FAMOUS would become widely known for the first use of manned submersibles on the Midocean Ridge, it actually employed every major technological tool then being used by marine geophysicists. For several years, scientists from the United States, Canada, England, and France conducted a comprehensive series of field programs preparing for the historic dives.

First and foremost, the mission required excellent topographic maps. This was particularly true for the Midocean Ridge, given its complex terrain, but also because its physical appearance reflected the underlying forces that were constantly reshaping it. The features that scientists saw on the seafloor would tell them precisely where it was lifting, sinking, shearing, or spreading.

Fortunately for Project FAMOUS, the U.S. Navy had developed a highly classified method of collecting precise topographic data in the deep sea. This mapping technology, called the Sing Around Sonar System (SASS), consisted of large sonar devices mounted on the hulls of two World War II Liberty ships. The sonar signal, a fan-shaped beam, traveled downward from these ships to the ocean floor and bounced off its various topographic features. The reflected energy then traveled back toward the surface, where computers processed these signals into ninety narrow beams (measuring one degree by one degree), each recording a separate depth. In effect, a contour line perpendicular to the direction in which the ship was traveling was divided into ninety separate dots. Connect the dots, and you had an accurate slice of ocean floor. Unfortunately, not all of the ninety beams could be used, because the ships rolled from side to side as they traveled over the ocean. For that reason an inertial navigation platform and other sensors had to select the central vertical beam, which varied from one slice to the next as the ship rolled, and thirty beams to either side. As the ship moved forward to record additional slices, computers combined them to produce a contour map.

Ships that used SASS could gather and process amazing amounts of data while ripping through the water at speeds of twenty or more knots. It was an extremely powerful tool, and the oceanographic community had never had access to it. For this historic expedition, however, the U.S. Navy surveyed the

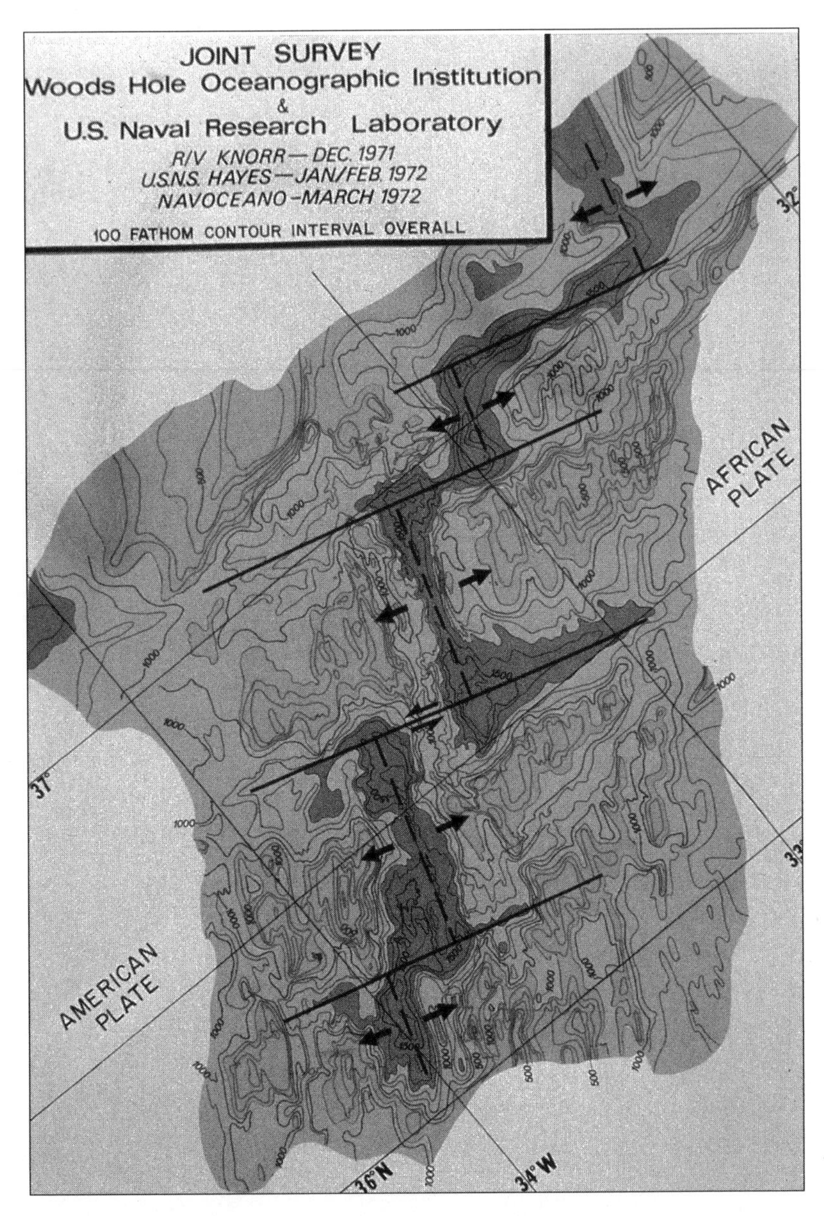

Bathymetric map of the Project FAMOUS study area, 100-fathom contour interval. (Source: Project FAMOUS.)

FAMOUS area to produce unclassified maps at contour intervals of five fathoms.

Other surface ships, meanwhile, crisscrossed the central rift valley using towed air guns that released large amounts of seismic energy. These explosive sounds penetrated deep into the crustal rock underlying the mountain range. After bouncing off its hidden internal structure, the returning signals were picked up by sensors at the surface. They too revealed a typical oceanic crustal structure beneath this representative segment of the ridge.

Such research ships also measured swarms of micro-earthquakes using seismographs placed on the ocean floor, and they recorded similar activity using arrays of floating hydrophones called sonobuoys. They found that the rift valley was almost constantly trembling. Meanwhile, geophysicists made precise heat-flow measurements showing that the seafloor was giving off more heat closer to the ridge, as would be expected given the magmatic upwelling occurring there. Scientists, in short, used every traditional sensor they could think of to ensure that they had chosen the right place for the study.

With these physical profiles established, the stage was set for detailed mapping. The emphasis now shifted from indirect geophysical sensing to more direct kinds of visual imaging. This phase of the project would produce data more familiar to field geologists, who had waited on the sidelines for years as geophysicists probed the seafloor. The scepter was being passed.

First came the side-scan sonar systems, riding on sleds. In particular, the large-area system called GLORIA (Geological Long-Range Inclined Asdic) from Great Britain (towed near the surface) and the Deep-Tow system from the Scripps Institution of Oceanography (towed near the bottom on a very long tether) added finer texture to the topographic maps. These surveys revealed the rift valley's rugged central zone, thought to be volcanic, and a complex network of fissures and fault scarps

Photograph taken by the towed camera sled ANGUS of small, haystack-like volcanic cones on the summit of central volcanic highs in the rift valley. (Source: Woods Hole Oceanographic Institution.)

reflecting how tectonic forces acted on the newly formed ocean floor. Diving teams would later zero in on those fissures and scarps.

Next came photography. Although the Deep-Tow system snapped a limited number of bottom photographs, the LIBEC (Light Behind the Camera) system from the Naval Research Laboratory brought back long film strips that formed continuous mosaics of the rift valley floor. Woods Hole's towed-camera system named ANGUS (Acoustic Navigated Geological

Undersea Surveyor), built specifically for Project FAMOUS, added thousands more photographs. These images began to give the American and French dive teams what they wanted: the first visual glimpse of the geology awaiting below. Scientists spent days walking around the pictures, laid out all over the floor of a navy gymnasium in Washington, D.C. At this point their excitement compared with that of space scientists receiving their first photographs from a new planetary mission.

In these images, researchers could clearly see extensive lava flows running down the central axis of the rift valley. The lava seemed unaltered by tectonic disturbances; it was fresh. Flanking this eruptive zone to either side were older, more sediment-covered volcanic terrains, deeply cut by open fissures. Those features suggested the earliest phase of seafloor spreading. Even farther away from the rift's central axis, but still within the valley floor, they could see extensive accumulations of rock debris, or talus, which had fallen to the base of vertical, inward-facing fault scarps. Those features could be viewed as the later effects of spreading, with larger displacements and complete fractures.

This primitive volcanic terrain had no equal anywhere on land. Glimpsed only through the lenses of black-and-white still cameras towed from the surface, it nevertheless looked spectacular. It also looked forbidding, as the Grand Canyon must have looked to its first geologist-explorers. Like the wooden boats of those river-running geologists, the heavy steel frames of the towed camera systems returned badly damaged by numerous collisions with an unforgiving, hard-rock terrain. They had banged their way through a valley, after all, hemmed in on two sides by steep cliffs. All around this canyon, and even inside it, mounds of craggy rubble and sharp, unweathered peaks awaited the approaching submersibles like rows of teeth in the mouth of a shark.

Mosaic photographs taken by the towed camera sled ANGUS of a fissure running down the axis of the rift valley of the Mid-Atlantic Ridge. (Source: Project FAMOUS.)

A Bathyscaph Leads the Way

It was August 2, 1973, when Xavier Le Pichon jumped aboard an inflatable rubber Zodiac for a short trip to the bathyscaph *Archimède,* wallowing in the gentle North Atlantic swell. Accompanying Le Pichon on this first dive to the ridge were two other crewmen: veteran pilot Gerard de Froberville, a naval officer and protégé of the *FNRS-3* pioneer Georges Houot, and Jean-Louis Michel, a CNEXO engineer. The French had decided that their bathyscaph should go first, in spite of its limited ability to maneuver through rugged terrain, because *Cyana* was still undergoing final development and testing.

Once the bathyscaph broke tow from its support ship, *Marcel Le Bihan,* it began to drift in the surface current. By the time it could submerge and start its one-hour free fall to the rift valley floor, de Froberville discovered they had floated a considerable distance from the expected dive site. Fighting a bottom current that almost equaled its maximum speed of two knots, the bathyscaph labored to get back on course. It finally entered the rift valley and came down near the foot of a vertical scarp more than 150 feet high. Le Pichon later described the landing:

> I must say I was very excited. The navigation was difficult and the pilot had to maintain the *Archimède,* which is fairly bulky, within three meters of the scarp against this fairly strong current. The pilot had to use simultaneously [and] nearly continuously our three propellers and this used up a lot of energy. We scraped the hull of the *Archimède* several times against the rocks [a dangerous, nerve-wracking tendency of bathyscaphs]. We landed a little bit downslope on a talus of broken pillow lavas, obviously fallen from the advancing flow fronts. There we sampled a fairly large pillow while fishes kept poking at our porthole.

For the next hour or so, Le Pichon explored a central high spot the French had named Mont de Venus. Since it was presumably a rift displacement giving birth to new seafloor, this French anatomical phrase meant, in approximate English

Photograph taken from the bathyscaph *Archimède* of a pillow lava form resembling a broken egg, from which more fluid lava has flowed. (Source: Jean Francheteau/CNEXO.)

translation, "womb of the earth." On this hill, and along the rift valley axis near it, fresh lava had flowed out of a subterranean magma chamber.

Although the central, volcanic part of the ridge had an overall linear shape, on closer inspection from the bathyscaph window Le Pichon found complex, interlocking volcanic flows. Many were very steep, almost resembling fault scarps rather than the usual rounded edges of lava flows. This rugged terrain made the going difficult. After spending two hours on the bottom, the bathyscaph had covered only a thousand horizontal feet. It spent much of its time rising up and over two lava flow fronts, then sinking down again behind them. But in traveling that short distance, its occupants made the first close inspection of new seafloor emerging between crustal plates.

The next dive into the rift valley proved even more difficult—in fact nearly fatal. As chance would have it, that was the day I finally realized my boyhood dream of diving to the bottom in a bathyscaph.

Sheer luck—bad luck, as it turned out—put me in this position. *Alvin* would not be joining Project FAMOUS until the following year. As the only American member of the dive team out on the ocean with the French that summer, I was high on the list of scientists scheduled to go down in the bathyscaph, for diplomatic reasons. Le Pichon asked me to make the second dive on August 5.

My journey, like his, began early in the morning with a splashing Zodiac ride to *Archimède,* rising and falling on each swell. Once aboard, I scampered down the access tunnel, through the gasoline float, and into the tiny steel gondola attached below. It was a tight fit as the pilot, Lieutenant Gilbert Harismendy, and a CNEXO engineer named Semac crushed in around me.

The descent was cold and wet thanks to the condensation that soon began forming on the inside of the cabin, and the chill I had gotten from the Zodiac trip never went away. That alone would have made the descent miserable, but I had also come down with a fever and the early signs of a strep throat, symptoms I was keeping to myself. Yet these discomforts barely dimmed my excitement at being the first American to dive on the Midocean Ridge and explore its spectacular terrain.

As we neared bottom, Lieutenant Harismendy began running *Archimède*'s propeller at full throttle to give us some upward thrust and help break our fall. This maneuver threw up a cloud of sediment. We landed with a sharp bump.

To see the bottom, I had to peer through a binocular device that allowed more than one of us to use the bathyscaph's single forward viewport. We had come to rest by a vertical fault scarp. The dimensions of this scarp showed up clearly on the bathyscaph's sonar; it extended farther than we could see through the darkness as a long, linear trace. As we rose up this cliff, a second parallel scarp, farther away, appeared on the sonar screen. We were climbing a giant staircase leading out of the rift valley on its eastern, or "African," plate.

Our job along this west-to-east traverse was to recover rock samples from the various lava flows or talus piles at the bases of these inward-facing scarps. Picking up samples was a slow and frustrating procedure, requiring close coordination between the pilot and the scientific observer. All three of us in the crew cabin had to remain as still as possible. If anyone moved or even shifted weight, the mechanical arm changed its position outside our capsule.

Finally, after many attempts, we collected our first sample. Just as we were about to continue our traverse, the bathyscaph lunged forward and then shot upward. A sudden dip in its electrical voltage had tripped an automatic safety feature, causing the bathyscaph to drop several tons of emergency ballast.

We were repeating the same kind of unexpected ascent that Houot and Willm had made in 1954, during the first French dive to the Atlantic seafloor in *FNRS-3.* At the time, I had no knowledge of this historic precedent. When I realized that we were suddenly going up, I felt strong disappointment that our journey in the rift valley had come to an abrupt end. I also felt feverish, and my throat hurt. But my thoughts and feelings quickly refocused as acrid fumes and smoke began to fill our tiny diving capsule.

Lieutenant Harismendy picked up the underwater telephone and began to report on our situation, which was growing worse. To kill the apparent electrical fire, Semac starting shutting down various systems in an attempt to isolate the problem. While he was working through this program, Harismendy turned off the bottles releasing oxygen into the pressure sphere. Denying the fire oxygen should cause it to die out. But at the same time, Harismendy was also denying our lungs the oxygen we needed. With a one- to two-hour ascent still ahead of us, holding our breath was not an option. We reached for the emergency breathing units beneath each of our seats.

Bathymetric profile made by *Archimède* as it traversed a series of volcanic flow fronts in the central volcanic axis. (Source: Robert D. Ballard/Odyssey Corp.)

The smoke in the air was making it hard to see, adding to the confusion. We first had to fill the breathing units with air from our lungs. In other words, we had to inhale the smoky air of our crew cabin, blow it into a bag, and then "rebreathe" this air. A mixture of chemicals in the bag would remove the carbon dioxide—or so the instructions said, in French.

I managed to do this correctly. My only problem was that Lieutenant Harismendy had failed to turn on the oxygen supply leading to my emergency unit. I tried to pull off the mask to tell him, but my two horrified companions thought I had panicked; they shoved it back onto my face. We struggled; I became dizzy. On the verge of passing out, I made a universal hand signal known among divers, drawing a finger across my throat: Not Getting Air. They figured it out, and finally turned on my oxygen.

So ended my first dive to the Midocean Ridge. Fortunately, many others lay ahead. Of all the ways in which disaster might strike in the future, being smothered to death by my own crewmates was at least one I had now rehearsed and would know how to avoid.

In all, *Archimède* made seven dives into the rift valley during the summer of 1973. Slow, clumsy, and difficult to maneuver, it nevertheless labored through this rugged terrain to investigate a swatch of the seafloor five or six miles long. To everyone's surprise, the central high spot—Mont de Venus—was found to be primarily volcanic, built by numerous lava eruptions. They must have been copious and recent, since the hill had not been significantly altered by subsequent tectonic faulting.

These preliminary dives also revealed that despite the tremendous size of the North American and African crustal plates, each thousands of miles wide, the actual zone of lava injection between them is extremely narrow. That made it ideally suited to submersible investigation. Had the boundary separating these spreading plates been as broad as intuition might have suggested, with lava oozing up for many

Divers photographing the submersible *Cyana* (foreground) and the bathyscaph *Archimède* (lights in background) during test dives off France prior to participating in Project FAMOUS, 1973. (Source: William R. Curtsinger/National Geographic Society Image Collection.)

miles on either side, our attempt to conduct a thorough study by manned submersible would have been an utter failure.

— *Alvin* Gains Strength without Gaining Weight

Project FAMOUS moved into its final phase in the following year, 1974. The submersibles *Alvin* and *Cyana* were to join the bathyscaph *Archimède* that summer to carry out the most comprehensive diving program ever attempted in the deep sea. But before they could start, *Alvin* had some catching up to do.

When Xavier Le Pichon first wrote to K. O. Emery proposing the joint expedition, *Alvin* had been certified to a depth limit of 6,000 feet. The average depth of the rift valley in the Mid-Atlantic Ridge was 9,000 to 10,000 feet. To reach that depth, *Alvin* would need a new pressure sphere. Bill Rainnie's Deep Submergence team had been maneuvering for years to get one, but the navy decided instead to cut back its spending on submersible programs. We found, however, a significant exception: the Naval Applied Science Laboratory in Brooklyn was planning to build two titanium pressure spheres similar to *Alvin*'s.

Titanium, an expensive metal that can be difficult to work, has a great advantage over steel: pound for pound, it is roughly twice as strong. A titanium sphere the same size and weight as *Alvin*'s steel sphere would therefore withstand twice as much pressure. It would allow *Alvin* to dive twice as deep, to 12,000 feet.

Navy engineers were hoping to use titanium in advanced submarine hulls. To learn more about its behavior under pressure, they planned to test two experimental spheres in a massive hydrostatic pressure chamber at Brooklyn, increasing pressure until the hulls imploded. That seemed like a waste of good diving capsules to Bill Rainnie; he finally convinced both the navy and Woods Hole to set aside one of those spheres for a "long-duration" test in *Alvin*.

This new sphere was still in Brooklyn, undergoing tests, when Le Pichon wrote to Emery in 1971. It was not ready to be installed in *Alvin* until the winter of 1972–73. The Woods Hole group then spent the following summer working on unexpected problems, such as stubborn small leaks around penetrators in the new hull. The deepest dive that year was only 3,822 feet. By the 1974 diving season, barely in time for Project FAMOUS, the new titanium sphere finally had penetrators that sealed properly, although no one completely understood why some had worked and others had not.

One more question, however, still troubled the *Alvin* group in the early 1970s. Could the submersible endure another fencing match with a swordfish? The potential threat from large fish was turning out to be a real problem. After the first attack on *Alvin* in 1967, I witnessed a similar event in 1969 while diving in *Ben Franklin* off the coast of Florida. It was my first dive in a submersible; we had been sitting several hours on the bottom. While the rest of the crew was occupied with other tasks, I saw a sailfish swimming rapidly in small tight circles, the way a shark will do just before attacking. It seemed aggravated by the submersible's presence. Suddenly the fish rushed toward the viewport through which I was looking. The only protective action I could take was to set off a powerful strobe light. I don't know if that light had any effect on the fish, but it did immediately swim away.

Two years later, in 1971, crew members diving in *Alvin* near Grand Bahama Island saw a streak flashing toward them. In this instance the attacker was a blue marlin. Before it hit, pilot Ed Bland had the quickness of mind to turn off the lights inside the pressure hull. However, he left on two external incandescent lights mounted high atop the sail. The crazed fish attacked those lights, blowing them out, and then sank dead to the bottom. That left the question still hanging: What if it had struck a viewing pane?

After several meetings with concerned pilots, the director of Woods Hole, Paul Fye, asked for engineering tests to determine the vulnerability of *Alvin*'s viewports to future attacks. Jim Mavor, the institution's safety officer, devised an experiment involving billfish swords being fired from an air gun at one of *Alvin*'s spare viewports, simulating an attack at sixty miles an hour. Fish traveling at top speed, not quite that fast, had reportedly rammed swords through thick wooden keels. From these air gun tests, however, it appeared that the sub's Plexiglas windows would survive a direct hit.

—— Hitting a Bull's-Eye from 10,000 Feet

On June 6, 1974, a new Woods Hole research ship named *Knorr* departed for the Mid-Atlantic Ridge. Filled with twenty-

The research vessel *Knorr* (background) towing the support ship *Lulu* (foreground) with *Alvin* aboard to the Project FAMOUS dive site, June 1974. (Source: Emory Kristof/National Geographic Society Image Collection.)

The bathyscaph *Archimède* resting next to its support ship *Marcel Le Bihan* (background) and the submersible *Cyana* resting aboard its support ship *Le Suroit* in the Azorian seaport of Punta Delgado, 1974. (Source: Emory Kristof/National Geographic Society Image Collection.)

four expedition participants—including scientists, technicians, students, and two journalists—*Knorr* was also carrying *Alvin* and towing *Lulu.* It joined the two French submersibles, *Archimède* and *Cyana,* and their support ships in the Azores. The diving began later in June.

Never before had three deep-diving submersibles carried out such a coordinated effort. Having more than one ship diving in the same area added to the overall level of safety; one team might conceivably come to the aid of another. But it also had its drawbacks. All three vehicles were commonly submerged at the same time. As they took soundings and navigated, a cacophony of acoustic signals bounced off the steep-sided bedrock surface of the inner rift valley. It became impossible at times to track every sub.

When doing fieldwork on dry land, geologists determine their location by visual surveying. The deep ocean is dark, however; sound, not light, is the medium surveyors must use. Tracking the position of subs underwater involves stopwatch-like timing. The time it takes an acoustic ping to travel from one point to another—say, from a moving sub to a support ship—reveals the distance between them. To avoid prematurely triggering the acoustic-response timers on board the various ships, we developed a method of timesharing so that each team could track its own submersible. Although we each got fewer tracking positions than if we had been operating on our own, the submersibles were usually moving slowly, or even sitting for long periods of time to gather samples. Our timesharing restraints, as it turned out, did not make a measurable difference in our ability to track the subs.

A precise knowledge of where each vehicle was at the time observations were made, photographs taken, or rock samples collected was critical for scientific accuracy. For that reason, each vehicle had its own network of battery-powered acoustic transponders moored to the ocean floor. These were part of the navigational system, developed at Woods Hole, that I had tested in 1971 during my work in the Gulf of Maine. Surface

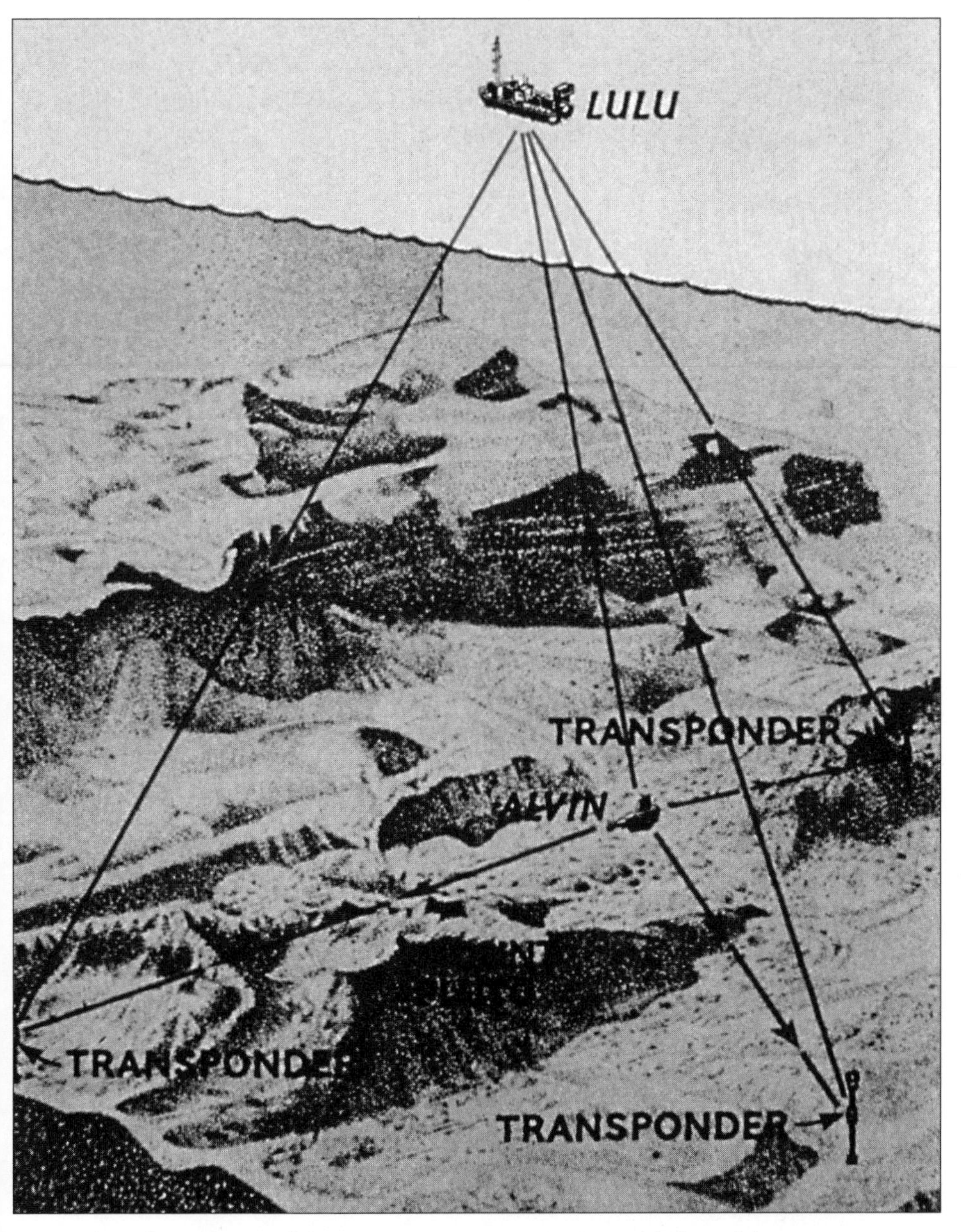

Artist's rendering of the acoustic navigation system used aboard the support ship *Lulu* to track *Alvin* during its dives in the rift valley of the Midocean Ridge. Water depth greatly reduced. (Source: Walter Hortens/National Geographic Society Image Collection.)

ships dropped each transponder carefully into position on a topographic high spot that looked down into the portion of the rift valley where its submersible would be working. It was important that no undersea terrain block the signaling path between the submersible and its transponders.

If we needed to obtain the position of, say, *Alvin,* we first had to know where *Alvin*'s support ship, *Lulu,* was located. Every ten to twenty seconds, *Lulu* transmitted an acoustic signal to the network of transponders. Each one responded with a unique signal. Given the responses from three transponders within the network, we were able to determine the exact location of *Lulu* using simple trigonometry. Next, *Lulu* sent a second interrogation signal to *Alvin.* On receiving this signal, *Alvin* would emit its own signal, evoking another round of responses from the transponder network. *Lulu* received these as well. From this latter set of data and *Lulu*'s previously determined position, we were able to determine *Alvin*'s position, once every thirty seconds or so. We displayed the positions on board *Lulu* by charting them on a navigational plot.

These navigating techniques had worked beautifully in the Gulf of Maine, and they served equally well during Project FAMOUS. Eventually they became standard operating procedure for precise deep-ocean fieldwork. Dolphins enjoyed the transponders as well. Being wonderful mimics, they sometimes learned to chirp at exactly the frequency that would evoke a transponder response.

To optimize the limited amount of time *Alvin* could spend on the bottom, we wanted to land the submersible as close as possible to its desired starting point for the day. While *Alvin*'s crew was preparing to dive, *Lulu*'s captain maneuvered the ship as if he were launching the sub. For the next twenty minutes, the science navigator tracked our surface position to determine where the winds and currents were pushing us. With this information and knowledge of deeper currents gained while the submersible had descended the previous day, we knew exactly where to launch. Several minutes after *Alvin*

The author boarding *Alvin* prior to its first dive to the rift valley during Project FAMOUS, June 1974. (Source: Emory Kristof/National Geographic Society Image Collection.)

A diver inspects *Alvin* prior to the flooding of its ballast tanks for a dive to the rift valley of the Mid-Atlantic Ridge. (Source: Emory Kristof/National Geographic Society Image Collection.)

submerged, the science navigator made additional adjustments by asking the pilot to drive in a particular direction during the lengthy descent.

These procedures proved very effective. After falling some 10,000 feet, the submersible usually touched down within 100 feet of the desired landing site. Often it came within 30 feet or sometimes even 15, less than the sub's length—a direct hit.

In all, *Alvin* made seventeen dives, while *Cyana* and the bathyscaph *Archimède* combined for twenty-seven. Despite the rugged terrain, the submersibles encountered few problems navigating through their assigned areas, a record that reflects the operating skills of the pilots. Scientists brought back from the bottom 3,000 pounds of rock samples, many water samples, some sediment cores, and more than 100,000 photographs.

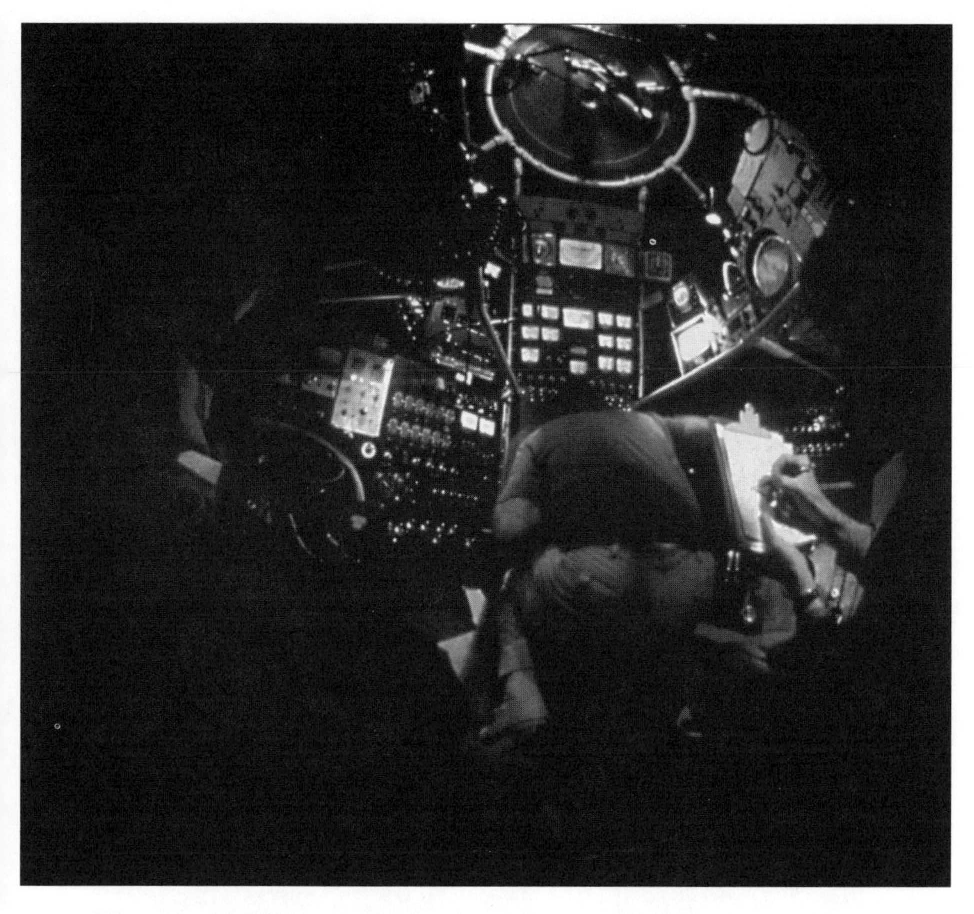

The author (left), pilot Val Wilson (center), and geologist Jim Moore (right) inside *Alvin*'s pressure sphere during the Project FAMOUS diving program. (Source: Robert D. Ballard/Odyssey Corp.)

These data would take years to analyze, but the general picture was already apparent when the expedition ended in August 1974.

The zone where new crust was forming was of course the main attraction. Although the press began calling this narrow rift the "crucible of creation," there were no volcanic fireworks. Instead, thick magma had oozed through cracks. It solidified quickly as it met the deep ocean's enormous pressure and near-

Project FAMOUS scientists Bill Bryan (left) and Jim Moore (right) processing rock samples collected by *Alvin.* (Source: Emory Kristof/National Geographic Society Image Collection.)

freezing cold, producing lava formations, collectively called pillow lava, unlike any seen before on land. Some of the weirder variations entered scientific literature with such descriptive names as toothpaste, trapdoor, breadcrust, and broken egg.

A much broader area of tectonic faulting surrounded the narrow volcanic zone. Within 1,000 to 1,600 feet of the rift center, most of the delicate extrusive lava forms had been destroyed. Sliced and offset by numerous faults, their surfaces were reduced to broken, jumbled blocks or extensive talus fans at the base of fault scarps. This faulting and deformation appeared to be a continuing process, whereas volcanic activity was episodic.

Artist's rendering of the inner rift valley of the Project FAMOUS study area showing the French submersible *Cyana* (foreground) and the bathyscaph *Archimède* (distant background) during the 1974 diving program. (Source: Davis Meltzer/National Geographic Society Image Collection.)

As simple and straightforward as those observations may seem, they confirmed the theory of seafloor spreading, providing the first systematic documentation of a crust-making process that has global significance. (It may also have extraterrestrial significance: these observations might help us understand features we already see or have yet to find on distant moons and planets, such as the recently discovered magnetic striping on Mars.) By the time the results of Project FAMOUS were published in two volumes of the *Geological Society of America Bulletin*, our detailed knowledge of plate tectonics had taken enormous strides forward.

Elongated lava tube flowing down the slope of a flow front in the central volcanic high of the rift valley. (Source: Project FAMOUS.)

Small Submersibles Drive Bathyscaphs to Extinction

Although the technology we used during Project FAMOUS proved highly effective, one historic tool soon fell by the wayside. The rigors of deep-ocean fieldwork sealed the fate of the bathyscaph *Archimède.* Before long, the French decided that it was simply too large, unwieldy, and expensive to continue operating. Their bathyscaph program had by then amassed a phenomenal record. In 1970 Georges Houot, the first French bathyscaph pilot, published a comprehensive history, *20 Ans de Bathyscaphe* (*20 Years with the Bathyscaph*). His book documents 93 dives by *FNRS-3* and 139 by its successor, *Archimède.* Researchers using these vessels carried out extensive studies in the Mediterranean, in the Puerto Rico and Japan

Photograph taken from *Alvin* of a fissure running down the axis of the rift valley in the Project FAMOUS study area. (Source: Project FAMOUS.)

Trenches, and of course in the deep Atlantic. The demise of the French program left *Trieste II* as the world's only operating bathyscaph. Before the end of the decade, the U.S. Navy would retire it, too.

Small submersibles, by contrast, had finally come of age. Both *Alvin* and *Cyana* would soon extend the French-American deep-seafloor studies around the globe. Early in 1976, *Alvin* dove in the Caribbean to investigate a region where two plates grind past each other in the Cayman Trough, a gash in the bottom four times deeper than the Grand Canyon. (The bathyscaph *Trieste II* completed those studies, diving to deeper parts of the canyon than *Alvin* could reach.)

The following year, *Alvin* moved on to examine a spreading rift valley in the Pacific Ocean. The researchers who made that

Photograph taken from *Alvin* of a vertical fault scarp in the inner rift valley, showing a cross section of truncated pillow lavas. (Source: Project FAMOUS.)

trip were looking for data to confirm a theory about heat flow; they would have been satisfied merely to advance the work of Project FAMOUS. Instead, to everyone's amazement, they touched off a profound upheaval in biology.

CHAPTER

6

HYDROTHERMAL VENTS
Exotic Oases

Isn't the deep ocean supposed to be like a desert? . . .
Well, there's all these animals down here.

—*Jack Corliss*

Although most dives during the busy second summer of Project FAMOUS went smoothly, *Alvin*'s dive on July 17 ran into a brutal trap. I was the surface navigator for *Alvin* that day. By midmorning, I had guided the sub's descent to the bottom and was checking its progress every few minutes in *Lulu*'s control room. I first became alarmed when I noticed a large blob of ink growing on our navigational plot. The pen was not moving; *Alvin* was spending an inordinate amount of time in once place.

I called down on our underwater acoustic telephone and urged Jack Donnelly, the pilot, to get under way. Mission time was running out, I told him.

"We're trying," he replied. "We don't seem to be able to rise."

More than the words themselves, the pilot's edgy tone of voice sent a chill through the control room. Immediately, we understood that something was wrong.

Donnelly asked for Val Wilson, *Alvin*'s chief pilot, who was not diving that day. Wilson took the microphone and began talking to Donnelly, some 8,400 feet beneath the ocean's calm and sunny surface.

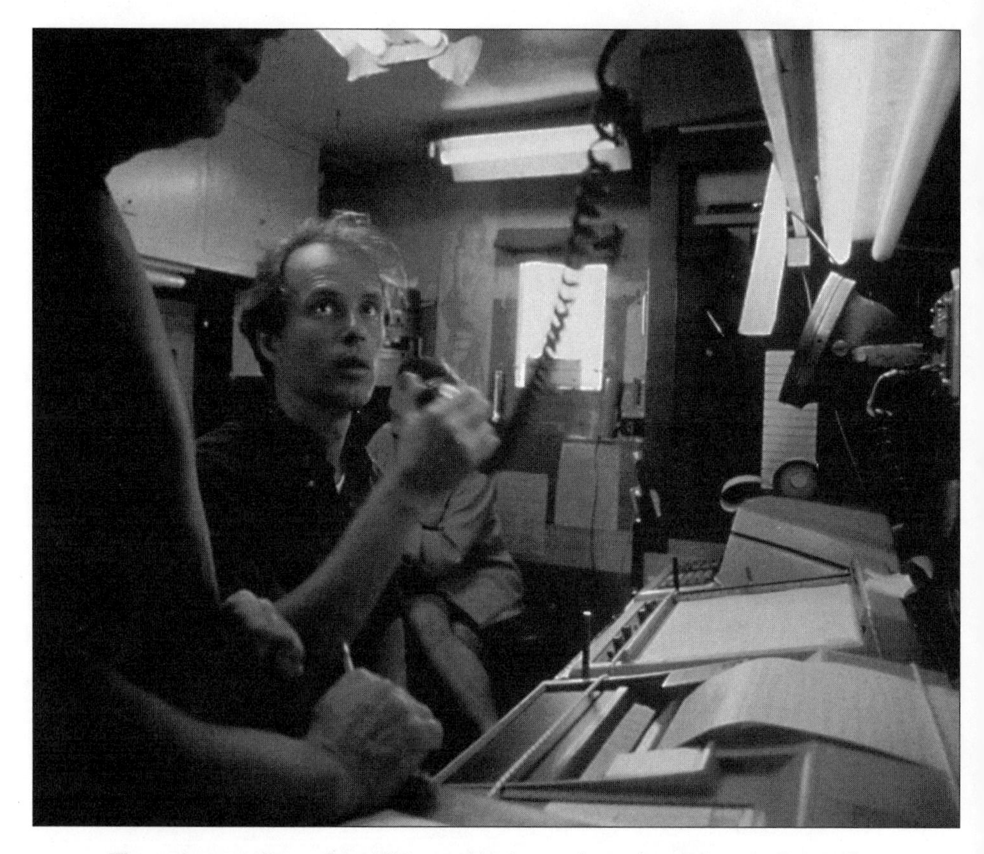

The author on the surface talking to *Alvin's* crew by underwater telephone after learning the craft has become trapped in a fissure. (Source: Emory Kristof/ National Geographic Society Image Collection.)

Scientific curiosity, it seemed, had gotten *Alvin* into a terrible predicament. The two scientist-observers onboard, petrologist Bill Bryan and volcanologist Jim Moore, had gone down with an endless series of unanswered questions dancing in their heads. They were keenly interested in the volcanic formations covering the rift valley floor. Submarine lava flows represented the most common rock type exposed on the earth's surface, yet they were the least understood. How thick were these flows? Did they come out all at once or in layers? Bryan and Moore could not answer those questions by looking down at the seafloor from above.

Looking down from above had also failed to answer another question: Where were the hot springs? In theory, ocean water should have been descending into the same cracks in the bottom that allowed magma to well up; the water should then have been rising again as hot springs. Yet no one had found any of these springs, which seemed strange. Perhaps the deep ocean was so cold that it instantly chilled any rising warm water. Bryan and Moore thought they might have better luck than other researchers looking for hot springs if they could sample the temperature of water confined in a deep fissure, where it might retain more heat.

As *Alvin* moved east on its dive that morning, the volcanic terrain grew older, fractured by wider and wider tensional cracks. At first they opened only a few feet, revealing limited views of beautifully truncated lava flows, one on top of another. Eventually the divers reached a gaping fissure, wider than the submersible, with no bottom in sight. The temptation was too great. They had to enter.

Slowly, *Alvin* inched downward, the scientists' noses pressed against its viewports. As the sub drove forward and deeper into the fissure, they dictated a steady stream of observations into their hand-held recorders. When *Alvin* reached bottom, Donnelly stopped to take a temperature reading. No sign of hot springs yet—the water was still ice cold. As he proceeded again, stopping frequently to let Bryan and Moore observe rock formations, no one noticed how the fissure walls were gradually drawing closer together. Suddenly, it seemed, they towered just inches from the viewports on both sides. By then, *Alvin* was already stuck.

Before I detected their lack of progress, the crew had been keeping the problem to themselves. Donnelly tried everything to extract the submersible—moving forward, wiggling from side to side, backing up. He could not, however, repeat in reverse the easy downward glide by which *Alvin* had maneuvered into this spot. Donnelly also tried to rise directly upward, but the opening was too narrow. *Alvin* had inserted itself into the fissure the way someone might insert and then turn a key in a lock.

Over the acoustic telephone, Donnelly and Wilson discussed their options. Releasing *Alvin*'s descent weights would not be wise. Although Donnelly could see only a small area around the sub, outcrops appeared to be looming above. Even if *Alvin* dropped weights to become more buoyant, this inwardly curved ceiling would probably block its escape. *Alvin* might need to descend again.

Meanwhile I radioed the French fleet, but I had to think carefully about what I would say. A writer for the *New York Times* on board *Knorr* was listening to our radio transmissions, as reporters will do; I tried to sound casual. I asked Commander de Froberville if the bathyscaph *Archimède* had finished its dive.

De Froberville paused. We did not usually chat like this in the middle of a busy day. Yes, he said. *Archimède* had returned to the surface.

When would it be ready for the next dive? I continued, as if making small talk—and as if it would be nothing out of the ordinary for a bathyscaph to make two deep dives in one day!

De Froberville paused again, even longer this time. No doubt my question made little sense to him. Then he caught on to my game. I was calling because *Alvin* was in trouble. Furthermore, I was trying not to say this directly, because if the news got out right away our radios might soon be jammed with calls, not only from the press but also from our sponsors—quite likely in the middle of an emergency rescue attempt. De Froberville finally replied that it "just so happened" his group would be ready to dive again *very* soon, and in fact they were thinking of diving near us.

His reply was reassuring, but we in *Lulu*'s control room knew that the chances of the bathyscaph actually helping to extract *Alvin* were remote. Surely it would not be able to enter the fissure, and what could it do from outside? Really, there was only one choice: Donnelly would have to try to knock his way out.

Until this dive, no one had considered using *Alvin* as a battering ram; pilots normally tried to minimize all banging and scraping. Yet here there was actually hope. The rock all around the trapped sub consisted of very young lava. It was brittle and covered with a glassy surface that forms when lava cools very quickly. By slamming his thrusters back and forth, over and over again, Donnelly began to chip pieces of this rock from the overhanging walls. They fell away into the open fissure below. Finally, after a couple of hours of violent thrashing, Donnelly's voice rose up from the depths: "We're clear and under way again." He sounded as if nothing had happened.

Everyone on *Lulu* sagged with relief; we had been tense and sweating the whole time. To our amazement, rather than surfacing right away, Donnelly and crew then continued their traverse to the east. But *Alvin* never entered a fissure again, and no one found the elusive hot springs.

— The Mystery of Missing Heat

For one group of scientists, the readings of consistently cold water came as a bitter disappointment. Dick Holland, a geochemist from Harvard, had led a team as eager to sample the water pushing up from deep, hot chambers as others were to examine lava. Would there be gushing deep-sea fountains or only blobs of moderately warm water welling up gently in the darkness? Any variation between those extremes might have been acceptable. But total absence raised potentially devastating questions. The submersibles' failure to find any such water was not a failure for Holland alone; it posed a problem in geophysics.

After Project FAMOUS, scientists needed to make a choice. They could either search harder to confirm the existence of submarine hot springs or else acknowledge a serious flaw in their basic understanding of gravity, water circulation, and heat flow.

Holland had had good reason to anticipate success. In the 1960s Clive Lister, a geophysicist at the University of Washington, had made a fascinating observation, which, he argued, supported the existence of hot springs. He and others had been speculating that the Midocean Ridge towers above the surrounding seafloor because it is swollen with heat energy. Unlike mountain ranges on land, the ridge, they believed, is essentially a heat blister on earth's surface. If the theory of seafloor spreading was true, then hot young crust emerging at the rift valley would cool as it moved away from the ridge. This cooling would cause the crust to contract and lose elevation. In other words, scientists were saying that the ridge's vertical profile represented a theoretical cooling curve. You could draw this curve by solving equations that describe heat transfer, and the actual topographic shape of the ridge should match it.

Well, this is exactly what scientists found. Heat and height seemed very well matched except, as Lister pointed out, along the rift valley axis itself. Heat-flow measurements made at portions of the seafloor near this axis were not as high as they should have been, according to theory. Apparently, some unknown process was stealing heat from the bottom of the ocean.

What could be removing heat? The only logical answer was hot springs. We all knew that magma chambers must underlie the ridge at a relatively shallow depth of one to two kilometers. We also knew that these chambers contain molten rock at temperatures of 1,200–1,400 degrees centigrade. Clearly, cold bottom waters within the rift valley—at three to four degrees centigrade—could easily enter the ocean floor through the cracks we had observed and penetrate to regions of hot rock surrounding the magma chambers. Once heated, this seawater would expand and force its way back to the surface, exiting as hot springs along the rift valley axis. The rising water would also, of course, carry heat away from the rocks—the missing heat.

Idealized model of the axis of the Midocean Ridge underlain by a magma chamber. As tectonic plates continue to separate, fissures formed in the newly crystallized volcanic crust provide seawater with access to the region around the magma chamber. This fluid, once altered and heated, rises back to the surface, forming hot springs. (Source: William H. Bond/National Geographic Society Image Collection.)

A second argument that predicted the existence of hot springs reinforced Lister's physical line of reasoning. It arose from geochemistry. In 1965 Jack Corliss, a graduate student at Scripps, was completing his doctoral work by analyzing basaltic rock samples dredged from the Mid-Atlantic Ridge. Those rocks showed chemical traces of hot-water circulation. His analysis clearly suggested that seawater was penetrating the hot rock surrounding magma chambers and leaching out various chemicals. This water became richer and richer in dissolved minerals as it sank and got hotter below the seafloor. Later, he argued, these minerals ended up as deposits on the newly formed rift valley floor when the superheated water flowed back to the surface and cooled once again.

Years after Corliss finished his chemical studies, French geologists who dove to the seafloor during Project FAMOUS indeed found mineral deposits. The samples they brought back supported the geochemical argument for hot springs. Yet try as they might, the French scientists, like their American counterparts, failed to find direct evidence of the hot springs themselves.

In 1975, the year after the completion of Project FAMOUS, two earth scientists joined forces to propose a new expedition, using manned submersibles, that would put the hot-spring theory to a thorough test. They were Richard von Herzen, formerly of Scripps but now at Woods Hole, and Jerry van Andel, also formerly of Scripps but now at Oregon State University. Von Herzen specialized in heat flow; he fully understood Lister's line of reasoning. Being at Woods Hole, he was also keenly aware of *Alvin*'s recent successful fieldwork. Van Andel, on the other hand, had been Corliss's adviser at Scripps; he knew the geochemical argument. Equally important, he had been one of the principal diving scientists during Project FAMOUS, and he knew firsthand that *Alvin* was up to the challenge. When van Andel accepted an appointment at Stanford University, however, Jack Corliss, by then a researcher at Oregon State, took over his leadership role in the effort to gain funding for a new expedition.

The two collaborating scientists sent a proposal to the National Science Foundation (NSF) to search for hot springs—not in the Atlantic Ocean but in the Pacific, along a segment of the Midocean Ridge called the Galápagos Rift. They had two main reasons for picking that site. First, Oregon State had already begun a large research program in the Pacific funded by the NSF. And second, the centers of seafloor spreading were

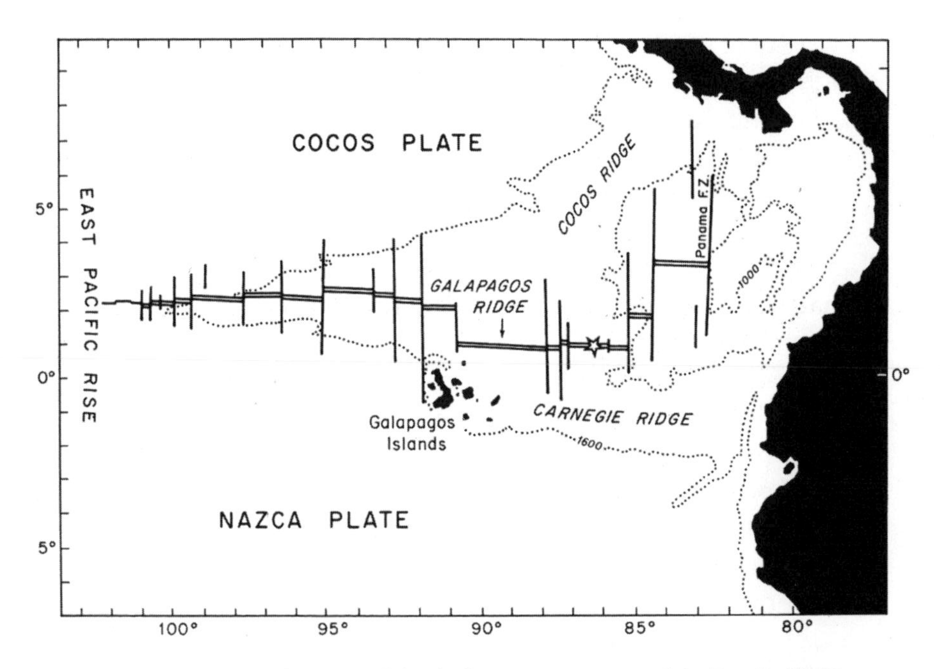

Regional map of the Galápagos Rift, which is an eastern spur of the East Pacific Rise formed as a result of the diverging Cocos and Nazca plates. Star indicates the location of the 1977 dive site where the first hydrothermal vents were discovered. (Source: William H. Bond/National Geographic Society Image Collection.)

pulling apart much faster in the Pacific than in the Atlantic. The faster spreading rate, they reasoned, meant that more heat energy was being released along the ridge axis there, raising the probability of finding hot springs. By the mid 1970s, a growing number of marine geologists were warming to the theoretical arguments that favored the existence of hot springs. But was it worth the expense to send humans down to look for them? To bolster his case, Corliss emphasized the potential practical benefits of the research. Understanding how hot, mineral-rich waters circulate through the seafloor, he argued, would someday help prospectors find ore deposits. Although no one believed that deep-sea mining would become practical anytime soon, NSF decided to fund the expedition. First, however, surveys from the surface would have to identify promising sites for a submersible to investigate.

During the summer of 1976, several research groups used a variety of instruments to probe the inner valley of the Galápagos Rift. What they found was indeed encouraging. Instruments aboard the Scripps Deep-Tow system provided evidence of heat anomalies—slight temperature rises in water near the bottom. That evidence convinced the NSF to fund the next phase of the search, a manned diving expedition. The Scripps team marked one promising area with two long-term acoustic transponders. They also noticed another kind of marker on the bottom: photographs taken by the Deep-Tow cameras revealed clusters of white clamshells, empty and dead. These shells looked like garbage thrown from a passing ship—remnants, perhaps, of a feast. Scripps scientists named that spot the Clambake site, not knowing its true significance.

— The Galápagos Hydrothermal Expedition

The stage was now set for the final phase of the program, a dive series by *Alvin* to pinpoint the suspected hot springs and observe them directly. This was scheduled to take place during the winter of 1977. By the time the Galápagos Hydrothermal Expedition got under way in February, I had been asked to be co–chief scientist with Richard von Herzen. Woods Hole and NSF wanted me to help lead the expedition not because of anything I knew about hydrothermal research but because of my experience conducting submersible programs in complex volcanic terrains.

The real investigating scientists on this cruise, in addition to von Herzen and Jack Corliss, included chemists Jack Dymond from Oregon State, John Edmond from MIT, and others. Jerry van Andel, though no longer a co-leader, came along primarily to help inexperienced divers cope with the rigors of deep-diving fieldwork. There were no biologists, because we expected to see nothing but fresh lava fields, as barren as a moonscape.

Woods Hole's research vessel *Knorr,* with *Lulu* in tow, began the expedition at Rodman Naval Base in the Panama Canal. Our destination was an area 400 miles west of Ecuador, along the rift in the Pacific seafloor that separates the fast-spreading Cocos and Nazca plates. We planned to concentrate on the sites where the earlier Scripps–Oregon State–Woods Hole expeditions had recorded temperature anomalies.

After several days at sea, we decided to break the tow and let the two ships proceed under separate power. That way the faster *Knorr* could arrive ahead of *Lulu,* install a network of acoustic transponders within the rift valley, and conduct some preliminary reconnaissance runs with ANGUS, our towed camera system. I felt that ANGUS would be essential to our success. The time that human observers can spend on the bottom is precious; we needed to make the best possible use of it. If we towed ANGUS all night, it would cover far more ground than *Alvin.* We could then use the sub as a precise observing and sampling instrument during the day, investigating spots that ANGUS had identified as interesting.

The year before, I had convinced the U.S. Navy to map our dive area on the Galápagos Rift using the same powerful SASS technology that had given us detailed maps for the FAMOUS expedition of 1973–74 and later for our diving program in the Cayman Trough. Those maps had set a new standard for bathymetric detail that all future submersible expeditions would seek to acquire.

When *Knorr* arrived at the diving area, we used its echo sounder to collect a series of bottom profiles perpendicular to the rift axis. From this information, our navy maps, the transponders left the previous year, and satellite navigation, we did our best to tie our location to the estimated locations of the thermal anomalies detected the year before. We then lowered the ANGUS camera sled into the rift and began towing it above the seafloor. ANGUS would snap a picture every ten seconds—this time with color film—giving us thousands of

ANGUS camera sled being brought back aboard *Knorr* after the first camera run down the axis of the Galápagos Rift, February 1977. (Source: Emory Kristof/ National Geographic Society Image Collection.)

overlapping images. In addition, for this cruise, Woods Hole had put a new instrument on the sled: a sensitive thermistor that could register temperature changes as minute as one five-hundredth of a degree Celsius.

At first, our temperature sensor recorded no variation in the near-freezing water just meters above the ocean floor. Then, as the first day's run neared its halfway point early in the evening of February 15, recorders onboard *Knorr* received an acoustic signal from ANGUS: a spike in water temperature. This anomaly lasted less than three minutes. Since we had keyed our time and temperature data precisely to the frames of film exposed by ANGUS's cameras, we would soon be able to review the pictures taken at the exact moment of this tem-

ANGUS photograph taken as the sled flew over an active hydrothermal vent, revealing large accumulations of white clams and brown mussels. (Source: Woods Hole Oceanographic Institution.)

perature spike. But first we had to finish the towing run, and then winch ANGUS back to the surface. By the time the film was developed, it was morning.

We were all eager to see the first detailed visual evidence of the hypothesized thermal vents, but nothing could have prepared us for what ANGUS had photographed one and a half miles beneath the surface. The 400-foot roll of color film revealed a bed of clams—hundreds of them, clustered in a small area on the lava floor of the rift. They were not empty shells, and not garbage. Instead, these remarkably large clams were clearly thriving, as if the volcanic, deep-sea environment was no more hostile than a sunny mudflat on the New England coast.

We couldn't help but wonder what so many animals were doing at that depth, in that eternal darkness. Where was their food supply? What source of energy could they possibly be tapping? Surely it was not photosynthesis, the harvesting of energy from sunlight. We remembered this much from basic biology: green plants, algae, and photosynthesizing bacteria ultimately nourish all animals on earth. And those plants and plantlike organisms—which form the basis of every significant food chain—require sunlight to grow. So how could clams be thriving in this dark and desolate place? What could be nourishing their animal flesh?

But we were not biologists. We were supposed to be finding warm water. The next step in our research plan called for *Alvin* to investigate whatever promising sites ANGUS had revealed. We could hardly wait. On that very afternoon, February 16, *Lulu* arrived with *Alvin* aboard. We lost no time in getting the submersible into the water at sunrise the following day.

After a descent lasting an hour and a half, pilot Jack Donnelly—the same pilot who had gotten stuck during the FAMOUS expedition—brought Jack Corliss and Jerry van Andel to the bottom at a point less than 300 yards from the clam beds. He then began driving along the lava floor toward that site. Along the way, the bottom looked as we all had expected: it consisted of fresh but relatively barren lava flows.

When *Alvin* reached its goal, however, the scientists observed a shockingly different scene. At the same time, *Alvin*'s temperature sensor began to beep. Warm water shimmered up from cracks in the lava flows. It was turning a cloudy blue as manganese and other minerals, carried from deep within the seafloor, precipitated out of solution to form a solid coating on the cooler surrounding rocks. But that was not all. The seafloor was teeming with life. Jack Corliss, talking to a graduate student on *Lulu* by acoustic telephone, sounded bewildered. "Isn't the deep ocean supposed to be like a desert?" he asked.

Corliss was looking at clams—giant specimens, measuring a foot or more in length. And that was only the beginning. He and van Andel stared in amazement as shrimp, crabs, fish, and small lobsterlike creatures passed their viewports. Corliss recognized a pale anemone. He could not identify the weird stuff growing on the bottom, like dandelions, or the wormy stuff attached to rocks.

Alvin's robotic arm, which had been expected to grasp only rock samples, now was pressed into service to collect outsized brown mussels that appeared to be bathed in the shimmering water. It also took a clam and a few rocks covered with a matlike, unidentified growth. Yet Corliss had no idea that many of these creatures were new to science. He spent the bulk of his time on this first dive as one very happy geochemist, making temperature readings and collecting mineral-stained rocks.

On the surface, we were equally ill prepared to handle biological specimens. When we had planned this cruise, our thoughts had been so far from biology that we had brought no preserving fluids along, except for a small amount of formaldehyde. Some of these amazing specimens, therefore, made the trip back to shore immersed in our strongest alcohol: Russian vodka we had purchased in Panama.

Over the next several days we scrambled to find more hot springs. They turned out to be warm rather than hot, with a maximum temperature of 63 degrees Fahrenheit—remarkably high, we thought, for the bottom of the ocean. At the same time, we became more and more excited about the creatures that invariably clustered around these hydrothermal vents. A suspicion dawned on us: these unexpected life forms might actually be a bigger discovery than the expected warm water.

Although geologists by training, the researchers on this expedition now eagerly took turns scouring the rift for similar signs of life. With continuing guidance from ANGUS, their precisely targeted dives in *Alvin* yielded rich rewards. We set

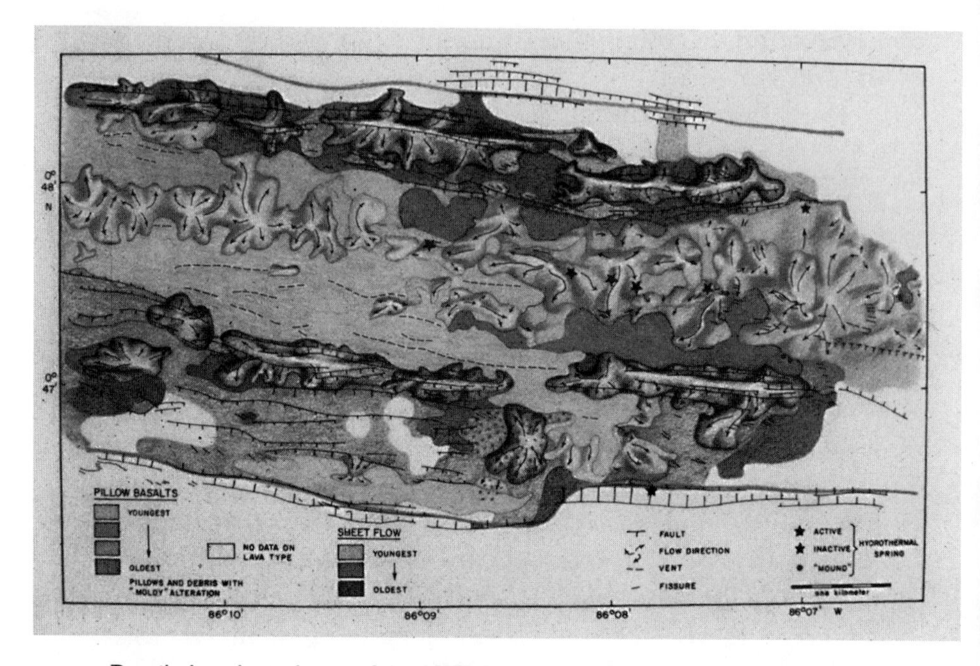

Detailed geological map of the 1977 Galápagos Rift study area compiled from numerous ANGUS camera lowerings. A central region of young lava flows is flanked to the north and south by old and more fractured volcanic terrain. The five stars running down the central axis are the active hydrothermal vents discovered by ANGUS. (Source: Robert D. Ballard.)

a record pace of eleven dives in as many days on the first leg of the cruise, then broke it on the second leg with thirteen dives in thirteen days. In the end, we identified five sites that teemed, or had recently teemed, with creatures as bizarre as they had been unexpected.

We termed our initial find Clambake I. At a second site, called Clambake II, a change in conditions had killed off the big bivalves and left only dissolving shells. This was probably the site that the Scripps Deep-Tow system had photographed the previous summer. Once apparently warm, the waters there had become cold again. Oyster Bed was our label for a patch of mussels we misidentified as oysters. (Our flawed attempts at taxonomy went temporarily unchallenged; we had no way to transmit photos to biologists on shore.) We dubbed another site Dandelion Patch because it was home to a population of hitherto unknown organisms resembling bright yellow dandelions.

Finally, we found a site called Garden of Eden, the lushest and most varied of these strange oases. Here again were the clams and dandelions, along with white crabs, limpets, small pink fish, and clusters of vivid red-tipped worms that protruded from stalklike white shells, or tubes. We called them, simply, tube worms. The largest tubes we observed measured a foot and a half in length. The animal itself filled more than half this elongated stalk. With no eyes, no mouth or any other obvious organs for ingesting food or secreting waste, and no means of locomotion, it was no worm, snake, or eel but no plant either—the strangest creature we had ever seen.

Midway through our cruise, we all became amateur biologists. And why not? We had discovered a new realm of life. We began to speculate that the animals in these deep-sea oases got their energy and nutrients not from the sun and photosynthesis but from the earth's hot volcanic depths. We felt as if we had glimpsed unknown, alien life on a new world, or at least an alternate version of life on our own. "We were dancing off the walls," John Edmond recalled afterward. He and most of the other researchers could not have been more excited if they had been sailing with Columbus.

—— Biologists Arrive on the Rift

After our third dive, we radioed biologists at Woods Hole to convey news about the strange life forms and ask for advice. They gave us many suggestions about how best to collect samples, what data to record, and how to preserve specimens. They could hardly wait to see the results, just as we could hardly wait to get *Alvin* started on its next dive. But inevitably a bit of tension developed. This was, after all, a geologists' expedition. The scientists who had come along had worked hard to secure funding for their own research and had already planned a full schedule of fieldwork. Now others back on shore were advising them to pursue different goals, ones that biologists would have pursued if they had organized the expedition.

Two years later the biologists got their chance. In January and February 1979, Woods Hole marine biologist Fred Grassle and I co-led a second expedition to the undersea oases of the Galápagos Rift. Fourteen other biologists accompanied us, ensuring that this time there would be no relying on vodka for preserving specimens. We also brought a film crew from the National Geographic Society, who chronicled our discoveries in the television special *Dive to the Edge of Creation.*

On this second trip, we faced a challenge quite different from our task two years earlier. During the first expedition, we were following the lead of remote sensors and had no idea what we would find. Any kind of hot eruption or warm spring would have satisfied us. Now we had a good idea of what we would find, but warm water in itself was not our target. Instead, we were trying to locate precisely the same exotic life forms we had visited two years ago, in a place with no prominent landmarks either above or beneath the surface. In that eternal darkness, on such a relatively featureless plain, a pilot might land the sub a hundred yards from a site we had visited and yet wander off blindly in a different direction, toward the lure of some other temperature spike.

As before, we deployed ANGUS as our eyes and our deep-sea thermometer prior to launching *Alvin.* Reviewing the thousands of frames exposed by the ANGUS cameras on the sled's first run, we began to resign ourselves to a long search. Then, with about four frames to go, we found what we were looking for: ANGUS had photographed a clutch of our mysterious dandelions. We knew we were in the right spot—if not one of the oases we had seen two years earlier, then at least a good alternative.

We had not rediscovered Clambake I. In fact, we would not relocate that particular site until the eighteenth dive, which

Alvin sits on its cradle aboard the support ship *Lulu* as members of the 1979 biology team prepare the sub for a morning launch. (Source: Emory Kristof/National Geographic Society Image Collection.)

was next to last. Yet all of us eagerly took turns in *Alvin* exploring the sites that we did find. One of the new oases proved to be the largest yet discovered—an otherworldly habitat for tube worms up to twelve feet long, eight times longer than any we had seen two years earlier. We called this new site the Rose Garden.

When my turn came again to dive and I entered the Rose Garden for the first time, I was overwhelmed. Never before or since have I seen so much strange and exotic life. While diving on the Mid-Atlantic Ridge several years earlier, I had grown used to its barren volcanic terrain. The freshest flows had a sugarlike coating of reflective glass that crunched like thousands of lightbulbs when the submersible landed. Life was sparse, with an occasional fish swimming by or a ghostly crab quickly disappearing into a tiny lava tube. At times the rift valley there resembled a Martian landscape—a lifeless desert filled only with the frozen shapes of violent events, created when huge rocks split apart and lava emerged from eruptive fissures.

In the Galápagos Rift we saw similar fresh lava, but here it was dotted with lush oases. Beyond a doubt, chemical nutrients coming from the vents made this dense life possible. The warmth of the water was not crucial in itself. On previous expeditions we had observed that many kinds of animals, from invisible microbes to scavenging crabs, can survive the near-freezing temperatures even at the deepest reaches of the sea. These bottom-dwellers feed on organic debris drifting down from the surface. Their numbers are sparse because this food is not plentiful. Here, though, the secret of abundant life was a cornucopia of locally derived nutrients. We geologists had guessed as much during our first expedition. On this second cruise, one biologist estimated that the water just above a hydrothermal spring contains three hundred to five hundred times more nutrients than ordinary seawater nearby.

During this second expedition, with our complement of biologists and biochemists, we were able to achieve a far more

The author removing a red tube worm from its outer protective casing. (Source: Emory Kristof/National Geographic Society Image Collection.)

sophisticated understanding of the deep oases—the creatures, their ecosystem, and how life sustains itself. Yet much remains to be learned, and research on these vent communities continues to this day.

From the moment we first saw a bed of gigantic, living clams in 1977, one obvious question engaged everyone's mind: How can these creatures flourish at such depths, in an environment totally devoid of sunlight? To find out, researchers needed to trace the food chain from top to bottom, back to the primary producers. On the earth's surface, the primary producers are photosynthesizing plants and microbes. Using solar energy to set off a chain of chemical reactions, they convert carbon dioxide and other nutrients into the organic molecules in living tissues. But what was the ultimate energy source in the deep, dark abyss, and what were the nutrients?

One key life-sustaining ingredient came to our attention quickly on the first expedition, when chemist John Edmond opened water samples *Alvin* had brought up from Clambake I. Although Edmond was working in a small laboratory on board *Knorr,* the odor of rotten eggs immediately pervaded the ship. No one could miss this powerful clue: we were smelling hydrogen sulfide. It is the same chemical that fouls the air around swamps, sewage plants, and natural hot springs; it meant that the water emerging from these deep hydrothermal vents was far from ordinary seawater.

At that moment we could forget about fancy instruments. Our wrinkled noses alone lent strong support to Lister's original hot-spring hypothesis. The source of this hydrogen sulfide had to be volcanic. Deep fissures in the rift were clearly allowing cold seawater to penetrate the ocean floor, down to the level of hot, newly formed layers of rock. As the temperature of the water rose, its chemical composition changed. The seawater left some of its dissolved components in the subsurface rock and leached out others. In addition, sulfates in the water were reduced to hydrogen sulfide. This could happen only at high temperatures. Finally, the heated water rose back

Electron microscopic image of the sulfide-oxidizing bacteria that form the basis of the unique communities of vent animals. (Source: Holger Jannasch/Woods Hole Oceanographic Institution.)

to the seafloor through other fissures in the crust, laden with enough hydrogen sulfide to kill most ordinary forms of life—but not the oasis creatures.

Living inside the clams and tube worms, as well as in water very near the vents, were bacteria that metabolize hydrogen sulfide. They were using it to energize a sort of biochemical battery that ran their molecular machinery. Through a series of reactions involving not only hydrogen sulfide but also the carbon dioxide and oxygen in seawater, these microbes were synthesizing the basic molecules needed for life without benefit of sunlight. Instead of photosynthesis, they were performing *chemosynthesis.*

Such bacteria had already been known to exist, but only in rare habitats like surface hot springs, where they remain fairly isolated from other living communities. Here in the deep-sea oases, however, they form the basis of a unique food chain. And the energy that makes the nutrients they need, most notably the hydrogen sulfide, comes ultimately from the earth's internal heat.

As it turned out, biologists learned that a nonvolcanic element also plays a significant role. Normally, these bacteria metabolize not only volcanic compounds that emerge from the vents but also the small amount of oxygen dissolved in cold bottom waters. They find it more efficient than relying on hydrogen sulfide alone. And since green plants and microbes produce virtually all the free oxygen on earth, life in the deep oases maintains an oxygen link with life on the surface. Yet life could survive in these deep oases without the help of oxygen-producing life on the surface. If the sun quit shining and oxygen stopped arriving, the microbes would still survive.

What would happen if the sun suddenly turned off? The eternal darkness would remain just as dark. Animals in the deep-sea vent communities would not immediately notice the death of the sun, or the subsequent death of life on the surface. They would continue to thrive until they exhausted the oxygen in seawater. But even after that, life in some form would persist. Sensing the lack of oxygen, bacteria at the bottom of the food chain would switch over to a basic, probably more primitive kind of metabolism that also relies on hydrogen sulfide. Their ancestors probably survived that way, without benefit of sunlight or oxygen, before green plants existed. And their descendants may yet return to that strategy if some catastrophe should destroy all life on the surface.

—— A Bizarre Natural History

During our exploratory dives in 1977 we knew almost nothing about these chemosynthesizing organisms, let alone

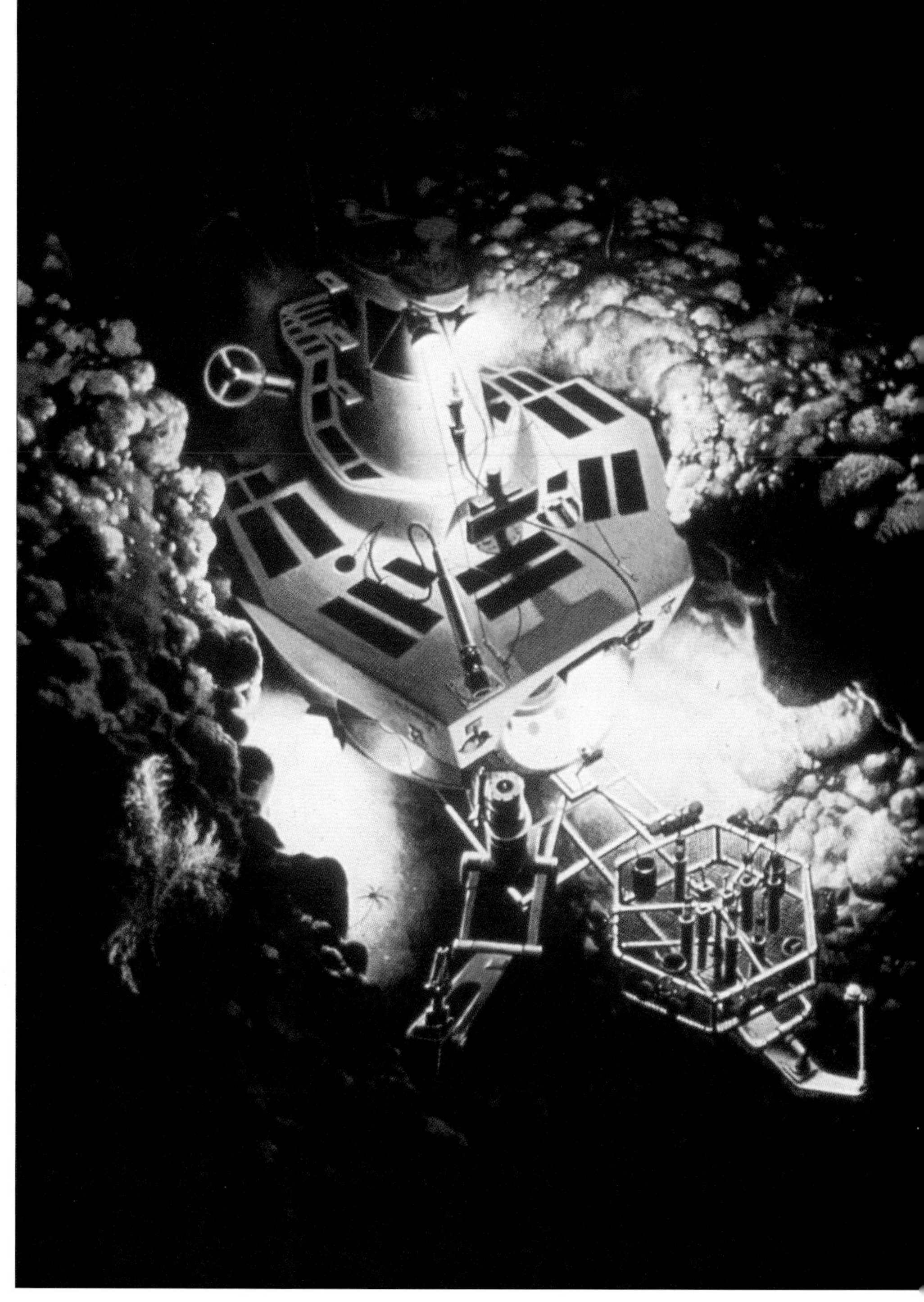

Artist's rendering of *Alvin* stuck in a fissure during its dive in the rift valley, July 17, 1974. (Source: Davis Meltzer/National Geographic Society Image Collection.)

Photograph taken from *Alvin*'s forward camera as the submersible enters an active hydrothermal vent. Large accumulations of clams and other vent animals can be seen. The object in the lower portion of the photo is *Alvin*'s science tray, loaded with various instruments. (Source: Woods Hole Oceanographic Institution.)

Close-up image of large red tube worms (*Riftia pachyptica*) characteristic of many active vents. Brown mussels (Mytilidae) and a white brachyuran crab (*Bythograea thermydron*) can also be seen. (Source: National Geographic Society Image Collection.)

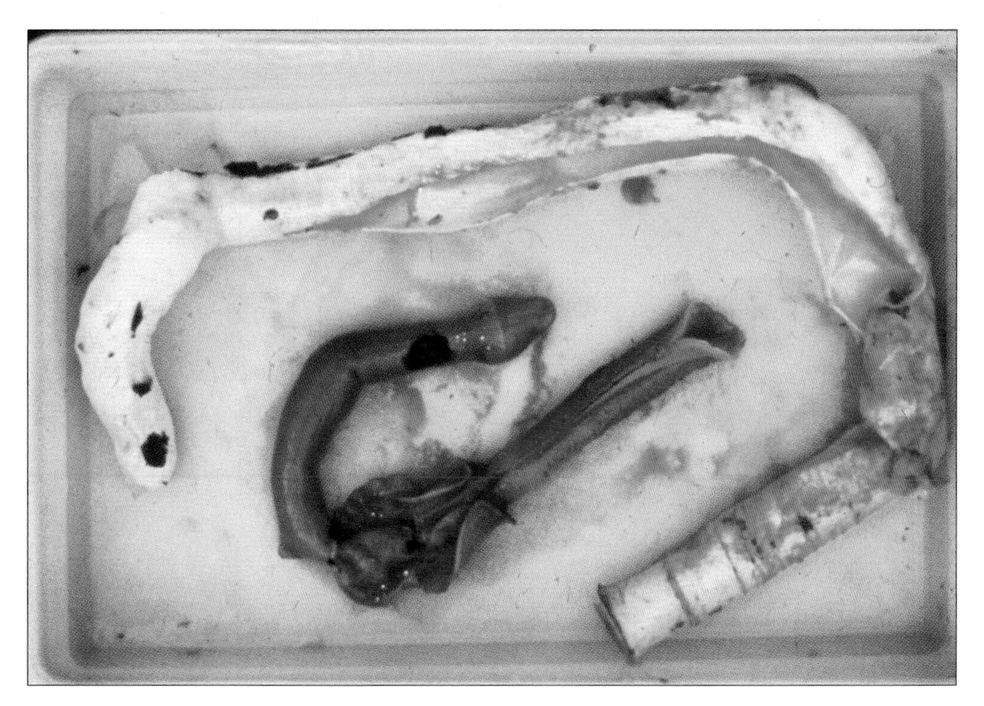

A red tube worm removed from its white casing reddens the science tray with its hemoglobin-laden blood. (Source: Emory Kristof/National Geographic Society Image Collection.)

Interior view of a large white clam, revealing its reddish flesh. (Source: Emory Kristof/National Geographic Society Image Collection.)

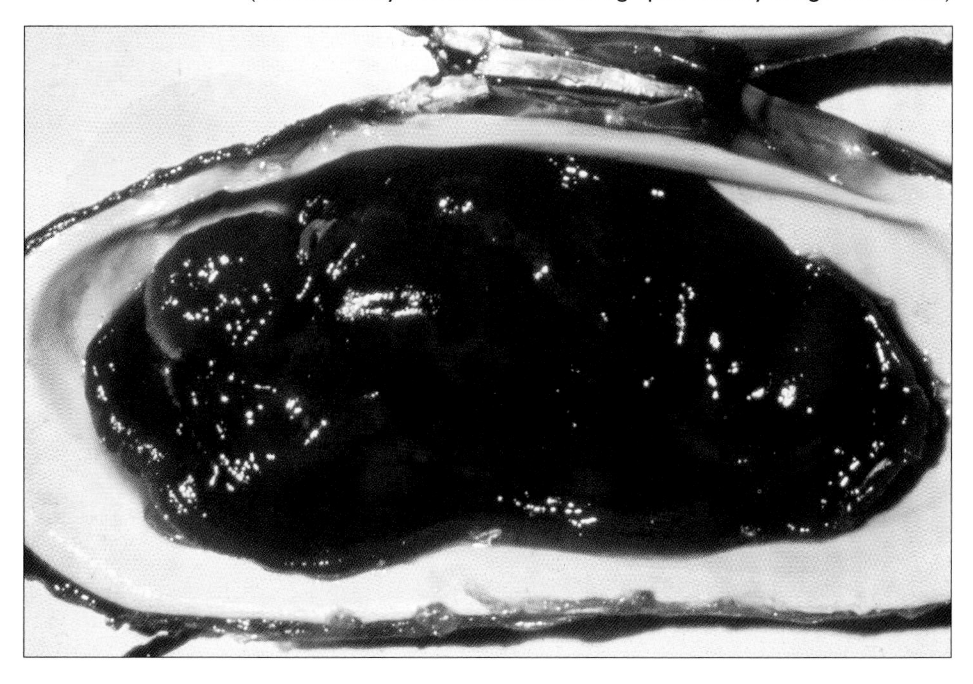

Photograph taken from the French submersible *Cyana* in 1978 of dissolving white clam shells at an inactive vent site on the East Pacific Rise at 21 degrees north. (Source: Jean Francheteau/CNEXO.)

Photograph taken by the author through *Alvin*'s viewport of a lava "pillar" surrounded by the rubble of a collapsed lava surface near an active hydrothermal vent. (Source: Robert D. Ballard.)

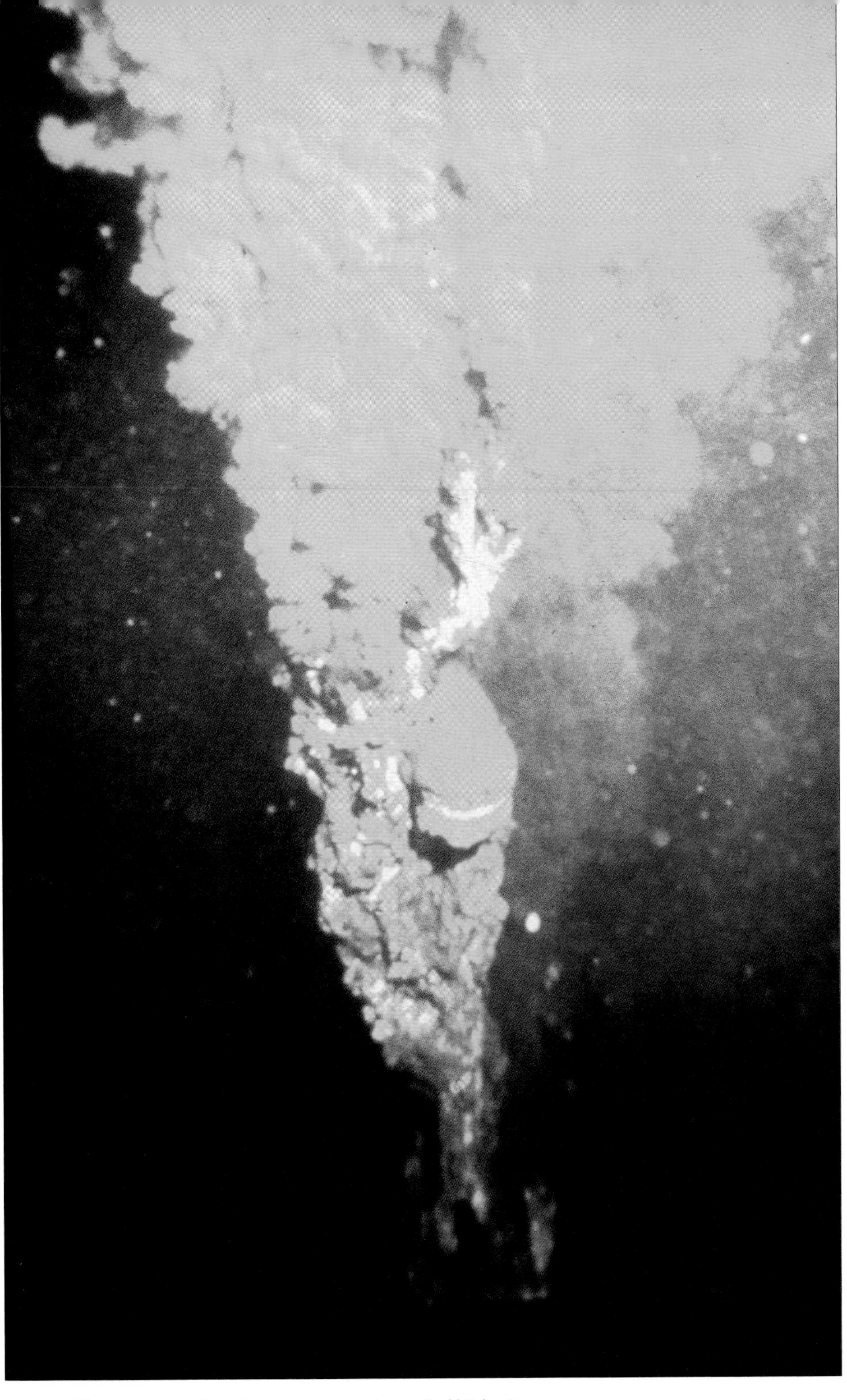

Photograph taken by the author through *Alvin*'s viewport of a "black smoker." (Source: Robert D. Ballard.)

The *Titanic*'s debris field contained the first artificial objects detected by *Argo*'s low-light video cameras. (Source: Woods Hole Oceanographic Institution.)

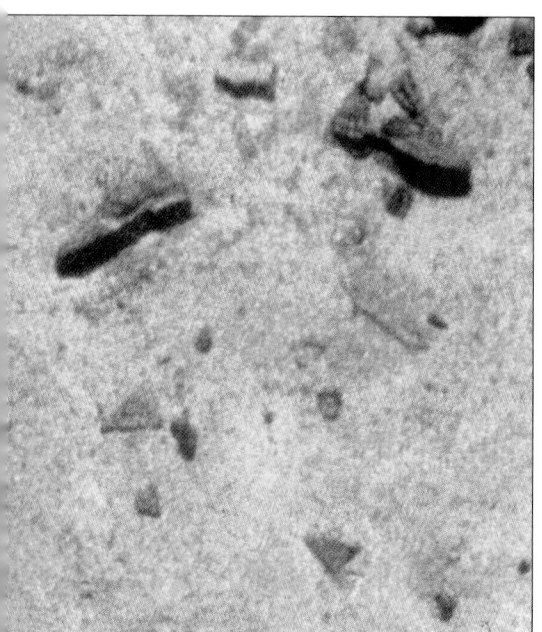

All that remains of what is thought to be a crew member of *Titanic* who perished during its tragic sinking. (Source: Woods Hole Oceanographic Institution.)

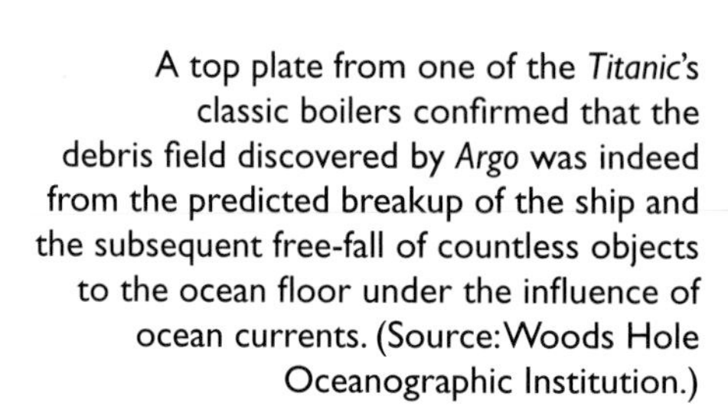

A top plate from one of the *Titanic*'s classic boilers confirmed that the debris field discovered by *Argo* was indeed from the predicted breakup of the ship and the subsequent free-fall of countless objects to the ocean floor under the influence of ocean currents. (Source: Woods Hole Oceanographic Institution.)

Rendering by artist Ken Marschall of the *Titanic*'s bow section resting on the bottom of the ocean. His research was based upon countless hours of video tape and still images collected by vehicles ANGUS, *Argo, Alvin,* and *Jason Junior* during the 1985 and 1986 expeditions. (Illustration by Ken Marschall © 1992 from *Titanic: An Illustrated History,* a Viking Studio/Madison Press Book.)

Photograph taken by the author from inside *Alvin* as *Jason Junior* peers into one of *Titanic*'s cabins on the ship's starboard boat deck. (Source: Robert D. Ballard.)

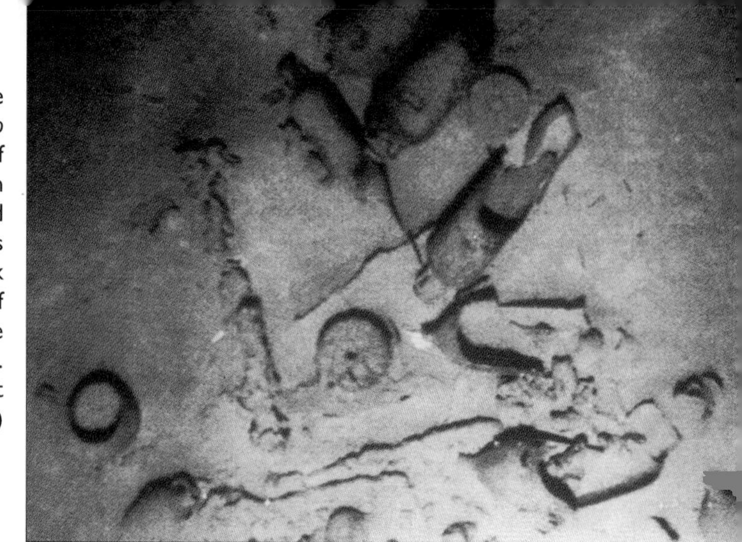

Electronic still image taken from *Argo* of the site of the wreck of an ancient ship, named *Isis* by the author's team, which sank in a storm north of Skerki Bank in the Mediterranean Sea. (Source: Quest Group Ltd.)

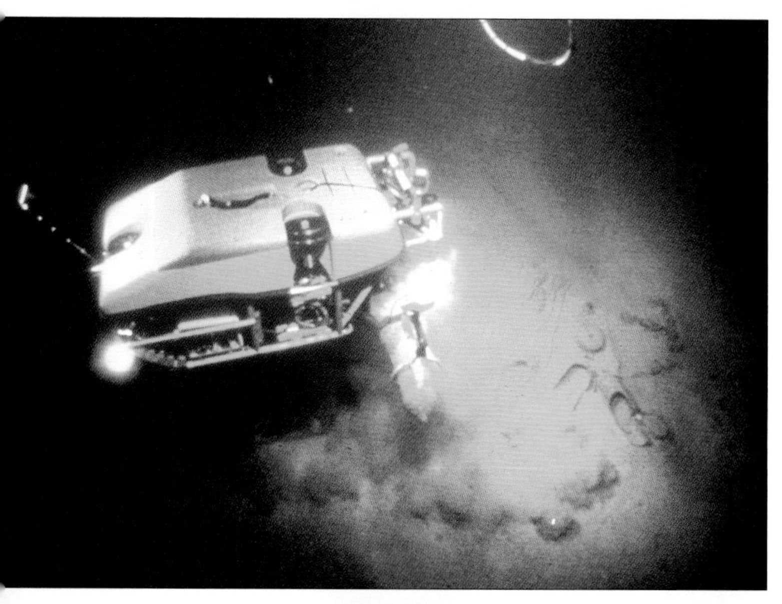

Image taken from *Medea* of *Jason* lifting an amphora from the *Isis* shipwreck site for transport to the elevator. (Source: Quest Group Ltd.)

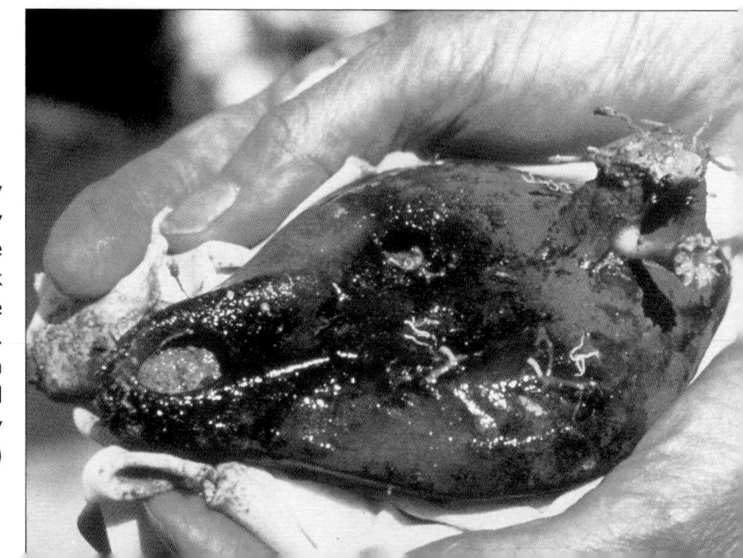

Small pottery lamp recovered by *Jason* from the *Isis* shipwreck site using the elevator system. (Source: Joseph H. Bailey/National Geographic Society Image Collection.)

what strategy some of them might follow in the barely imaginable end game of life on earth. Just the same, many of us enjoyed speculating about such matters. The second Galápagos Rift expedition filled in many of the gaps in our knowledge.

What became apparent soon after it ended is that many forms of bacteria contribute to the chemosynthesis at the bottom of the food chain. They are active in three basic environments. The first lies within the subterranean vents cutting deep into the volcanic terrain; they harbor heat-loving microbes that thrive in water hot enough to kill other kinds of microbes. The second chemosynthetic environment is the surface of rocks right next to the vents, where large microbial mats grow. The third is within the bodies of larger, symbiotic organisms such as tube worms that cluster around the vent openings.

The larger organisms we observed within the oases also inhabit various environments, from an inner zone where the water is warmest and most saturated with volcanic compounds to outer zones where the water is successively cooler and less toxic to ordinary life forms. In the Galápagos rift valley, the warmest water that anyone measured exiting from the vents was 73 degrees Fahrenheit, and the dominant macroorganism living near the vent openings was the giant red tube worm, or *Riftia pachyptila.* These spectacular organisms form large clusters, or hedges, standing up to ten feet high. All around them we could see shimmering flows of milky white, mineral-rich water.

Without eyes, mouth, or a digestive tract, the worms rely on bacteria living inside them to survive. They absorb oxygen and other inorganic compounds from the water through the exposed red tips of their bodies, which look something like heads poking out of the tubes. Actually they are more like gills, with hundreds of thousands of tiny tentacles arranged on flaps lining the exposed red tip. The bacteria living inside them use the absorbed compounds for chemosynthesis. Some of the ingredients, such as hydrogen sulfide, well up from the vent fluids, while others, such as oxygen, exist in the ambient bottom water. The worms position themselves in the area of mixing just above the vent openings, clustering in thickets to direct the exiting fluids up past the tips of their tubes. Being immobile and yet sexually differentiated, male or female, tube worms most probably reproduce by broadcasting eggs and sperm into the water.

Even closer to the warm springs, living in some cases inside the vent openings, a variety of simple-looking limpets (Archaeogastropoda) greatly excited the biologists. They had never seen anything like them except as fossils, which suggested that these creatures might be akin to species considered extinct. Large mussels (Mytilidae) also lived nearby, either on other organisms such as tube worms or in beds on the volcanic rocks, attached by strong threads.

A little farther from the vent openings, we commonly saw giant white clams (*Calyptogena magnifica*) wedged inside small fissures cutting across the volcanic terrain. As the clams grew, their shells conformed to the jagged outlines of the fissure openings, wedging them all the more tightly in place. Their anterior ends pointed down and their hinges up, an ideal feeding position as the hydrothermal fluids flowed past them. On my first dive in 1977, I found these clams, often a foot or more long, as impressive as the tube worms. What surprised us was not only their size but also the intense blood-red color of their flesh. Biologists on the expedition of 1979 confirmed that the clams, as well as the tube worms, contain unusually high amounts of hemoglobin, the oxygen-binding molecule also found in mammals' blood. They seem to use this extra hemoglobin to store a backup oxygen supply, which comes in handy when shifting bottom currents bring them less than the ideal amount.

The giant clams turned out to be good indicators of hydrothermal history. Unlike the remains of most other vent organisms, which quickly vanish after the flow of warm water turns off, the large white clamshells persist for years before dissolving in the bottom water. Thanks in part to such clamshell

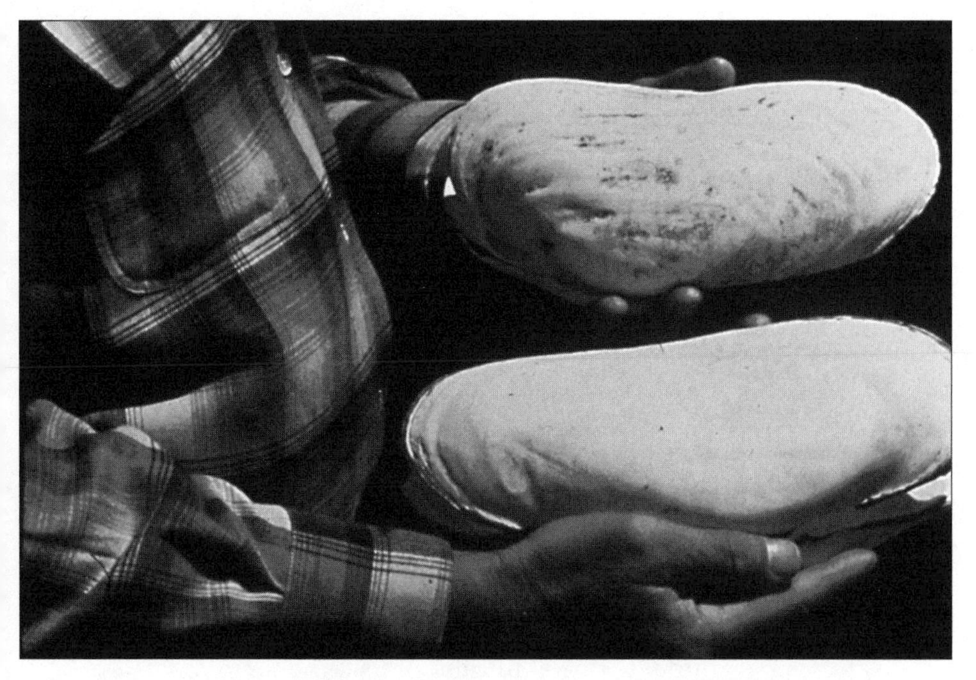

A photo comparison of large white clams (*Calyptogena magnifica*) collected at the Galápagos Rift (upper) and East Pacific Rise at 21 degrees north (lower). (Source: Emory Kristof/National Geographic Society Image Collection.)

evidence, we now know that hydrothermal vents are highly ephemeral, or short-lived. They turn on and off in a matter of a few years or tens of years as the deep plumbing that feeds them opens or closes with tectonic movements around the rift.

Oasis animals have evolved a fascinating strategy for surviving these local catastrophes. The microbes most likely produce dormant, seedlike spores, and the larger organisms, larvae, which are able to settle quickly out of the water when a vent turns on, grow fast and reproduce early. Before the warm and nourishing fluids stop flowing, the colonizing pioneers can easily disperse new spores and larvae into the water column, which may transport a lucky few to new vent environments.

Other important organisms that biologists found living in and around the Galápagos vents include a variety of anemones (Actinarians), brachyuran crabs (*Bythograea thermydron*), galatheid crabs (*Munidopsis*), and a highly unusual worm (enteropneust) clustered in spaghetti-like piles. The blind white crabs that frequent the oases and feed on dead mussels and clams are apparently members of a heretofore unknown crustacean family.

And the strange organisms we called dandelions? Despite their plantlike appearance, these turned out to be animals, a new kind of jellyfish (rhodaliid siphonophores) related to the Portuguese man-of-war. Unlike that ocean-roaming variety, however, these jellies spend their lives attached by threadlike filaments to rock formations on the bottom. Each of the creatures' "petals" has a different purpose. Some capture microorganisms, others digest them, and still others are involved in reproduction. All surround a buoyant pocket of gas, which allows the animal to bob at the end of its tethers.

During the second Galápagos Rift expedition in 1979, we also collected new species of leeches, more conventional-looking worms, barnacles, and whelks. Biologists even took away some two hundred strains of bacteria, brought back alive to Woods Hole for whatever clues they might offer to the basis of the remarkable food chain. Then, in November 1979, a third expedition returned to retrieve instruments and trays left on the bottom in February. Biologists on this expedition found that bacteria had grown in amazing profusion, and they actually measured the respiration of living deep-sea mussels by plopping them into instrumented containers.

—— Earth's Oldest Ecosystem?

Clearly, we had discovered something fundamentally new to science. Before the first Galápagos Rift expedition, biologists had assumed that all major ecosystems depend on photosynthesis. Even the holothurian, or sea cucumber, well adapted to living at great depths in a sunless world, depends for its survival on organic material that drifts down from the sunlit surface thousands of feet above. But within the Galápagos vent fields we found, for the first time, a community of animals capable of subsisting on chemosynthesis. They need no sunlight at all for survival, owing their existence instead to the internal warmth and chemical bounty of the planet itself.

The experience of visiting these strange oases permanently changed many of us. Jack Corliss, especially, seemed transfixed. He was no longer a narrowly focused geochemist; instead he began musing on a cosmic theme: the secret of life itself. During the first expedition, Corliss suggested a profound idea that literally turned conventional notions about the origin of life upside down.

According to a common theory, life began on earth in a shallow sunlit pool, a just-right environment stocked by chance with all the needed raw materials. From that comfortable cradle, the earliest microbes evolved to colonize other, harsher environments. According to this theory, those harsher environments would have included, as we now know, the volcanic deep-sea oases. Life started at the surface, in other words, and then one or more weird offshoots evolved to survive in the darkest deeps.

Corliss, however, proposed the reverse. Life started at some primitive version of these vents, he suggested. The water flowing through them, after all, was filled with ready-made nutrients, physically churning and chemically recombining. Just as important, perhaps, the enormous mass of ocean above these oases could have buffered emerging life from disasters such as meteorite bombardment, common in the early years of the solar system. Here, in the deep abyss, according to Corliss, we might be seeing the direct offspring—four billion years later—of earth's original Eden.

With the help of several colleagues, Corliss developed these ideas. In 1980 he, John Baross, and Sarah Hoffman published a modest, ten-page "Hypothesis Concerning the Relationship

Between Submarine Hot Springs and the Origin of Life on Earth." Over the years, after much further research, this bold hypothesis has been developed into a substantial theory. The idea that life originated at hydrothermal vents has gained support, but the matter is far from settled.

We all got a little giddy then, in February 1977. No matter what strange ideas tickled our imaginations, a very strong suspicion consistently tantalized the explorers among us: surely such organisms could not have been confined by evolution only to one obscure stretch of the Galápagos Rift. At how many other places on the bottom of the ocean, we wondered, do such communities thrive? And how many other as-yet-unknown species draw life from the steaming, mineral-rich depths of the earth's emerging crust?

CHAPTER

7

BLACK SMOKERS
Recipe for a Salty Ocean

One day we may learn that [life on earth] was sparked by . . . heat from an undersea volcano. . . . Our saline blood, the salty sweat . . . all betray man's ocean genesis.

—*Jacques Piccard*

While biologists were studying deep oases, another group of scientists awaited their turn to use *Alvin.* In April 1979, our Woods Hole diving team moved north to join them: a group of geologists doing research at 21 degrees north latitude off the west coast of Mexico. Oceanographers had been coming to the site since the 1950s. The reasons were the usual ones: this was yet another section of the Midocean Ridge, called the East Pacific Rise, where the seafloor is spreading—and it was close to Scripps.

Some of us now had another reason: we considered the East Pacific Rise a likely place to find more hydrothermal vents, new life forms, and perhaps interesting mineral deposits. However, the Scripps geophysicists in charge of the cruise had their own agenda. They were not about to abandon carefully made plans and take up a chase after shimmering oases. Instead, when *Lulu* and *Alvin* arrived, the expedition leaders pressed them into a comprehensive geophysical survey. They were gathering magnetic data to trace local seafloor spreading; placing seismographs on the bottom to study tremors; and making extremely difficult gravity measurements to indicate the location of magma chambers.

Deep-Tow, the Scripps sled, played a leading role in this program. Among other variables, it was equipped to record temperature spikes, which would suggest the presence of active hydrothermal vents. But finding such vents was not the primary purpose of the expedition. The Scripps team kept their sled outside the local rift valley, surveying a broad region that had not yet been studied, and no evidence for hot springs turned up during any of the runs. After about a week of this, Deep-Tow was laid up for repairs. Finally, ANGUS—the Woods Hole sled—had access to the towing winch.

Immediately we devised a towing pattern that would focus exclusively on the narrow rift valley. During its very first run, ANGUS recorded temperature anomalies and photographed giant white clams, which we knew marked the location of warm springs. The two geologists scheduled to make the next dive in *Alvin* now had to decide whether or not to visit those sites. One of them, Bill Normark of the U.S. Geological Survey, had ruined a previous gravity reading after his glasses slid down his nose. He sat perfectly still in *Alvin* for half an hour but then pushed his glasses back up, and that was enough to muddle the delicate instrument. He clearly needed but little persuasion to postpone such tedious work.

On the morning of April 21, Normark and Thierry Juteau, a French volcanologist, dropped into the water in *Alvin.* The pilot was Dudley Foster. As targets for this dive, we had given him the coordinates of three likely hydrothermal vents.

At the first site a few sulfide deposits, obviously the work of hot water, excited the two geologists. Foster did not want to linger; he had seen better examples on previous dives. As he drove over fresh lava flows toward the next site, he began to notice small white crabs on the horizon. They reminded him of similar scenes he had witnessed in the Galápagos Rift. Yet as he entered this second vent field, it didn't feel the same as others he had seen. He passed a cluster of dead clams; there were not as many living creatures as usual. The water seemed cloudier than usual. Suddenly, a chimneylike spire about six feet tall came into view, with a dense black fluid billowing from its top. Foster told the surface crew it looked like a locomotive "blasting out" stuff, which resembled black smoke. Clearly the water nearby was warm, and maybe hot; it shimmered. But how hot?

As Foster brought *Alvin* closer, he found the maneuvering difficult. Something was pulling him into the roiling black cloud.

That pulling force proved to be the updraft, or chimney effect, caused by the rising black fluid. As this heated water gushed upward, the "black smoker" was pulling cooler ocean water in from all sides. And since *Alvin* was neutrally buoyant, it was also going along for the ride—straight toward the center of this strange volcanic outburst. Foster had no choice.

Lava Lakes and Superhot Water

Once again, a submersible probing the dark abyss had turned up something fundamentally new to science. Unlike the deep-sea oases, however, this latest discovery seemed sinister, perhaps even deadly. Yet if *Alvin* could emerge from the chimney fluids unscathed, scientists might collect some remarkable data. That would be a fitting payoff for all the work leading up to this find.

As with many "accidental" discoveries, careful planning and preparations had made this one possible. That same preliminary research had also hinted that dangerous hot springs might erupt in these depths. But the thought of it happening while *Alvin* passed by—and that a searing hot fountain might suck in whatever moved past it—had never crossed the threshold of our imagination. Like a meteor strike or a bolt from the blue, it had simply never figured into our plans.

The plans that sent *Alvin* to the East Pacific Rise began to take shape soon after the conclusion of Project FAMOUS in 1974. Impressed with their first visit to a midocean rift valley,

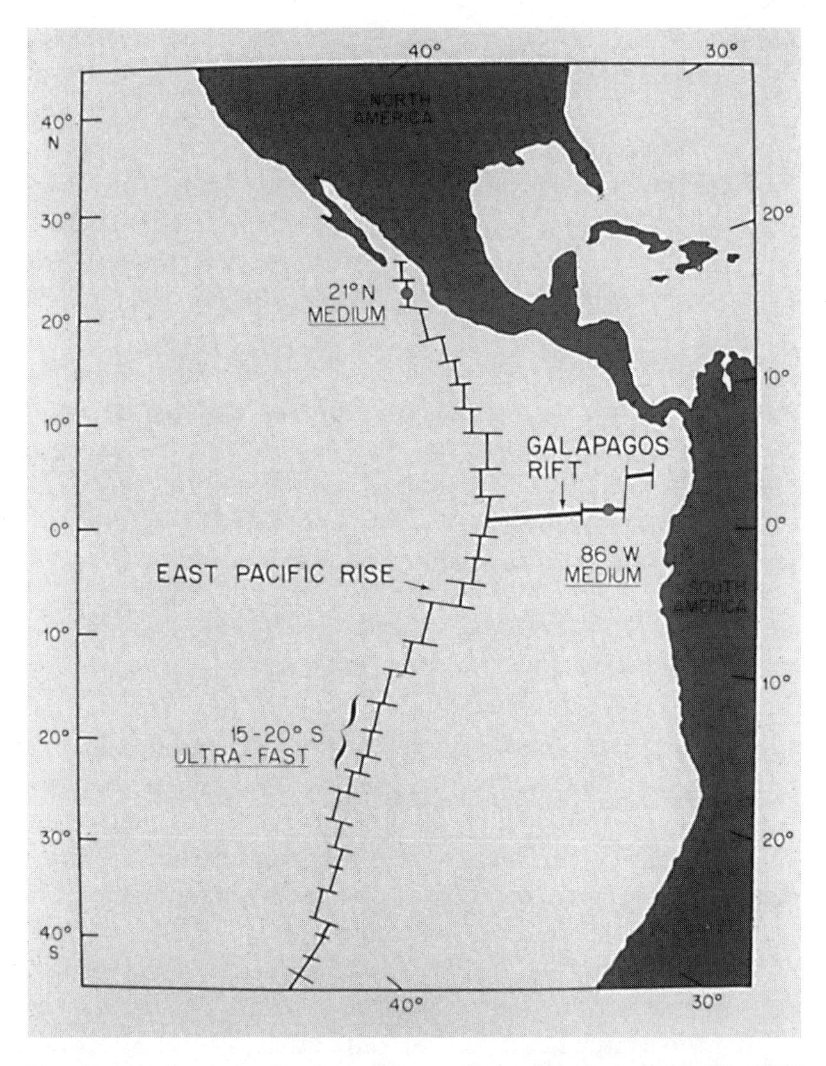

Map showing the area of various diving programs conducted between 1977 and 1979 in the Galápagos Rift and on the East Pacific Rise at 21 degrees north. *Medium* and *Ultra-fast* refer to the spreading rates of the crustal plates along these segments of the Midocean Ridge. (Source: Robert D. Ballard.)

the French participants were eager to conduct another large diving program with U.S. scientists and engineers. As with any successful expedition, new questions had popped up as quickly as old ones were answered. One obvious question was this: If the geologists in Project FAMOUS had found such spectacular terrain in an "average" segment of the Mid-Atlantic Ridge, how would the same volcanic and tectonic processes play out in a more active, faster-spreading part of the seafloor? Would there simply be more of the same tortured, frozen volcanic landscape, or might researchers find something entirely different?

Jean Francheteau, a geophysicist who had participated in Project FAMOUS, spearheaded the French effort to launch a new expedition. His group chose the East Pacific Rise at 21 degrees north latitude, where crustal plates are spreading at a rate of two to five inches per year, several times faster than in the FAMOUS area. As the program developed, the French asked a number of U.S. scientists, including me, if we would be interested in a joint investigation.

Woods Hole had played the leading U.S. role in Project FAMOUS because the research was conducted in the Atlantic Ocean and involved the use of *Alvin*. In the Pacific, however, Woods Hole was not the logical choice to lead a program. Since scientists at the Scripps Institution of Oceanography had carried out a great deal of prior research on the East Pacific Rise, Scripps became the leading U.S. institution for this second French-American program. Fred Spiess, a Scripps geophysicist, would play the role Jim Heirtzler had played during Project FAMOUS.

The French wanted to use their submersible *Cyana* to carry out the first series of dives, and they invited me to participate in the initial phase of our joint program. We arrived at the dive site in February 1978, and over the next several weeks Francheteau's team completed a total of twenty-one dives. Their traverses over varying terrain gave us vivid pictures of this faster-spreading seafloor.

Photograph taken by the ANGUS camera sled of a "sheet lava" flow in the central portion of a lava lake along the axis of the Galápagos Rift. (Source: Woods Hole Oceanographic Institution.)

As in the Mid-Atlantic Ridge, the East Pacific Rise had a relatively narrow central rift, roughly one-quarter to half a mile wide, flanked on either side by older, tectonically altered terrain. Fissures and small fault scarps had broken up the lava there. Within the rift valley, however, the young seafloor looked different than in the mid-Atlantic. Here we saw pahoehoe, or "sheet lava" flows, a type of deep-sea lava first seen a year earlier in the Galápagos Rift. We also saw large depressions that had clearly been lava lakes. When filled with molten rock making contact with cold bottom water, they must have

Artist's rendering of a large lava lake discovered in the Galápagos Rift in 1977 and delineated by the ANGUS camera system. (Source: Robert D. Ballard.)

been spectacular, seething cauldrons. Rings around their sides, like rings around a bathtub, and isolated pillars suggested that lava had drained away into deeper chambers after filling these lakes.

Clearly the faster rates of seafloor spreading at the East Pacific Rise, as well as in the Galápagos Rift, were producing high volumes of lava that flooded large areas, as opposed to the toothpaste-like dabs, or pillow lava, that we had seen in the Atlantic.

Lacking a towed camera sled, the French group could not scan large areas before choosing their targets for diving. Human observers aboard *Cyana,* with its limited horizontal range, were the only eyes on the bottom. It is therefore not surprising that geologists on the 1978 expedition did not discover any active hydrothermal vents. Yet Francheteau's group saw indirect evidence of them within the rift valley: badly dis-

Artist's rendering of a lava "pillar" situated near the outer perimeter of a drained lava lake. The pillar supports a remnant of the "quenched" surface of the once-molten lake that did not collapse into the lake when it drained. "Bathtub rings" documenting the drainage event are seen on the sides of the pillar and adjacent walls. (Source: Robert D. Ballard.)

solved white clamshells, colored deposits on rocks, tall cones of highly oxidized mineral deposits. In hindsight, it is easy to label some of this as evidence for substantial eruptions of hot water, in keeping with the large flows of lava.

During one dive a Mexican scientist aboard *Cyana* came across chimneylike structures on older volcanic terrain flanking the rift valley. Using the submersible's mechanical arm, his diving crew collected samples. Having no idea how unusual they were, expedition members stored the colorful, glistening rocks among hundreds of others taken from the bottom. Months later, analysis on shore showed those chimney rocks to be pure sphalerite (an ore that is mainly zinc sulfide) containing 50 percent zinc, 10 percent iron, and 1 percent copper, with trace concentrations of lead and silver.

This was a remarkable finding. Such an ore deposit could have been carried to the seafloor only by scalding hot water, emerging at hundreds of degrees Fahrenheit. Along the Galápagos Rift, by contrast, the highest temperature that anyone had measured in vent fluids was 73 degrees Fahrenheit. That temperature was not, however, the whole story. Analyses of water samples from those lukewarm vents showed that the fluids had started their journey at much higher temperatures as they left the reactive zone around a magma chamber: between 660 and 750 degrees. The discovery in 1978 of sphalerite on the ocean bottom suggested that similarly hot water—unbelievably hot, capable of melting some metals—might actually have erupted directly onto the seafloor.

Although we had seen only moderately warm springs in the Galápagos Rift, we now knew—or should have known—that searing fountains of heated fluids could actually reach the surface. Would such eruptions be fairly continuous? Somehow it seemed unlikely. In theory it might be possible, but in practice we found it difficult to find even tepid water, let alone hot. Perhaps the escape of superhot water was extremely rare, so that no human would be likely ever to witness an outburst. No one really knew.

Rivalry on the Rise

The American phase of the U.S.-French expedition to the East Pacific Rise took place in the spring of 1979. Scientists in the American group, headed by Fred Spiess, wanted to understand how the seafloor was spreading on a time scale of months to years. Previous research, carried out since the 1950s, had confirmed the theory of seafloor spreading over hundreds of millions of years—the time scale of continental drift, during which crustal plates traveled thousands of miles. The Scripps geologists now wanted the most recent details, inch by gritty inch.

To understand how forces active right now were ripping the seafloor apart, these scientists needed to know the physical characteristics of the upper ocean crust: its composition, density, porosity, and so forth. That meant using *Alvin* primarily to deploy instruments such as seismographs and carry out experiments such as hammering the bottom by dropping a weight, rather than as a vehicle for making qualitative, visual field observations.

One key to understanding the very recent local geology, we all agreed, was to pinpoint the sites where new seafloor was actually emerging—the active volcanic zones. At that time, only a few of us were beginning to understand how narrowly confined the central, active volcanic axis of the Midocean Ridge truly was. Other participating geologists, particularly those from Scripps, were convinced that the zone of volcanic extrusion was much wider. They believed they would find significant areas of eruption several kilometers beyond the central rift valley, and they ran the first search. They focused on a reasonably flat area to the west of the central axis known as Tortilla Flats, where they hoped to locate active volcanoes and fresh lava flows using their Deep-Tow system. This was also where they wanted *Alvin* to dive.

The primary sensors on Deep-Tow were a side-scan sonar, a temperature probe, and a magnetometer. These remote-sensing devices painted a broad, regional picture of the bottom terrain and some of its physical properties. Although Deep-Tow also had two black-and-white cameras, for slow-scan television and still imaging, it spent little time in close visual contact with the bottom. The Scripps team had to rely mainly on remote sensing to find clues to volcanic activity, rather than specific visual markers like clamshells or fresh glassy lava. Day after day passed as the Scripps research vessel *Melville* pulled Deep-Tow over the area. We saw no signs of active venting; Tortilla Flats proved to be older than the rift valley seafloor, and it was covered by a thick blanket of sediments.

Another team aboard *Melville,* which included Jean Francheteau and me, then got a chance to survey the bottom. We were convinced that we had to look elsewhere to find volcanic activity. During both the 1977 and 1979 expeditions to the Galápagos Rift, I had noticed that active hydrothermal vents lay along a straight line near the youngest lava flows in the rift valley. Once we had found a vent, it became relatively easy to find additional vents along any particular fissure system simply by driving along the strike of the fissure, which generally paralleled the rift axis. Now, if our team here on the East Pacific Rise could find warm water the same way, we should also find what the Scripps group wanted—fresh lava, active volcanoes—in addition to the warm springs.

Our tool for the search would be ANGUS, the Woods Hole towed-camera sled. Unlike the Deep-Tow system, which needed to steer clear of suspected obstacles in order to protect fragile components, ANGUS was intended by field geologists to remain within constant visual range of the bottom. It had therefore been designed to survive head-on collisions with rugged volcanic terrain—"Takes a Lickin' But Keeps On Clickin'," as crew members had painted on the sled. Part of ANGUS's uncanny resilience was due to what it had—a rugged frame—and part to what it did not have—whenever possible, delicate sensors were left off. That explained ANGUS's unofficial name: Dope on a Rope.

This strategy—keeping it simple and robust—allowed us, unlike the Scripps geophysicists, to lower our camera sled directly into the narrow rift running down the axis of the ridge crest.

But how should it scan the bottom? Unlike traverses made by submersibles, which were usually perpendicular to the rift, we decided to run ANGUS's traverses more or less parallel to the rift axis. This was not desirable for any geological reason. Traverses perpendicular to the rift axis, from younger

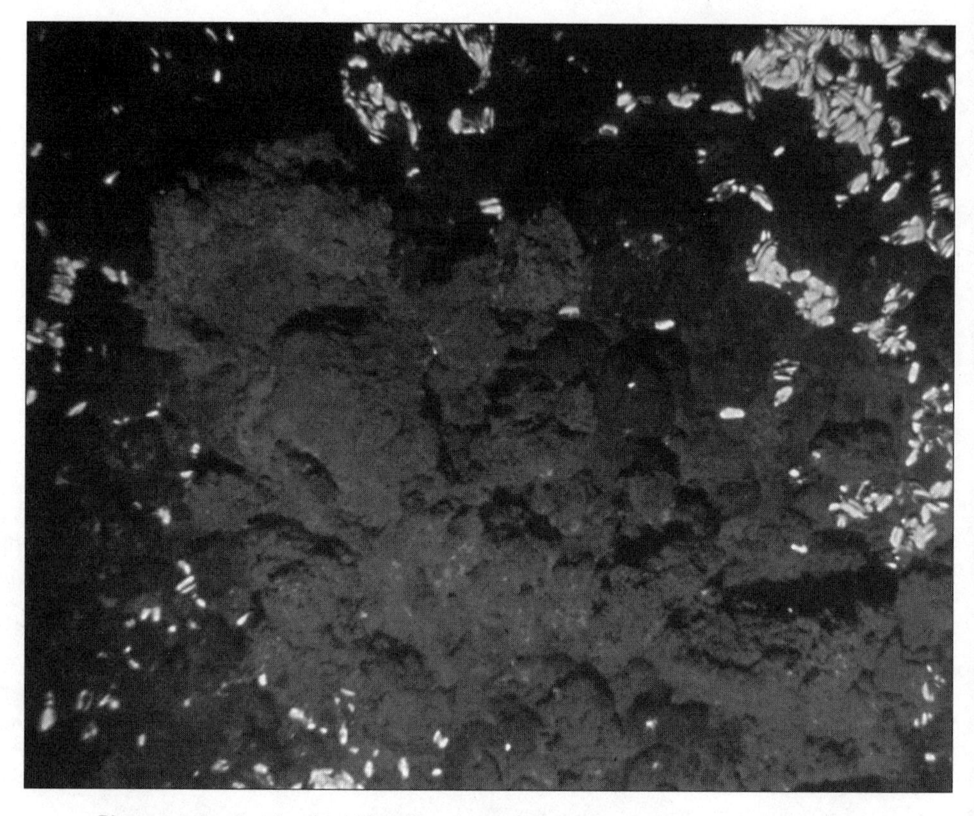

Photograph taken by the ANGUS camera sled as it crossed an active hydrothermal vent at 21 degrees north in 1979. A large deposit of sulfide minerals formed by "black smoker" precipitation is seen in the middle of the vent field, surrounded by large white clams and other vent-community organisms. (Source: Woods Hole Oceanographic Institution.)

to older seafloor, were actually more useful scientifically. But since fault scarps bounded the rift to either side, we decided on a compromise to avoid countless collisions. Even ANGUS had limits.

Our first trackline resembled a slalom run, as we towed the vehicle from side to side down the strike of what we hoped was the central volcanic axis. No sooner had ANGUS begun its unusual traverse than its temperature sensor alerted us to a warm spike. Repeating procedures we had followed on the

Galápagos Rift, we recovered the sled and processed its color film as quickly as possible.

Later, as we reviewed the photos, we saw frame after frame of young volcanic terrain covered by a fresh glassy surface, just as we had seen in the Galápagos Rift. Then came the first indication of an approaching warm spring: it was not a remotely sensed rise in temperature, but the appearance of small white galletea crabs dotting the otherwise barren flows. This gradient of crabs quickly increased, giving way to larger and more densely packed sessile organisms, in particular the white clams so typical of the Galápagos Rift oases. As the center of this vent site approached, we could see the familiar milky water, laden with minerals, along with increasing amounts of suspended particles. But then the scene began to look odd. Unlike the Galápagos oases, the center of this vent site was not covered with tube worms and clams. Instead, we saw large, yellowish-brown deposits, for the most part devoid of life.

We radioed the coordinates of this vent site to Francheteau, who had transferred to *Alvin*'s support ship, *Lulu*. Since *Alvin* and ANGUS shared a network of bottom transponders, it would be easy to vector *Alvin* to any site discovered by ANGUS. The geologists scheduled for the next dive, on April 21, decided to visit the site along with two others whose coordinates we also transmitted.

— *Alvin* Begins to Melt

When Dudley Foster, *Alvin*'s pilot, arrived the next morning at the unusual site we had seen in the photos, he too noticed something odd about it. And then he saw what we had not seen: the black smoker. The fluid rushing out appeared to be hot, but only a measurement could verify his hunch. Moving closer, Foster felt the submersible being pulled toward the chimney.

As *Alvin* entered the billowing thick cloud, Foster lost all visibility. That was unfortunate, because steering and driving were already difficult. During the struggle to regain control, *Alvin* bumped into the chimney, which toppled like a huffing house of cards. For a moment all was confusion and swirling black smoke. Fortunately, the collision improved the situation. Black fluid now flowed from the broken base of the chimney instead of its top, releasing *Alvin* from the updraft. Foster activated his variable ballast system, making the sub less buoyant, and landed slowly on the bottom, fully in control.

Using the vertical lift propellers, Foster climbed a gentle mound surrounding the fallen chimney. Obviously these structures were fragile; the entire mound, which was some thirty feet in diameter, consisted of numerous broken chimneys that had fallen on their own, without the help of submersibles. As he approached the latest toppled chimney, Foster saw that it was hollow, lined with mineral crystals that reflected the submersible's lights. With *Alvin* resting firmly on the bottom, he could bring its mechanical arm into play. He moved it to *Alvin*'s science tray, groping for a temperature probe with an easily grasped T handle.

Lifting this probe from the tray, Foster rotated it to the right and positioned it above the vent opening, which was still gushing black fluid. The temperature readout, displayed inside *Alvin*'s pressure hull, shot up. Next, Foster inserted the probe inside the vent. Its reading went off the scale.

Now Foster grew nervous; he urged the two geologists to report this peculiar event to the surface. The same probe had been used in the Galápagos Rift to measure the temperature of water exiting from vents. Its readings had never exceeded 23 degrees Celsius (73 degrees Fahrenheit), comfortably within its 33–degree Celsius range.

When Jim Akens, the engineer who had built the probe, heard what was going on, he knew something was wrong. He had designed the probe to handle higher temperatures than anyone expected. Time after time, water temperatures in the Galápagos Rift had borne out those expectations. Akens therefore doubted the off-scale reading; he suspected that the instrument had malfunctioned. He urged the crew in *Alvin* to try another measurement.

Less worried now, Foster again nudged *Alvin*'s mechanical arm into the black fluid. Again the temperature reading shot off the scale. Another malfunction? Foster tried to move closer to make a better placement but lost his way in the thick black smoke. By now it seemed obvious that he would not get an accurate reading; he pulled back, eased away, and drove on to the third vent site.

Thankfully, this last site resembled others Foster had seen at the Galápagos Rift: benign and teeming with life. Thinking the temperature probe broken, he stopped using it. There was still plenty to look at and sample, especially with two awed passengers aboard who had never seen a deep-sea oasis before this dive. Foster and the two geologist-observers finally ascended without ever making any gravity measurements.

Later that day, with *Alvin* safely stowed on *Lulu*'s deck, Jim Akens removed the troublesome temperature probe to see what was wrong. He found plenty. The tip of its plastic holder had completely melted; the rest had charred. Stunned, the engineers on *Lulu* began searching their manuals for the melting point of this particular plastic. Finally they found it: polyvinyl chloride, 356 degrees Fahrenheit.

Alvin's forward viewport, which had been sitting in leisurely repose only few feet from the vent opening, was made of a similar plastic. If that Plexiglas window had also melted, or even softened. . . .

Before he could call it a day, Dudley Foster felt compelled to inspect *Alvin*'s exterior skin near the lower viewports, which had hovered directly over the chimney top. He found melted fiberglass.

—— Hell Itself

Jean Francheteau and I were scheduled to dive together in *Alvin* the next day. We were professional colleagues, and also close friends. Over the years, we had participated in numerous expeditions together, diving in *Archimède,* in *Cyana,* and now in *Alvin.* Of all the scientists on this cruise, the two of us were the most interested in finding active hydrothermal vents. The year before, during our dives in *Cyana,* we had come close. Now that we had found one—in superhot spades—it was wonderful that we could see it together firsthand, through the viewport of a submersible.

Before we dove, Ralph Hollis, our pilot, made sure we had a temperature probe that could stand up to high heat.

Reaching bottom, we approached the black smoker cautiously. The surprise of being the first humans to see it was no longer possible, but the thrill was the same. And knowing what to look for had its advantages. On this dive we saw several chimneys, some belching black and others white smoke. Francheteau said it best in his wonderful English: "They seem connected to hell itself."

As we explored this deep inferno, looking for all the world like a dark industrial landscape, we were able this time to take accurate temperatures. It meant bringing *Alvin* close to the searing hot smokestacks, very cautiously—taking readings along the way. Luckily for us, the temperature gradient turned out to be steep: icy cold within inches of a vent, then shooting sky high with every nudge closer. The highest reading from inside a vent showed an incredible 350 degrees Celsius, or 662 degrees Fahrenheit—more than hot enough to melt lead, let alone our Plexiglas viewports.

Out of which we stared in utter amazement.

— Salt of the Earth

Here, in 9,000 feet of water, we were looking at the first direct, visual proof of what geophysicists and geochemists had only theorized. Here was the actual superhot water, rushing up from deep chambers below the seafloor, blackened by the high concentrations of mineral salts it had leached from within the earth's crust.

Here also was a crystal-clear demonstration of what had eluded marine scientists for centuries: a logical explanation of the oceans' chemistry. As John Edmond had put it during our first Galápagos expedition, the question seemed embarrassingly simple: Why is the ocean salty?

Before our discovery of hydrothermal vents, first in the Galápagos Rift and now here on the East Pacific Rise, the composition of seawater had baffled scientists. We knew about the water cycle on the earth's surface—how water heated by the sun evaporates from oceans, condenses into clouds, and precipitates into rivers that return it to the sea. Initially, geochemists thought that compounds dissolved in seawater came from the global river system. But they found instead that the compounds flowing into the sea from rivers were different from certain mineral salts that scientists found uniformly throughout the world's oceans. There were chemicals present in the oceans that could not be found in rivers, and vice versa.

Now we had finally found the compounds missing from the equations. They were coming from hot springs on the seafloor. In addition to the water cycle on the earth's surface, here was a second kind of water cycle, through the crust—driven not by solar heat and evaporation, but by the earth's own internal heat.

What Francheteau and I were watching was part of the same deep recycling of seawater that nourished the oases of the Galápagos Rift, providing hydrogen sulfide to the bottom of the food chain. Here, on the East Pacific Rise, we found hotter springs: black or white smokers. The color of the exiting water depended on its temperature and mineral content. The chimneys themselves consisted mainly of polymetallic sulfides—pyrite, anhydrite, and chalcopyrite rich in sulfur, zinc, iron, copper, lead, and other metals, with traces of gold and silver. Their rapid growth testified to the mineral richness of this subterranean broth. As the fountains of hot fluids cooled, they left material that built up the chimney flue pipes ever taller: in the 1990s, one group of researchers saw a chimney grow from zero to sixteen feet tall in less than a year. Some, the "pagoda"

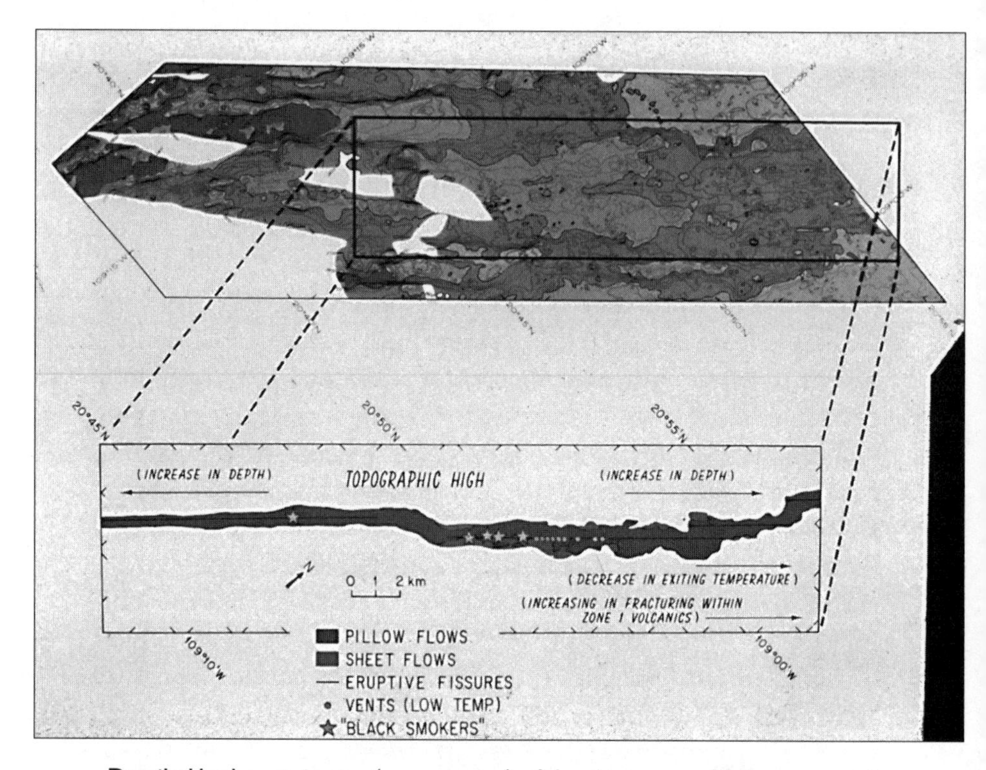

Detailed bathymetric map (upper image) of the rise crest at 21 degrees north compiled from SEABEAM sonar data collected by France's *Jean Charcot.* Five-meter contour interval. The lower image is a simple geological map based upon ANGUS camera runs in 1979 that delineates the young central volcanic axis and active vent sites. (Source: Jean Francheteau and Robert D. Ballard.)

chimneys, were grotesque constructions up to thirty feet high, with mushroom shapes and flanges projecting from their sides.

Other researchers on the East Pacific Rise expedition of 1979, although they were not biologists, continued to discover important vent animals. One of the most impressive was dubbed "Pompeii worm," or *Alvinella pompejana.* These creatures live in a mass of honeycomb-like tubes near high-temperature vents. The worms, which move freely into and out of their tubes, attach themselves even to the sulfide chimney

Fragment of "Black Smoker" chimney lining revealing interior crystalline assemblage of sulfide minerals. (Source: Emory Kristof/National Geographic Society Image Collection.)

walls of the hottest black smokers. The water flowing past their point of attachment must be highly mixed, and therefore at a lower temperature than the hottest exiting fluids. Even so, the tiny tubes seem precariously close to disaster. They are constantly being covered by fine mineral grains, which precipitate from the rapidly cooling waters all around them.

During dives on the Galápagos Rift, we had already seen the nourishing effects of nutrient-rich vents on animal communities. Now, observing the eruption of much hotter fluids through the perforated juncture between two crustal plates, we began to speculate on the broader relationship between water and the earth's crust, which the oceans largely conceal.

A "Pompeii worm" (*Alvinella pompejana*) collected at a high-temperature hydrothermal vent at 21 degrees north on the East Pacific Rise. (Source: Emory Kristof/National Geographic Society Image Collection.)

Staggering amounts of seawater apparently cycle through these deep-ocean springs. Scientists have since estimated that all the water in the seas may seep into the hot, lower crust and rise back up through such vents every eight to ten million years. This is the planet's lifeblood. As it circulates, it supplies vital elements—such as iron, copper, zinc, and manganese—necessary for sustaining life, even on the surface. Our own cells contain trace amounts of these minerals. In addition, minerals carried up to the seafloor precipitate and harden into ore deposits—one explanation, perhaps, for some rich deposits on dry land that was once covered by the oceans. In this sense the chimneys not only resemble industrial factories; they actually are natural refineries. Valuable minerals exist in seawater at extremely low concentrations. The cycling of water through vents builds up far more concentrated deposits.

Suspecting all this, geochemists Rachel Hayman and Randy Koski searched an old copper mine on the Arabian peninsula in 1983. The rocks inside had been created 95 million years ago, on a midocean ridge. And knowing now what to look for, those same scientists found in the dusty ore what no one else had recognized: tube worm fossils.

Hydrothermal Bonanzas

Jean Francheteau had suggested a connection between the high-temperature hot springs and hell itself, and after their discovery all hell certainly did break loose. It was beginning to look as if hydrothermal vents might be common rather than rare, and that scientists would have to take account of a second, previously unknown, water cycle through a little-known part of the earth's crust. The implications had a profound impact on many fields of research—biology, chemistry, geology, and geophysics. Just as the theory of plate tectonics had mobilized earth scientists in the early 1960s, so the discovery of hydrothermal vents now mobilized researchers from a broad range of disciplines.

All of a sudden, scientists who had never set foot in a manned submersible or expressed much interest in the spreading axis of the Midocean Ridge were submitting deep-sea proposals to the various funding agencies. They considered it vital to investigate vents. To justify their requests, some emphasized the importance of the recent discoveries to basic science. Others cited the possibility of mining the mineral deposits—a familiar incantation to the gods of funding. Still others argued the importance of such deep-sea resources, especially "strategic metals," to national interests.

The rush to the vents began full force in 1981, with an expedition to the Galápagos Rift funded by the National Oceanic and Atmospheric Administration (NOAA). Led by geologist

Alexander Malahoff, researchers used *Alvin* to probe further into the area where geologists in 1977 had first discovered hydrothermal vents. The NOAA group eventually reported finding more than twenty chimneys atop a huge mass of mineral deposits. Malahoff estimated the vein of ore lying under them to be more than 100 feet thick, 600 feet wide, and 3,000 feet long—consisting of millions of tons of minerals. If those metallic sulfides could be mined, he claimed, their copper alone might be worth $2 billion.

Needless to say, Malahoff's estimates raised eyebrows, some in disbelief. Nevertheless, a few venturesome capitalists, as well as governments—notably that of the United States—expressed a keen interest in finding out how many vent sites might exist, and how many of those might offer similar pay dirt.

Before Malahoff went about his ore-finding business, Jean Francheteau invited me to participate in an explorer's dream: a three-month journey, in mid-1980, along the East Pacific Rise. We would be cruising aboard *Jean Charcot,* the premier French research ship. Taking advantage of the latest American technology in bottom mapping, the French had purchased the first unclassified multi-narrow-beam sonar system, called SEA-BEAM, and mounted it on the hull of *Jean Charcot*. For the first time, the world scientific community could survey potential diving sites along the Midocean Ridge quickly and in great detail without having to rely on the secretive whims of the U.S. Navy, as we had done during earlier programs.

The timing could not have been better; all kinds of people were clamoring to find vents. By now, it was clear to Jean and me that several factors controlled the distribution of hydrothermal vents. They were situated in the youngest volcanic terrain. And the faster the rate of seafloor spreading, the more likely that vents would be found. Yet we knew more.

Our studies along the Galápagos Rift and the East Pacific Rise—and even the Mid-Atlantic Ridge, where hydrothermal venting had not yet been found—had revealed similar geo-

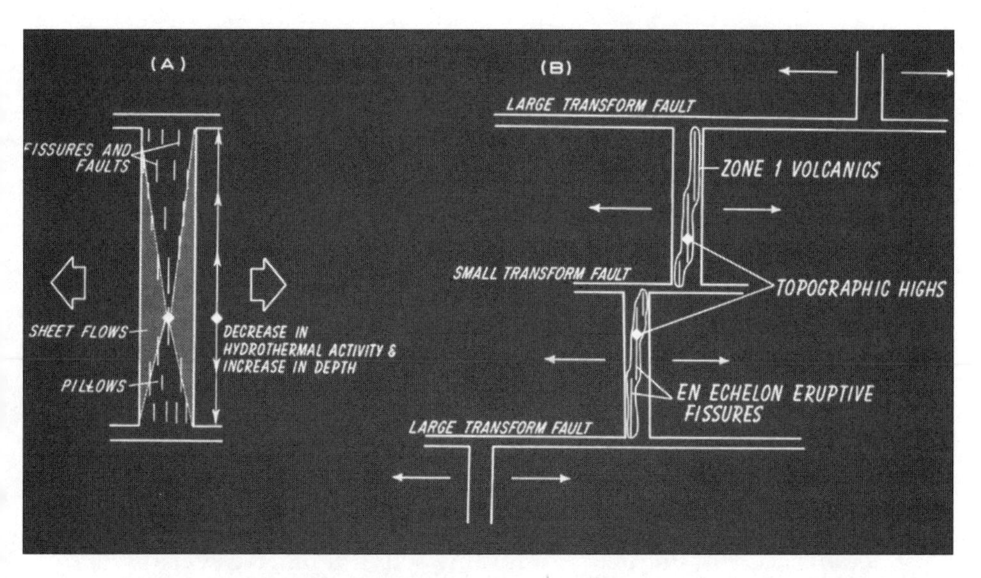

An early model developed by Jean Francheteau and the author to assist in finding active hydrothermal vents and to better explain the presence of "sheet lava" flows and young unfaulted volcanic terrain at topographic highs within individual spreading segments of the Midocean Ridge. (Source: Jean Francheteau and Robert D. Ballard.)

logical patterns. Each ridge was sliced into distinct segments, bound at each end by faults perpendicular to the axis of the ridge. As the crust on each side of these faults slipped one way or the other, it offset the ridge to either side. When we approached the intersections between a ridge axis and these numerous transform faults, the height of the ridge began to decrease. Since the Midocean Ridge is essentially a heat blister on the seafloor—a result of thermal expansion—we reasoned that the higher its elevation, the hotter the crust there, and the more likely we would be to find hydrothermal activity.

We decided to test our model by mapping a long stretch of the East Pacific Rise and then searching for vent sites at the high spots between transform faults. From May until July 1980, *Jean Charcot* zigzagged slowly over the East Pacific Rise, starting at 22 degrees north latitude and moving toward its fastest-spreading segment at 22 degrees south, near Easter Island. From our SEABEAM survey, we could clearly see individual spreading segments. We hoped to investigate the topographic high points on as many segments as possible whenever we returned.

Our first chance came in April 1981 during a cruise aboard *Melville,* the Scripps research vehicle, to the East Pacific Rise at 20 degrees south. Using *Jean Charcot*'s SEABEAM maps to guide us, we conducted a series of ANGUS camera runs down the rift valley axis of a fast-spreading segment near its high point. That strategy quickly found active hydrothermal vents.

We had another chance in January 1982 when Jean brought the submersible *Cyana* aboard *Le Suroit* to dive at 13 degrees north on the East Pacific Rise. Once more, our model proved to be an excellent predictor for finding active vents. We even dove in *Cyana* where we did not expect to find vents, near the intersection of a transform fault with the ridge axis. Sure enough, we failed to find any, and this further strengthened our confidence in the model.

— From Hot to Cold to Extraterrestrial

By then, of course, we were not the only team searching for new vent sites. Peter Lonsdale from Scripps, who had played a major role in the discovery of warm springs in the Galápagos Rift, was using *Alvin,* and his considerable field-research skills, to search for vent sites in the Gulf of California. He focused on a series of small spreading segments in the Guaymas Basin, and in January 1982 he located a number of active vents. What made them unique was their setting. They were not in barren fresh lava fields but in an area covered by thick sediments, often thousands of feet deep.

For millions of years, the Colorado River has deposited a tremendous volume of rich organic sediments into the Gulf, covering even the young, actively spreading volcanic seafloor with its thick, mucky accumulation. Hydrothermal fluids must rise through this sediment cap before reaching the bottom waters of the Guaymas Basin. During this final vertical journey, superheated fluids essentially cook the organic sediments, producing many kinds of petroleum compounds. Even the surface water smells like fuel oil.

On the soft bottom sediments, Lonsdale's group found extensive green, bright yellow, and white bacterial mats. They turned out to be previously unknown species living on hydrogen sulfide, producing methane gas, and growing to such jumbo size that individual "microbes" could be seen without a microscope!

During and after 1981 the study of hydrothermal circulation intensified and spread throughout the world. A team headed by Peter Rona of NOAA discovered the first hydrothermal vents on the Mid-Atlantic Ridge in 1986. Both French and American researchers found additional vents along the East Pacific Rise at 10, 11, and 13 degrees north. Similar sites turned up in 1984 on the Juan de Fuca and Discovery Ridges off the coast of Washington and British Columbia. Across the Pacific, in 1987, geologists tried a different setting. They dove in *Alvin* west of the Mariana Islands, near a zone where one crustal plate is burrowing under another, producing magma. In this active volcanic area, where slabs of seafloor are converging rather than spreading, the scientists found shimmering vents and spectacular, beautiful oases.

Not to be outdone, other expeditions to the Mid-Atlantic Ridge located active vents at 26 and 23 degrees north. More recently, hydrothermal vents have been found between 37 and 38 degrees north, near the Azores crustal hot spot. Large lava lakes dot that seascape as well, proving, as we suspected, that not all of the Mid-Atlantic Ridge is "average."

At the other temperature extreme, scientists on a geology expedition in 1984 found oasis creatures in near-freezing water. Just as their astonished predecessors had done after finding the first warm oases, these scientists instantly expanded their mission. Instead of rocks, they began grabbing biological specimens.

It started when two researchers diving in *Alvin,* biologist Barbara Hecker and geologist Ray Freeman-Lynde, reached the base of the West Florida Escarpment in the Gulf of Mexico. This huge limestone cliff rises half a mile from the seafloor. Scientists on the previous dive had noticed dark stains beneath it, left by fluids seeping from its base, and large clamshells—which had fallen, they assumed, a very long way from the top of the cliff. Now, everywhere that Hecker and Freeman-Lynde looked, they could see animals in the cold bottom water, living all around these equally cold black seeps.

Further investigation showed that the black sediments were rich in hydrocarbons—petroleum compounds—and that the water was rich in hydrogen sulfide. Apparently, nutrients seeping out from the limestone supported chemosynthetic bacteria in these cold waters off Florida. As in the warm oases, such bacteria also lived inside the gills of animals nearby: large mussels, white clams, and red tube worms. The densely populated cold-seep communities also included galatheid crabs, gastropods, sea anemones, serpulid worms, and other organisms typical of warm-water oases.

Farther west in the Gulf of Mexico, off Louisiana and Texas, additional cold-water seeps provided habitats for similar deep-sea communities. More recently, cold hydrocarbon seeps off the coast of California, in the North Sea, and in the Sea of Okhotsk have been found to support similar ecosystems. The same organisms have even been found sprouting on and around the oily bones of a decomposing whale in the deep Pacific off California.

In the late 1970s, when we first tried to envision how such organisms might find new homes after fluids stop flowing at their vent sites, we thought of spores and larvae drifting on long and perilous journeys. The odds seemed slim that many would survive. We never imagined that such creatures might easily hopscotch all around globe, lighting on any potential home from warm vents to cold seeps to large carcasses. Obviously, these ecosystems are not confined to fresh volcanic terrain at the boundaries of shifting crustal plates. And they are winking on and off in surprising numbers. In years to come, chemosynthetic animal communities will probably be found throughout the world's oceans and lakes, wherever nutrients exist and conditions arise to spur their growth.

Such habitats may not be confined to planet earth. Recently, the vent mania has even influenced space exploration.

In 1975, when the *Viking* lander touched down on Mars, its robotic arm sifted the dusty, reddish-brown soil for signs of microbial life. No conclusive evidence emerged. These days, however, the search no longer focuses on the surface. As this book is being written, astrobiologists, who speculate about extraterrestrial life, are turning their thoughts to other environments on Mars, once considered impossibly hostile but now thought to be potentially suitable for chemosynthesis. In coming years, life-seeking missions may probe volcanic terrains, for instance, below the surface of Mars, where hydrothermal fluids might have circulated in the past and perhaps still do.

Farther afield, a robotic probe may someday visit the ice-covered surface of Jupiter's moon Europa. Scientists believe that this moon has internal heat, and that a liquid ocean sloshes beneath the ice. Volcanoes have been seen erupting on another Jovian moon, so who knows? Maybe distant Europa has black smokers, a salty ocean, and vents just right for alien life.

For the time being we can still be certain of this: most of the earth remains unexplored—never seen by human eyes. Even the seafloor viewed indirectly, by cameras towed over the bottom, adds little more to the minuscule total amount we have seen, probably less than 1 percent. Scientists have barely looked into our own dark abyss.

PART III
DETACHMENT

CHAPTER

8

A TETHERED EYEBALL RACES TO FIND THE *TITANIC*

The endeavor between France and America has been technically and spiritually difficult, yet the actual *Titanic* site was explored with her dignity in mind at all times.

—*Jean-Louis Michel*

More and more in the 1980s, I thought about death in the black abyss. One reason was common sense—irrepressible at times, when one is slowly falling inside a potential coffin, surrounded by a cold, dark, and relentless pressure. Another reason was the changing nature of my deep-sea exploration. More and more often, I was leading searches for sunken ships. Their silent presence on the seafloor always reminded me of the humans who had gone down with them. During the summer of 1984, in particular, my mind kept returning to a tragic accident with much loss of life. I was leading a crew from Woods Hole on a mission to examine the wreckage of the USS *Thresher.* The assignment required me to picture how the huge submarine had broken apart and sunk to the bottom, so I couldn't help but imagine the final scene.

It began with *Thresher* gliding silently in the deep Atlantic, a thousand feet below the surface. In abrupt response to a leak and short circuit, its nuclear reactor had shut down. Gradually the sub slowed and halted; then it started to sink. As pressure squeezed tighter on the hull, thick steel plates started bending imperceptibly inward. Suddenly, at around 1,500 feet, a big section of the hull behind the engine room collapsed. All 129 men aboard would have heard this hammer blow—felt it—as a visceral, loud thud. Instantly, a massive slug of seawater roared into the gaping hole. Ramming ahead like a huge boxer's fist, it exploded through heavy sealed bulkheads, shredded interior compartments, ripped out walls, and expelled debris. The new combat submarine, nearly 300 feet long, weighed 4,000 tons in water. The ocean demolished it in about two seconds.

U.S. Navy personnel on *Skylark,* a rescue ship cruising on the surface, heard the brief implosion. Then nothing. It was 9:18 on the morning of April 10, 1963.

Deciphering a Debris Field

Twenty-one years later, I visited *Thresher*'s graveyard. The navy had asked me to document and map the debris, and also to locate and monitor the vessel's nuclear reactor for any trace of leaking radiation. Despite all the time that had passed, the navy still did not have an accurate picture of the site. And there remained some doubt about just what had gone wrong during the last few minutes of *Thresher*'s fatal test dive. Knowing which parts of the structure had failed, and the sequence of the breakup, would help investigators reconstruct the accident. Even now, something useful might be learned that would help to improve future designs or procedures.

Thresher had disappeared in 8,500 feet of water off Georges Bank, southeast of Cape Cod. The bottom there consists of rolling hills cut by small rounded gullies and covered with soft gray sediments. Imagine too a generous sprinkling of large and small rocks on this surface, dropped by melting icebergs that have drifted for millennia in the water overhead. A few more objects landing in scattershot fashion would hardly be noticed in the overall jumble, especially when the bottom was scanned by sonar. In fact, the shattered remnants of a submarine might be difficult to locate by any means unless you were practically on top of them, staring down at exactly the right spot.

The task of finding wreckage in this lumpy and glacially littered landscape had frustrated the navy. After a long search, the bathyscaph *Trieste* eventually discovered big chunks of twisted metal late in the summer of 1963. They were fragments of *Thresher*'s hull. *Trieste* retrieved a piece of twisted, broken pipe, but it made no wider survey. A bathyscaph was not suited for a large debris-mapping project. Nor was *Alvin,* which poked around inside parts of *Thresher*'s hull twenty years later. Although more mobile than a bathyscaph, on a typical dive *Alvin* still traveled much farther vertically (from surface to bottom and back) than horizontally across the seafloor. Diving in *Alvin* would not be a practical way to scrutinize broad patches of seafloor. Yet the navy needed a thorough survey. Bewildering clumps of debris lay everywhere around this site, trailing off an unknown distance into the darkness.

On July 11, 1984, a year after *Alvin* came and went, my team arrived on site to conduct the survey. This time I felt confident that we had the right tool for the job. We were traveling on *Knorr,* the Woods Hole research ship, and bringing along a new kind of vehicle. We were all a bit anxious, for this would be its first deep trial in the open ocean.

In many respects, our new vessel looked uninspiring. Although we had named it *Argo,* after the legendary ship that had carried Jason to the Golden Fleece, any passengers who tried to board it would clearly need help from the gods. For one thing it was smaller than *Alvin;* for another there was nothing to climb into. *Argo* contained no pressure sphere for passengers. In fact it looked skeletal, like a camera sled—just a simple framework of steel tubing, painted white. Fastened to this frame, *Argo*'s equipment—powerful floodlights, sensitive cameras, various electronic controls—lay exposed to seawater. And finally, in a throwback to the days of Beebe and Barton's bathysphere, a tether connected *Argo* to a winch on the deck of *Knorr.* We had designed our little sled to be dragged

The *Argo/Jason* system as it was initially envisioned and published in *National Geographic* magazine in 1981. The ship deploying the system uses a hull-mounted SEABEAM sonar to digitize the bottom terrain in real-time minutes before the *Argo* vehicle crosses over the bottom. The *Argo* towed vehicle is outfitted with passively through deep waters, with no motors or propellers.

Anyone not familiar with *Argo* might easily have dismissed it as a crude platform for towed instruments. That would have been a mistake. True, it was towed, and yes, it carried instruments rather than humans. Yet I preferred to think of *Argo* as the leading component of a new kind of diving *system*—one part of a larger whole. In fact, if you knew where to look, you would find near *Argo* a very large crew cabin: a control van the size of a small mobile home, firmly attached to the deck of *Knorr*.

Our scheme for connecting this cabin to *Argo* set our new system apart from all other deep-diving craft, whether towed-instrument sleds or manned submersibles. When *Argo* dove to the bottom, its tether—which contained a high-capacity data link—would transmit instantaneous, real-time imagery to the surface. That rich stream of visual information, impossible to obtain before the 1980s, would allow *Argo*'s crew to forgo the discomfort and danger of squeezing inside it and actually descending to the seafloor. Instead, while *Argo* glided far below like a distant eyeball, its "passengers" would remain inside the control van aboard *Knorr*, observing the seafloor by means of video monitors.

On the first day's towing run, we experienced an uncanny sensation. While munching hot buttered popcorn, sipping soft drinks, and listening to background music inside the van on *Knorr*'s deck, we could see the bottom passing by as if we were actually inside *Argo*, looking out through impossibly large viewports. We also watched *Argo*'s pilot at work. Sitting beside us while controlling the winch, he "flew" us over the boulder-

---

an array of digital acoustic and visual sensors that create a "virtual reality" for its scientific operator aboard the ship. Once an object of interest is found, the ship uses its dynamic positioning system to keep station while *Jason* is deployed from inside *Argo*. A satellite link provides real-time access to participating scientists ashore. (Source: Davis Meltzer/National Geographic Society Image Collection.)

strewn seafloor, maintaining an altitude of roughly forty-five feet. Using a sophisticated control system, he could decide which aspects of the vehicle he wanted to manage and which he wanted a computer to manage. In effect, he flew with a computer as copilot. Meanwhile the captain of *Knorr* could leave the bridge at any time, come down to *Argo*'s control van, and drive *Knorr* from a chair inside our van; he could stop or change direction in response to what we were seeing on the bottom.

Working this way as a team inside our control van, we could immediately absorb and react to the clear views we saw. Although our bodies remained on the surface, our eyes, nerves, and minds had gone down with *Argo*. No longer shivering and confined inside a tiny pressure sphere, we were nevertheless still exploring the abyss.

On that first day, like a person mowing a lawn, we cruised back and forth in an expanding pattern that would eventually cover the entire area where *Thresher* had gone down. Everyone enjoyed the illusion of deep-sea flying, and we were thrilled that *Argo* was working as planned. But then sobering pieces of wreckage began to appear on the monitors: a rubber glove, a bit of insulation, pieces of pipe, electrical cable. Before long we could see big chunks of *Thresher*'s hull lying helter-skelter on the bottom. It was an appalling scene of near-total destruction.

Once we had mapped the largest debris and recorded every square foot of the site on videotape, we expanded the survey. Our next series of runs, on subsequent days, took us farther from the main wreckage. In almost every direction we cruised—north, east, and south—we found less and less debris, until finally the debris field ended abruptly. That made sense. We were moving farther and farther away from the site of *Thresher*'s implosion. As the distance increased, less and less ejected debris would have tumbled that far. It looked as if our work would soon be finished. In only a few days, we had examined more of this debris field than all other missions combined over the past twenty-one years.

But then something spoiled our neat—and premature—conclusion. As we probed toward the west, we kept crossing over more debris. We would find a long, continuous patch, double back, cruise over more of that patch farther west, and so on—with no end in sight. Way over here, for example, lay a rubber boot and clenched glove from the engine room. Yet most of that room itself—big metallic chunks of it—had landed way back there, at the extreme eastern end of the debris field. Something had rearranged all these fragments, mixing and mismatching pieces. They lay now on the bottom in a new assortment, not at all according to their original location on the sub.

This weird scattering over great distances, and the even weirder redistribution, baffled me. The wreckage no longer resembled the compact debris fields I had seen at shallower sites, with most of the pieces smack in the middle of the field. My puzzlement forced me to rethink what had happened twenty-one years ago. I had to imagine, in more vivid detail, how deep-ocean forces would demolish a submarine. It was the last thing I wanted to do. Pieces of *Thresher* were all that we saw now, but human beings had once been inside it.

At first, when I replayed the sinking in my mind, the dreadful implosion dominated everything. With awesome power and merciful quickness, the ocean must have instantly killed every living person on board. The dying was over, I supposed, before anyone could suffer. And after that? Silence, darkness, falling wreckage—and suddenly I realized what else had happened. I had imagined most of the scene correctly, but with too little emphasis on the silent epilogue. The story did not end at 9:18, with the loud implosion. As terrible as that sounded, it meant only that *Thresher* was gone—obliterated—untraceable from above. And yet its sinking had barely begun. The submarine had vanished relatively near the surface. Fully five-sixths of its final descent still lay ahead.

In silence and in utter darkness, as I now reimagined the scene, parts of the ship and the bodies of its crew were tumbling down into the abyss. The pull of gravity, and the gentle tug of deep-ocean currents, now finished the grim work of demolition. Gradually the wreckage dispersed. As pieces of the broken hull headed for the bottom, a mile and a half below, debris continued to spill out. Each item set off on a separate journey. The most massive and compact debris hit first and hardest. Plowing into the ooze at more than a hundred miles an hour, the heavily shielded nuclear reactor buried itself in a crater. Within seconds, large fragments of the hull crashed down nearby, followed by a rain of trailing debris. These lagging pieces made ever more gentle landings as smaller and lighter objects arrived. Meanwhile the very lightest debris was barely moving from the site of the implosion. A mechanic's glove, a nautical chart, a piece of insulation—each found its own peculiar trajectory and speed, as the seafloor waited below with infinite patience.

The reason we were seeing such a distorted pattern of wreckage now seemed clear. *Thresher*'s lighter debris, instead of spreading evenly around the hull, would have been carried by currents as it fell, predominantly toward the west. Along the way, with the bottom so far below, the ocean had plenty of time to re-sort these lagging pieces. It did so with fanatical precision, and then laid them out in a new, yet logical, progression on the seafloor. The process reminded me of the way farmers used to winnow wheat. In a gentle wind, seeds fell almost vertically. Stems, being lighter, drifted to one side. The lightest chaff simply blew away. The result in *Thresher*'s case was a cometlike trail of debris that stretched out from the largest impact craters over a distance of more than a mile. The lighter the object, the farther it drifted before reaching the seafloor.

I had expected a small and circular field of debris. Instead, I now realized, we had been crossing and recrossing a linear trail. With our navigator's help, I quickly laid out a new search track. Returning to the lightest debris we had seen, we changed course and headed due east. We were now traveling along the length of the trail, not cutting back and forth across it. At last, the pattern made sense. On our monitors, small and light pieces gradually gave way to larger chunks of twisted metal. We followed this trail like a well-marked highway, directly back to *Thresher*'s main wreckage.

During our very first search using *Argo,* we had learned a valuable lesson. We now had a key to unlock other secrets that the world's vast oceans had long been hiding. Many lost ships, I thought, might now be found. The key was knowing how a debris trail forms in deep water, and which way its arrow is pointing.

But the key was more than knowing. It was also seeing. With *Argo,* we had eyes good enough to recognize small bits of debris far away from the biggest pieces. We could miss the main site by a mile, strike debris instead, and then still find everything, without ever descending to the bottom. We knew how great this advantage was because we had also installed on *Argo* the latest side-scan sonar technology, which was often used to hunt for sunken ships. The difference in what we saw using video and sonar was striking. At the far end of the debris trail, *Argo*'s low-light video cameras gave us clear views of electrical insulation, gloves, fiberglass, and so on, mixed among a background of glacial rocks. The sonar traces, however, revealed only the rocks, dropped erratically by glaciers; the finer pieces of man-made debris remained invisible.

Our new knowledge and tools, I thought, could lead us to spectacular new finds. Using *Argo,* we might even find the *Titanic.*

—— A New Paradigm

I had been waiting a long time for a chance to find that singular wreck. Even while pursuing my research as a geologist, and while working with other scientists to find undersea volcanoes, hot springs, and deep oases, I had dreamed of searching for the lost British luxury liner. Given the right technology, I thought, finding it should not be a matter of blind luck. Now, having tested *Argo* on *Thresher*'s wreckage, I felt an irresistible urge to use it on something bigger. This had to be the right technology.

But why that particular target? My motive was in part sheer fascination—some at Woods Hole used the word "obsession"—but it was not a morbid fascination. There was a strong practical component. Finding the most famous wreck of all time—especially after others had failed—would prove the new *Argo* concept in a way that a mapping of *Thresher*'s debris never could. And a spectacular success seemed important, for I fully believed that *Argo* (and its later companion craft *Jason*) represented the next major advance in deep submergence technology.

For half a century, pioneers of deep exploration had invented clever new ways to cram humans into hollow capsules and send them plummeting into eternal darkness. Each new strategy—from Barton's bathysphere to Piccard's bathyscaph to *Alvin* and other submersibles—improved on previous methods. Now our group at Woods Hole was making a radical break with that trend: we wanted to detach the crew cabin from the diving craft—permanently. *Argo,* our tethered eyeball, represented the first part of that plan. A more advanced tethered vehicle would follow, carrying a motorized remotely operated vehicle (ROV) named *Jason* as its passenger—a robot that could sally out to explore, take pictures, and collect samples while its human "crew" remained on the surface. Detached from the actual discomfort and danger, they would still be virtually present down below, still probing the black abyss.

Once again, new technology was changing the nature of deep-sea exploration. Surely such a potent robotic combination would outperform the current generation of submersibles. Or so I believed in 1984, as my thoughts turned from *Thresher* to *Titanic.* Perhaps foolishly, I declared my confidence ahead of time. "You'll never see much in *Alvin,*" I told an interviewer for the *Cape Cod Times.* "Manned submersibles are doomed."

My enthusiasm for new, unmanned vehicles irritated some colleagues, especially among the *Alvin* group at Woods Hole. *Alvin* was, after all, a dear old friend. Remarks that its days were numbered seemed especially unwelcome coming from me—a marine geologist who had built his career on the ability of manned submersibles to open a new frontier, and who had originally joined the team as *Alvin*'s cheerleader. Now that I was advancing a new paradigm for deep-sea exploration, the skeptical and sometimes angry responses coming from this group were understandable. Yet I could not believe any differently. By the end of the 1980s I would log more diving time inside *Alvin* than any other scientist-observer. And that experience was leading me to this conclusion: Manned submersibles, despite their many successes, have inherent problems that will always limit their efficiency. Ultimately, those limits will become intolerable.

The chief advantage and glory of manned submersibles is clearly the human intelligence on board. Allyn Vine, who championed the idea of submersibles in the 1950s, viewed this human presence—constantly sensing, probing, adjusting, and guiding—as irreplaceable. Think back a century, he urged his colleagues, to the great age of surface exploration. What would be the best scientific instrument you could take on an observing and sampling expedition—that you could, say, load aboard the *Beagle*, a British ship that traveled the world in the 1830s collecting biological specimens and copious amounts of data on habitats and climates? The best possible instrument aboard the *Beagle?* Why, Charles Darwin, of course.

Submersibles, Vine argued, could be this century's *Beagle*s. Oceanographers had long been dragging nets and dredges through the deep ocean; they would eventually add remote-sensing techniques and instruments on sleds. But if a few modern Darwins actually went down to explore in submersibles, Vine insisted, the results would be incomparably richer. Whole fields of science might take off in profoundly different directions.

The truth of this argument, however, depended on a small technicality: humans must stay alive to exercise their wonderful intelligence. On the surface, most of the time, that's easy. In the deep sea, however, we are fragile trespassers. The environment we need in order to survive there—a roomful of air, constantly refreshed—must be surrounded by thick, heavy walls. The vessel that houses this fortified chamber requires extraordinary design, construction, and maintenance. None of this comes cheap. *Alvin*, for example, would cost $25 million to build today. But *Jason*—a miniature, unmanned, remote-controlled version of *Alvin* without a pressure sphere and other technology needed to keep humans alive—would cost about $5 million. In light of such enormous differences in cost, it has always been a debatable question whether human presence in a deep-diving craft is more a liability or an asset.

Consider the working space, for example. Because of technical challenges and cost, the amount of space inside a pressure sphere must be limited. Such tight quarters greatly reduce the number of instruments and the amount of documentation a scientist can have there. The small space also reduces, of course, the number of scientists—often to just one or two. From the human point of view, then, the crew space inside a submersible is cramped. But from a budgetary point of view it's bloated. Human presence requires even a "small" submersible to be relatively large and expensive, with backup systems for life support and safety. And once the diving craft gets big, its operating costs escalate as well. A typical manned vehicle weighing twenty tons requires a large, sophisticated support ship with matching crew. The much larger bathyscaphs that preceded small submersibles required even more surface support, which was one reason why small submersibles replaced them.

Now consider removing the humans. You immediately achieve another reduction in size, comparable to the reduction from bathyscaphs to submersibles—this time from "small" submersibles to even smaller, unmanned tethered eyeballs and robots. That, once again, reduces operating costs. At the same time, the space available to participating scientists increases. The number of researchers who can join a remotely controlled "dive" is now large: the number who can fit into the control van on the surface ship. Eventually it will be even larger: the number who can establish a virtual presence—through the Internet, say—from anywhere on the planet.

In addition to creating far more space for crew members, ROVs give each participant much more observing time. A typical dive to 12,000 feet in a manned submersible requires two and a half hours in the morning for the descent and a similar amount of time in the late afternoon for the return to the surface. Such a dive yields only three to four hours on the bottom, with one or two scientists exploring, at best, approximately one mile of terrain. This short period of time on the bottom, and the small observing area, come with a very high price tag—currently $25,000 or more per day at sea, or roughly $8,000 per hour of observing time on the bottom. ROVs, by contrast, can remain underwater for weeks at a time without resurfacing at a cost per observing hour of $1,000 or less—which, when combined with the larger number of scientists who can observe, greatly increases the payoff. Equally important, ROVs can be deployed from a large number of support ships around the world with dynamic positioning capabilities. As a result, they are far more portable, working one month from a ship in the Mediterranean Sea and the next from a different ship in the north Pacific, while submersibles like *Alvin* can be operated only from a single specialized support ship, greatly limiting their range.

Finally, a more subtle notion began to shape my thinking in the late 1970s, turning me toward new technology. I realized that "manned" operations in submersibles like *Alvin* are never completely manned. That is to say, they never use the whole person. Most of the human body that's sealed inside goes along as extra baggage.

Unlike an astronaut on the moon, or Darwin on a tropical island, a scientist in a submersible cannot get out to explore. He or she cannot walk around on the bottom, feeling brittle rocks crunch underfoot or soft mud sucking; cannot bend over to inspect a burrowing creature; cannot use feeling in the fingers to pick just-right samples like fruit in a market; cannot tweak or fine-tune instruments. All that freedom of motion, seeing, and sensing contributes in subtle ways to data gathering and discovery. But instead, an aquanaut must always remain crouched inside a tiny capsule. He or she must peer through small windows with restricted views and must fumble with a remote mechanical arm to accomplish anything outside. In other words, "manned" submersibles are in large part remotely operated—and thus unmanned. They are really a hybrid of the human intellect and mechanical, robotic extensions.

Despite these inherent limitations, a large part of the scientific community decided in the late 1970s and throughout the 1980s that sending a researcher to the bottom of the ocean was worth the expense, given the unique contribution a controlling human intelligence could make on site. That decision proved wise at the time. A steady stream of humans diving in submersibles produced some of the most important discoveries ever made by marine scientists seeking to understand the geology, geophysics, biology, and chemistry of the deep sea.

But history has often shown that just when a new paradigm begins to receive broad acceptance, its replacement is already close at hand. By the time that diving on the Midocean Ridge was becoming routine, new technology was making it possible to consider a different way for humans to explore the abyss. Vehicles that could respond immediately to a controlling human intelligence, yet be operated even more remotely, were becoming possible. With the new technology, it would make little practical difference whether the person exercising commands was situated just a few feet away from the robotic arms of a diving craft, inside a pressure sphere, or many thousands of feet away, on the surface. Manipulator commands can move back and forth through a telecommunications cable at the speed of light. In theory, a person could perform every mechanical function just as well from the surface.

There was still, however, the matter of visual inspection. Submersibles often take passengers right up to lava flows, rocky outcrops, warm springs, wreckage, and other features. Crew members stare intently while deciding what to do—make a measurement, take a sample, creep closer to some unexpected feature, or move on. In the end, my thinking about the two paradigms came down to one question: Was an observer's view of the deep-sea environment superior from inside a submersible, or could we replicate that view in a system controlled from the surface? Was it possible, in other words, to achieve the illusion of *telepresence?*

The answer had always been no—we could not replicate the view. Neither ANGUS, the Woods Hole camera sled, nor Deep-Tow, the Scripps system, achieved anything remotely resembling telepresence. ANGUS could not transmit images; it had to be hauled back to the surface and its film developed. Deep-Tow had a data link embedded in the tether, but its small capacity permitted only a slow-scan black-and-white image to be transmitted to the surface. These time-delayed, snapshot-like views, if they showed much at all, could not be used to control the sled from the surface. In theory, telepresence could make deep-sea operations far more efficient. In practice, into the 1980s, our technology was not up to the challenge.

—— *Argo, Jason,* and *Hugo*

In 1979 I left Woods Hole for a yearlong sabbatical at Stanford University. During that quiet, reflective time I wrote scientific papers on the geology of the Midocean Ridge and planned new research objectives. I also visited local laboratories, where I saw to my amazement that the last barriers to telepresence were swiftly crumbling. There, in the heart of Silicon Valley, researchers and engineers were unleashing a dramatic technological revolution that is still in progress. Integrated circuits were shrinking the size and cost of computers, with no immediate limits in sight. Marvelous new electronic devices, such as digital low-light video cameras, were becoming smaller and cheaper, yet more powerful. Meanwhile advances in robotics were achieving the precision required for remote, free-floating dynamic tracking and control. It now appeared possible that, without even watching on a monitor, we could automatically position and hold a gripping claw within an inch of its target—even though it drifted miles below in inky blackness while we, on the surface, drifted in our own wind and currents. But as impressive as all those developments seemed, to my mind fiber optics was the key innovation. It made the return of tethers worthwhile and telepresence possible.

Telephone companies were just beginning to install fiber optic cables for long-distance communication in the late 1970s. In those networks, tiny solid-state laser diodes (a remarkable innovation in themselves) sent pulses of light through exceptionally pure strands of glass. Compared with their efficiency, the standard copper wires that transmitted electrical signals seemed like stone-age relics. In Beebe and Barton's bathysphere, a pair of wires embedded in thick rubber had carried telephone conversations back and forth between the bathysphere and its support ship on the surface. Now a single glass fiber, roughly the diameter of a human hair, could transmit millions of such conversations simultaneously, or the equivalent of the entire thirty-volume *Encyclopaedia Britannica* every second. With signals riding on beams of light, one fiber optic cable could handle every aspect of deep-diving telepresence. It could transmit richly detailed, continuous video signals along with data from other instruments, plus all the robotic manipulator commands needed to control mechanical systems.

While still at Stanford, I began sketching my ideas for a pair of integrated diving vehicles that would take advantage of fiber optics and the other new technologies. Before long a familiar feeling began to take hold. I had felt the same way when the bathyscaph *Trieste* arrived in San Diego, and again when I first saw plans for a small submersible, a prototype of *Alvin.* Those pioneering vessels had excited me as a student and young professional. And both had fulfilled their promise. They had opened new frontiers to human observation, and the knowledge they had helped gather was still profoundly changing our view of the planet. Once more, I realized while working at Stanford, a new kind of deep-diving craft was becoming possible. What new frontiers might it open?

Sooner or later, however, a dream needs funding. By the end of 1982, I had it. One primary source was the Office of Naval Research, the same office that had supported the development and use of the bathyscaph *Trieste* and the submersible *Alvin* in the 1950s and 1960s. Additional support came from the National Science Foundation, to help develop the fiber optic technology and a dynamic positioning system for *Knorr,* both of which were needed to permit a tethered robot to work in 20,000 feet of water. Meanwhile at Woods Hole I formed a new group, the Deep Submergence Laboratory (DSL), to develop and then operate the world's first deep-water teleoperated exploration system. Like *Alvin,* the new vehicles would be used most often, we expected, by the marine science community.

Although we were developing new technology, the idea of ROVs was not new. Years earlier, the navy had become involved in a series of top-secret "black programs" to develop specialized robotic vehicles and the surface platforms to control them. The CURV robot, for example, helped a surface ship snag and lift a hydrogen bomb from the Mediterranean seafloor in 1966, after *Alvin*'s crew had found the missing bomb. The millions of dollars poured into such programs led to many advances in deep submergence technology. Eventually the military declassified some of that technology. Long before the scientific community showed much interest, oil companies began using it to develop robots that would inspect and maintain underwater drilling equipment. At Woods Hole, my new deep submergence group also began adapting these control techniques.

Our short-term goal was to place human operators—a pilot, a navigator, and other members of our crew—in an advanced control center aboard our research ships. A high-capacity fiber optic tether would connect them to ROVs exploring the ocean below. Our long-term objective was to extend the illusion of telepresence to a large network of scientists. We wanted them to participate in deep-sea operations from shore-based satellite receiving centers. They would not only see exactly what we saw on board our ship, they would also have, at appropriate times, full control of the submerged vehicles. When a scientist stepped into such a center on shore, he or she would, for all practical purposes, be stepping inside our control center at sea.

We called the pair of vehicles we intended to build the *Argo/Jason* system. Their design incorporated the best qualities of both towed vehicles and manned submersibles. *Argo,* the advance-scouting eyeball, would conduct the search or mapping phase of a mission. It would hang passively at the end of a long tether while being towed by *Knorr.* After surveying an area with *Argo,* we planned to lower the motorized, remotely operated *Jason* robot—equipped with its own fiber optic tether, cameras, and a mechanical arm—to inspect and sample interesting finds. This strategy paralleled the technique we had developed earlier of surveying an area with ANGUS, the towed camera sled, and only then diving in *Alvin.* But our new scheme improved upon that older technique. If both *Argo* and *Jason* performed as expected, we would then construct a second, larger *Argo* that would actually house *Jason.* We called it *Hugo,* for huge *Argo.* With *Jason* deployed in *Hugo* and always ready, we would eliminate a long series of separate launches, descents, and recoveries. This would maximize our exploring range and minimize the endless hauling of vehicles up and down that eats away at expedition time.

Our DSL engineers developed *Argo* first. Like ANGUS, *Argo* had to be towed, or "flown" as we said, just above the bottom, allowing constant visual contact with the ever-changing terrain. As a result, we placed major emphasis on its visual imaging capability. Three silicon-intensified target cameras gave *Argo* its video eyes. The lenses in these cameras, originally developed for night-vision devices, electronically amplified available light by a factor of ten thousand. Thus *Argo*'s forward-looking wide-angle camera, a similar downward-looking camera, and another downward-looking zoom camera made optimum use of every photon to pierce the eternal darkness. (In the first version of *Argo,* these were black-and-white television cameras; the latest version, *Argo II,* has high-quality three-chip color cameras as well.) Incandescent running lights provided constant illumination for this video imaging, while strobe lights flashed for a 35 mm color camera, which carried film for later processing. Also mounted within *Argo*'s heavy-duty frame were an echo sounder and a 100-kHz side-scan sonar system; these allowed a much wider area to be surveyed, in swaths up to 300 meters wide. Of course, the price we paid when using the wider sonar as opposed to visual scanning was a great loss of detail.

A new fiber optic cable was still being designed for the academic community in 1983, when *Argo* was first developed. While waiting for it, we decided to tow the vehicle temporarily on a 0.68-inch coaxial steel cable. Since the coaxial cable was electrical rather than fiber optic, its carrying capacity would be lower but still sufficient for black-and-white transmission.

By the summer of 1984, *Argo* was ready for sea trials. These would be funded by the office of the Deputy Chief of Naval Operations for Submarine Warfare. In fact, that office had agreed to support a monthlong expedition using *Argo* every year for four years. Remembering the role *Alvin* had played in finding a lost hydrogen bomb off Spain in 1966, the navy wanted to make sure that the *Argo/Jason* system was used often enough to be kept in good working condition. For the first annual navy mission—*Argo*'s maiden voyage—we were asked to conduct a complete visual survey of the USS *Thresher* wreckage. The second mission, we were told, would be a survey the following summer at the site where the USS *Scorpion* had gone down in 1968.

From the moment *Thresher's* first piece of debris came into view, Argo worked beautifully. The navy got all the documentation it needed, and our DSL group learned how to find and follow a debris trail. Yet none of this early success heralded a new era in deep-sea exploration. Our mission, after all, was secret. The scientific community remained uninformed. We had demonstrated *Argo*'s exciting capabilities to no one besides ourselves and the ghosts of *Thresher.*

—— Another Summer, Another Submarine

When we began to build *Argo,* and even before, I made no secret of the prize I was seeking. In one part of the deep submergence community—a part that included engineers who craved a technical challenge, marine historians, and treasure hunters, some of whom were less reputable—discovering the final resting place of the *Titanic* had become a tantalizing lure, like the descent into Challenger Deep decades earlier. Several expeditions had already gone after the Big Prize, and the thought of someone else finding it before we finished *Argo* added tension to all our efforts. But entering that contest would bring new risks. As with Beebe's or Piccard's early descents, the intense public scrutiny would set off reactions beyond our control. If we failed, we could expect harsh treatment by the press, scorn from our peers, and hard times for the *Argo/Jason* program.

The *Titanic* itself made matters even more difficult. It had gone down in a cold, deep, and stormy part of the ocean, and it had somehow managed to disappear completely on the bottom. Its sinking had also become a legend—tragic and symbolic. To many minds, that enormous, "practically unsinkable" boat remained a powerful synonym for hubris—an overweening pride in technology—and perhaps, in a larger sense, for the approaching fate of our modern civilization. Yet none of that made the historic ship a *scientific* prize. For our group at Woods Hole, a direct approach was not possible. Finding the *Titanic* simply because it was there did not, in the minds of the oceanographic community, merit the search. Yet in the summer of 1985 we received indirect backing. Again, as expected, the navy asked us to map submarine wreckage. This time they wanted complete visual documentation of the site where the USS *Scorpion* had been lost, seventeen years earlier, in 11,500 feet of water southwest of the Azores. It was understood that once we completed this survey, which had been allotted three weeks, we had permission to use *Argo, Knorr,* and the remaining navy-funded time as we saw fit.

Knorr arrived at the site of *Scorpion*'s wreckage in mid-August. We knew that this submarine, unlike *Thresher,* had not been crushed and shredded in deep water by a major implosion. As best the navy could determine, it was instead destroyed at or near the surface. Either the battery compartment had exploded or *Scorpion* had been sunk by its own torpedo, ejected from the forward torpedo tube during an accidental "hot" (armed) run. Once released, the torpedo would have actively sought a target; it would have turned back on the submarine itself. In either case—internal explosion or torpedo strike amidships—the sub broke in half. Each part then filled with water and sank to the bottom.

Towing *Argo* over the site, we found two large impact craters on the ocean floor. One contained the forward part of the sub and the other the stern part, which included the heavy nuclear reactor compartment. The stern fragment had telescoped into a crumpled heap. Altogether, *Scorpion*'s hull looked far more intact than *Thresher*'s; yet debris still lay scattered all around. After the sub broke in half near the surface, debris must have fallen out as the two main pieces descended. Although this wreckage told a tale of accidental sinking completely different from *Thresher*'s fate, it had left a similar debris field on the ocean floor. We traced this new pattern with great interest. Again we saw a long trail on our monitors, with lighter items giving way to heavier ones, and again we could easily follow that trail back to the main wreckage.

Working day and night in rotating shifts, we mapped the debris field thoroughly. In four days, we had completed the job to the satisfaction of our naval observers. Immediately, we headed toward colder Atlantic waters and *Titanic*'s last reported position.

Rivals and Partners

In fact we were following on the heels of several recent attempts to find the great ship. A multimillionaire Texan named Jack Grimm had financed three of them, and his failures proved instructive. His first expedition, in the summer of 1980, ran into trouble from high seas, damaged equipment, and confusion about where to search. His second attempt, the following summer, searched a new area and came up with the location of what Grimm announced to be a propeller. His third expedition, in 1983 (again hampered by foul weather), found no propeller or anything else that a magnetometer could identify as metallic. A fourth search—conducted just weeks before my group's arrival—was led by Jean-Louis Michel of the Institut Français de Recherche pour l'Exploitation de la Mer (IFREMER). Michel's team represented half of a joint French-American effort. My Woods Hole group was the other half.

I had arranged our partnership the previous year. After watching Grimm's escapades, which pitted his determination to follow wild hunches against the judgment of expert oceanographers he had hired, I became convinced that we would have to conduct a more patient, systematic survey. That meant, among other things, spending a longer time at sea than Grimm had been willing to support. Yet it seemed doubtful that my Woods Hole group could shake itself free to look for the *Titanic* for any longer than one or two weeks after mapping the *Scor-*

pion wreckage. We would obviously need help, and the French had long been my exploring allies. During a trip to Paris, I met with colleagues in IFREMER (the successor to the French group CNEXO, which had initiated the Project FAMOUS dives to the Mid-Atlantic Ridge). The French looked forward to a transfer of our latest deep-sea technology, as had happened after Project FAMOUS. They also saw glory in the mission; they seemed to view finding the *Titanic* as a deep-sea equivalent of landing astronauts on the moon. We quickly agreed on a joint expedition.

When Michel and I planned our strategy, we decided that his group would survey a broad area using his research vessel *Le Suroit.* A remarkably sensitive new side-scan sonar, the Sonar Acoustique Remorqué (SAR), which Michel had spent years developing, could produce images that looked almost as good as black-and-white photographs—if towed back and forth with absolute precision, at exactly the right height above the bottom. We knew that Michel's group could achieve the required precision. Stormy weather would slow their search, but if given enough time we felt certain they could find a wreck as large as *Titanic,* unless their search pattern had holes in it. On their sonar traces, the *Titanic*'s huge silhouette would be hard to mistake, but it might look questionable if seen from an unusual angle, or indeterminate if the ship had broken apart. Our job would be to inspect such likely targets visually after Michel's group found them. The French would probably find the main wreckage, and we Americans, using *Argo,* would merely confirm its identity.

In principle, this strategy would allow the unique capabilities of our groups to complement each other. In fact, however, the *Titanic* proved just as mysteriously elusive as ever.

Every group looking for the *Titanic* had access to the same historical information, and all of us came up with the same search area of roughly 100 square miles. Such a plot of seafloor may seem small, but when a ship tows a cable in 12,000 feet of water it can't move very fast. The enormous drag against the cable forces the towing ship to move at a speed that is commonly less than 1 knot. All searches proceed at a snail's pace, and every team has a limited number of days at sea.

Michel's group, which I later joined, arrived in midsummer and persevered through extremely bad weather to maintain accurate towing patterns. The rough seas, however, prevented them from covering the entire search area. In the zone they did manage to survey, they found no convincing trace of the great ship. Michel and I were beginning to think that something might be obscuring our view of it. Huge mudslides, for instance, set off by an earthquake in 1929, might have buried a large part of the wreckage. By the time my group arrived on site, I thought that the odds were running against us. We were now facing a less focused search with *Argo* than we had hoped to undertake, and time was running terribly short.

What else might have caused everyone to miss the *Titanic?* All previous efforts, we knew, had used side-scan sonar to search for the ship's main hull. These systems were state of the art. Grimm had tried both the Scripps Deep-Tow system and a new one, belonging to Columbia University's Lamont-Doherty Geological Observatory, called Sea Marc I. The chief scientists in charge of their use, Fred Spiess and Bill Ryan, had accompanied Grimm while at sea. Both were accomplished veterans, as was our partner Jean-Louis Michel. None of these expeditions, however, had been able to survey its entire search area. So there were common threads: excellent sonar technologies, yet incomplete searches—plus nagging doubts about whether the wreckage remained on the seafloor surface in good enough shape to be identified at all. The combined failures and unanswered questions had left a residue of frustration, jealousy, and rumor floating about the quest as *Knorr* and *Argo* arrived on the scene.

Scientists, like everyone else, easily become rivals as they struggle for recognition and funding. In particular, there has always been a strong sense of competition between researchers at the three largest and oldest oceanographic institutions in the United States: Scripps, Woods Hole, and Lamont. Scripps, the oldest, remains the largest. It was the first to develop tethered vehicle systems for deep-sea exploration, most notably Deep-Tow. Woods Hole, on the other hand, was the first to obtain a manned submersible. For years the *Alvin* group at Woods Hole and the Deep-Tow group at Scripps had been subtly competing for primacy while arguing the merits of manned versus unmanned vehicles. Each group had its share of successes and supporters. Now, with the development of the *Argo/Jason* system, this delicate balance seemed threatened. Here was a new kind of system claiming to achieve the best of both vehicle types—towed and manned—through telepresence.

Grimm's failure to find the *Titanic* using Deep-Tow opened the way for us to demonstrate our new capabilities by finding it with *Argo*. The competition between my Woods Hole group and the Deep-Tow group had intensified years before, when we had used our ANGUS camera sled and then follow-up dives by *Alvin* to find the first "black smokers" on the East Pacific Rise (after Deep-Tow found no active vents). Now it seemed we were again going head to head. *Argo* was much more sophisticated than ANGUS, and *Jason* would be even more advanced once its development got into full swing. But the status quo would remain unchanged if no one found the *Titanic*—much was riding on our ability to find it, and to find it fast.

—— A New Search Strategy

During our transit from the *Scorpion* site, it seemed absolutely clear to me that we should not continue looking for the *Titanic* with side-scan sonar. Although we had assumed that its immense hulk would stand as tall on the bottom as an eleven-story building, somehow everyone had missed it. How could we be sure that the hulk remained intact? And even if it had, what if it had dropped into a canyon or behind an obscuring ridge? In some terrains, if it fell just right, even a large sunken ship might not show up on sonar.

I knew that we could always keep trying with *Argo*'s side-scan sonar. We could start by towing it over the area the French had not been able to cover. But I was chafing to abandon our two-step strategy and skip directly to using *Argo*'s cameras. That would involve a major change in search techniques. Since we had no likely targets from the preliminary sonar survey, *Argo* would have to scan large areas visually. And since we could imagine parts of the wreck being hidden from view—under a mudslide, for instance—I thought we should concentrate on finding the broadest possible target: an entire debris field.

Argo was, of course, uniquely capable of finding debris. Yet I knew I would have to consult with the French before changing our strategy, and perhaps defer to a hunch or two. A few targets that Michel's group had found during their SAR scans, although not definitely wreckage, could not be ruled out. I couldn't just discard their work. Even if those targets turned out to be rock formations, suggestive mounds, or other mirages—as others had always been—I knew we would have to proceed with all due respect. Michel had not been with us when we were finding and mapping *Thresher*'s debris; some convincing would be needed. But I hardly had time to do it.

When we arrived in the area the French had been searching—on August 24, 1985—we had only eleven days to find the *Titanic.* Fortunately, Michel's group had eliminated 70 percent of the primary search area. But all the previous failures now caused us to question whether the original search area was large enough. For several days, as we cruised north from the *Scorpion* site, Michel and I re-examined every aspect of the historical data, over and over and over again. This time we allowed for greater uncertainties; as a result, our search area grew. The French had covered only half of this larger area during more than a month at sea. We somehow needed to search the remaining half in far less time.

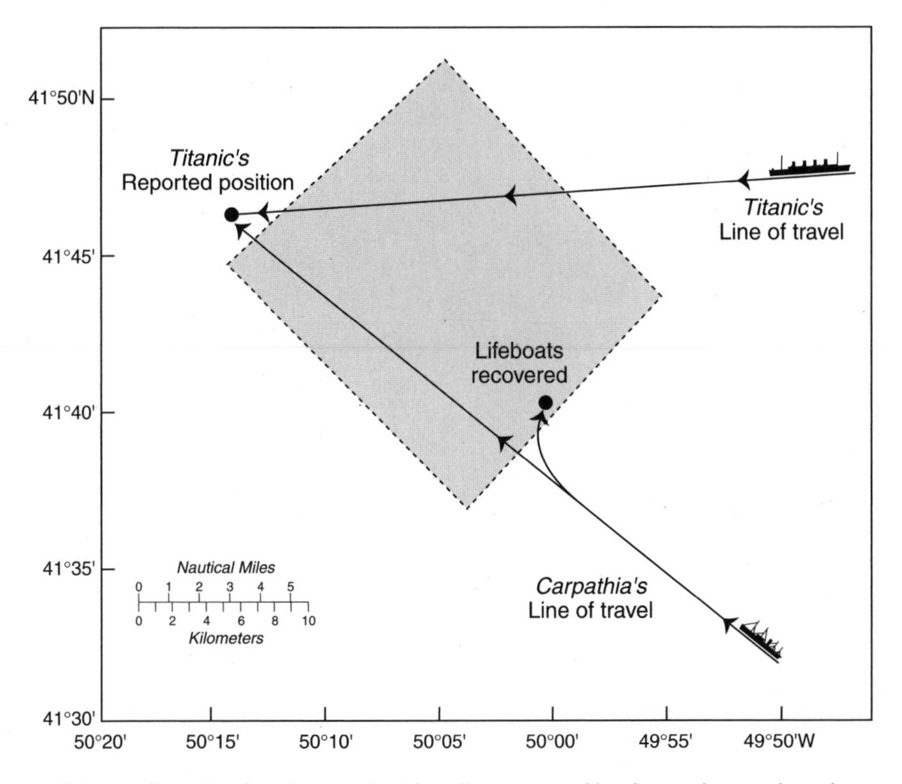

Primary (large box) and secondary (smaller screened box) search areas based upon an analysis of the historical database associated with the sinking of the *Titanic*. (Source: Robert D. Ballard/Odyssey Corp.)

During the French survey, I had been a restless co-leader aboard *Le Suroit*. Michel and I had become preoccupied with a prominent feature on the seafloor, discovered and named years after the famous sinking: Titanic Canyon. It ran through the heart of the search area, from the northeast to the southwest. Could we have missed *Titanic* down inside that canyon, resting up against its walls? Or might the canyon be a conduit for debris from the wreck, transported into it from surrounding tributaries?

Those questions, and the desire to investigate some of the targets identified on the previous French survey, led us straight

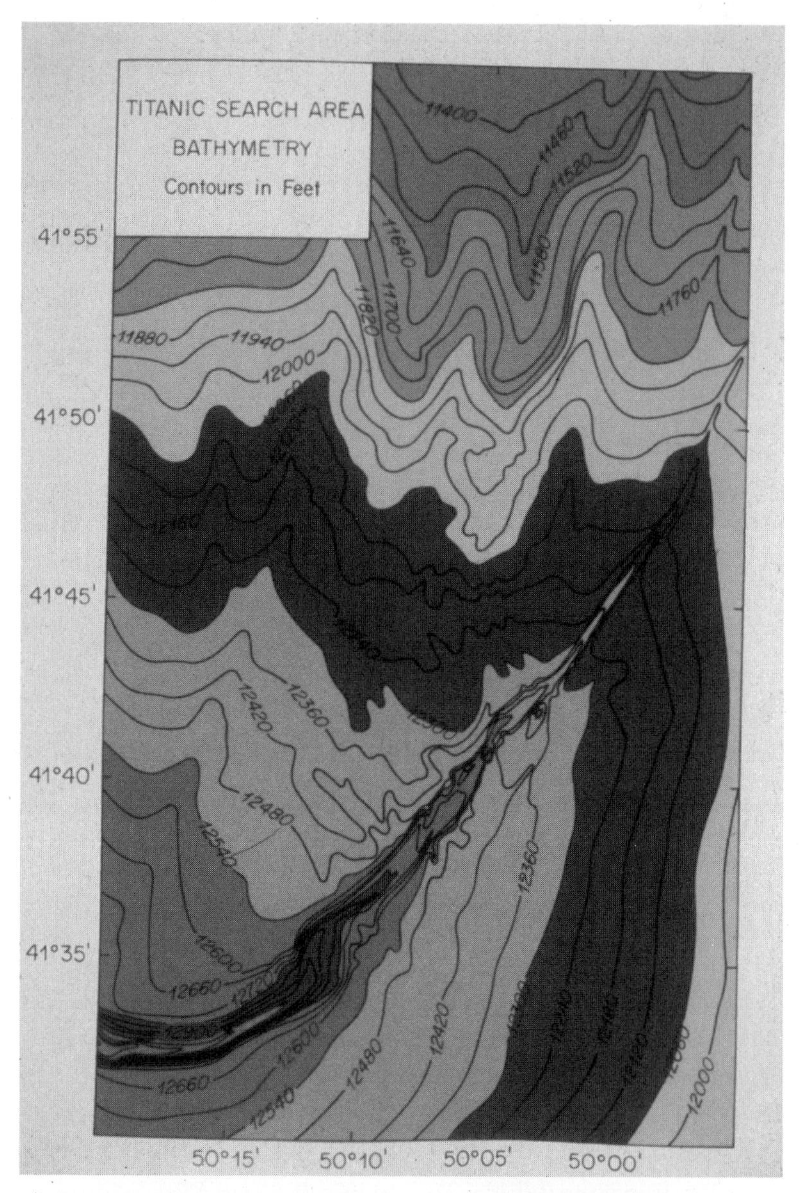

The bathymetry in the area where the *Titanic* sank is dominated by Titanic Canyon, a large submarine canyon that dissects the continental rise from the northeast to the southwest. (Source: Robert D. Ballard/Odyssey Corp.)

to Titanic Canyon. After installing our network of acoustic transponders, we made our first *Argo* lowering. Hours and hours passed as we watched a barren seafloor slide beneath us—or rather appear to slide past on our monitors in the control van. We traveled the length of the canyon without seeing any signs of the ship or debris.

It was time now to try our new search strategy, based on what we had learned while mapping the wreckage of *Thresher* and *Scorpion.* Yet we had reason to doubt that those lessons would apply. Both submarines, we knew, had been torn apart before sinking, badly enough to start shedding debris. In *Titanic*'s case, however, eyewitnesses disagreed on the manner of sinking. The most popular account had the ship going down in one piece, with the iceberg-damaged bow dipping below water while the undamaged stern rose skyward. Eventually, in this view, the entire ship slipped beneath the gentle waves. The stories of several survivors—Second Officer Charles Lightoller, Colonel Archibald Gracie, and Third Officer Herbert Pitman—strongly supported this account. There were others, however, who argued that the ship broke in half at the surface. Notable among them was Jack Thayer, a passenger who had jumped off the sinking ship and who gave a detailed account of the breakup shortly after the sinking.

Since Thayer was closest to the ship when it sank in near total darkness, we based our new search strategy on his account. We gambled that the *Titanic* had indeed broken in half and scattered debris. We assumed, further, that the lightest debris had formed a very long trail on the bottom, like the debris from *Thresher* and *Scorpion,* and that we would be able to see it with *Argo.* We decided to plan our surveying runs to intersect that line. Michel, who had a logical, precise engineer's mind, agreed to this plan once we had explained our experience to him. He understood quickly, in fact, how this new technique might just beat the odds that seemed stacked so heavily against us: too large an area to search in too little time.

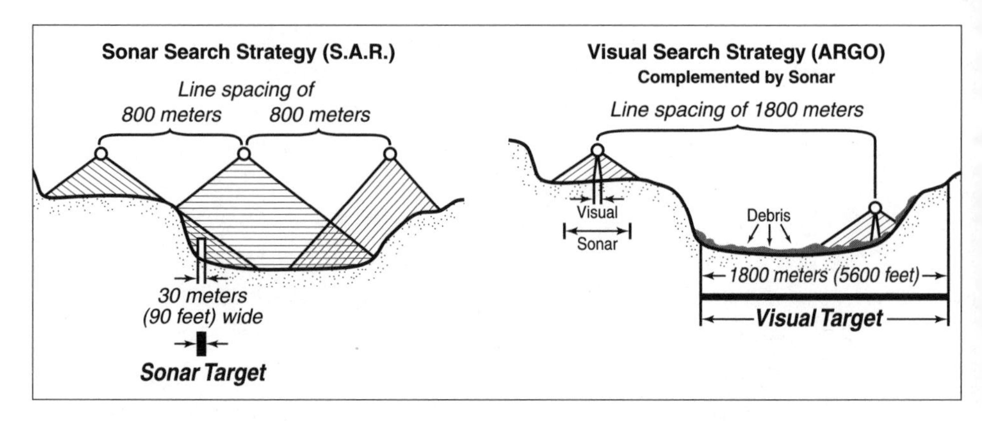

Comparison of two different search strategies used to locate the sunken British luxury liner RMS *Titanic*. The first (left) is an acoustical search strategy using a side-scan sonar; the other relies upon a visual imaging system. An acoustical search in rugged terrain requires overlapping coverage, whereas a visual search allows for more widely spaced survey lines. (Source: Robert D. Ballard/Odyssey Corp.)

The main question was where to begin. We decided to start searching away from what seemed the most likely location of the ship's remains and work toward the likeliest spot. The night the *Titanic* sank, another ship had been nearby: the steamship *California*, ironically also owned by the White Star Line. The *California*, also traveling westward, was cruising ahead of the *Titanic* the night of April 14, 1912. When it encountered an ice field running north-south across shipping lanes, its captain, Stanley Lord, wisely came to all-stop at 10:40 P.M. and began drifting south.

An hour later, the watch aboard *California* saw a luxury liner, which they thought was the *Titanic*, enter the ice field farther south. Later, that same watch observed rockets rising above this mysterious liner (many believe they were distress signals fired by *Titanic*). Unfortunately, Captain Lord misinterpreted the importance of the rockets—part of a celebration, perhaps?—and remained stationary with his radio operator fast asleep. The next morning, when he finally learned of *Titanic*'s fate, Lord got his ship under way to the south. By then, *Titanic* had already disappeared, along with 1,522 souls who could not get into lifeboats. The *California* was unable even to pick up survivors; the steamship *Carpathia* had already completed that task. Yet despite the tragic lateness of *California*'s arrival, its long presence near the stricken *Titanic* proved critical to our search seventy-three years later.

From *California*'s log, we were able to determine that the surface current that night was running at approximately 0.7 knot on a bearing of 170 degrees. In other words, debris falling out of the *Titanic* once it broke in half—*if* it broke in half—would have drifted south. The question was, how far south? Clearly, the lightest of all debris from the *Titanic* were its lifeboats, which did not sink. They were spotted, along with 700 or so survivors, by rescuers on *Carpathia* as it steamed northeast toward *Titanic*'s estimated sinking position, arriving two hours after the ship sank.

With this critical information, our search strategy was set. First, we would begin approximately five nautical miles south of where *Carpathia* had reported picking up the lifeboats. This would allow for a reasonable amount of error owing to the fixing of ships' positions in those days by celestial navigation. Even the lightest pieces of *Titanic*'s hull would then lie north of our starting point. Second, we decided to run our *Argo* search lines east-west to maximize our chances of intersecting a north-south debris trail. Third, we would space our lines slightly less than one nautical mile apart—considerably wider than we could have spaced them using a sonar scanner—since the debris trails of *Thresher* and *Scorpion* were both more than a mile long and both had sunk in shallower water. And finally, we would continue pushing our search lines northward until we reached the limit of coverage our French colleagues had attained using their side-scan sonar.

We intended to tow *Argo* over a huge tract of seafloor, roughly 100 square miles in extent. The French had faced an equally daunting task a few weeks before on *Le Suroit* and had not been able to scan their whole area. But our new visual strategy allowed us to sample relatively narrow swaths with wide gaps in between, and thus cover a large area two or three times faster than would have been possible when doing a complete sonar survey. Since we had less than half the time of any previous search effort, therein lay our only hope.

— Technical and Spiritual Difficulties

On August 28 we reached the southern limit of our search area and lowered *Argo* to the bottom. Eight and a half days remained in the expedition. Hours later, as *Knorr* approached the end of this first search line, the watch leader prepared to execute a turn to the north. In an effort to make the turn easier, he slowed the forward progress of *Knorr.* Drag on the cable lessened, as planned, while *Argo*'s momentum carried it forward and down like a pendulum. As the winch operator pulled back on controls to take in the slack, however, the cable take-up drum could not keep up. Seeing a dangerous loop, the winch operator reversed the controls. Suddenly, the cable jumped off the traction unit and became wedged in the winch axle. The winch then began cutting through the cable's protective steel-armored jacket. Quickly, we cut power to *Argo,* and the winch engine stalled. *Argo,* meanwhile, continued its pendulum swing toward the bottom. Plunging into the mud, it began to act like an anchor, with its cable ready to snap should the tension rise too high. We had to stop and let *Knorr* drift in the gentle current.

If we lost *Argo* here, so far from our tracking transponders, we would probably never find it again. Worse, a heavy steel cable that snaps under tension can whip through the air with incredible speed. It can easily slice through any part of a human body unlucky enough to be in its path. Yet many of us were now swarming around the winch, working frantically to save the mission.

The *Argo* search system as it is launched from the R/V *Knorr* during the 1985 phase of the *Titanic* project. (Source: Emory Kristof/National Geographic Society Image Collection.)

Quickly, we attached a set of "Chinese fingers" (a flexible, braided cable) to the taut towing cable, outboard of where it was wedged in the axle. The fingers, secured to a scaffold on the stern, gave us a second line we could use to bypass the tangled part of the towing cable. Slowly, we transferred tension to the fingers. This finally allowed us to unspool the towing cable, untangle it from the axle, and carefully respool it. Once we had wheeled the damaged part of the towing cable back onto the take-up drum, we could remove the fingers and begin winching *Argo* back aboard.

With *Argo* finally recovered, we inspected the damage to the cable's armored jacket. Clearly it would break if ever asked to bear tension again. Then we considered the internal coaxial

Inside *Argo's* control van. Co–chief scientist Jean-Louis Michel (left) stands next to the author (center). (Source: Emory Kristof/National Geographic Society Image Collection.)

electrical wires. If they were damaged, our quest to find the *Titanic* would surely be over. To our joyful surprise, when we turned the power back on, *Argo* came to life. We were still in the game as long as we worked in waters shallower than the point at which our cable had snagged. As luck would have it, our first search line had traversed the deepest portion of the search area. As we moved north, the bottom would grow shallower; our search could continue. We had eight days left.

Turning gingerly now at the end of each run, *Knorr* towed *Argo* on its damaged cable back and forth across the search area. We worked around the clock, with three watch crews rotating shifts in the control van every four hours. As we moved steadily northward, our spirits sank. We should have seen something within three days, most of us thought, if *Titanic* had gone down anywhere near our calculated most-likely spot, and if it had indeed broken apart and left a debris trail. If not, it had probably escaped notice between our search lines. We were all becoming exhausted, and bickering had broken out. Some of the crew were second-guessing our search strategy, challenging Michel and me. They were hatching new schemes that ranged, I thought, from desperate to crazy. By the ninth east-west run, our expedition time had dwindled to four days.

The End of the Trail

Shortly after midnight on September 1, a sense of failure was slowly taking hold as I prepared to catch a few hours sleep. Jean-Louis Michel's watch had just taken over in the control van. Nearing the end of our ninth search line, in the northwestern corner of our search area, we had approached within a few hundred yards the zone where the French had begun their sonar search. The *Titanic* was not likely to be in the remaining gap between our two areas—a place that our two separate expeditions had judged to be marginal. Our gamble had probably failed.

At twelve minutes before 1:00 A.M., something that looked metallic showed up on a monitor screen. Michel's crew sprang to life. All kinds of metallic objects were soon flowing by, including a riveted hull plate. *Argo* had begun to cross a debris-covered seafloor. But what kind of debris? Since the *Titanic* had sunk in well-traveled shipping lanes, this wreckage might belong to any number of ships. German U-boats, for example, had sunk scores of Allied cargo carriers in these waters during World War II.

At 1:05, *Argo* passed over a large circular metallic plate. "A boiler?" Michel asked softly. At the same moment Bill Lange, a video technician, shouted his recognition: "It's a boiler!" Deep inside its hull, the *Titanic* had been lugging twenty-nine huge boilers. Every one of them was needed for churning out steam to power its massive engines, which pushed the heavy, double-plated ship at a respectable 21 knots. That was a signature clue. Everyone was jumping, pointing at screens, and shouting. Almost everyone. Michel, with incredible patience, began flipping through pictures in a book, looking for *Titanic*'s boilers, cast—and photographed—in a Belfast foundry. He calmly kept turning his pages. And there it was—the top plate from one of those boilers. The picture in the book matched the image that glowed from the video screen.

Shouldn't someone tell Bob? This refrain had started a little before 1:00. But no one wanted to leave. The flow of debris petered out, then started again. A minute or two later: "Let's go get Bob." Finally the ship's cook walked into the van, and those on watch sent him to get me. I returned to find the Watch of Quiet Excellence—as they had named themselves—exploding in celebration. They were shouting something about boilers.

The queen of all lost ships had been found, in 12,460 feet of water. Just as we had surmised, the debris field ran north to south over a length of a little more than a mile.

All the time we had passed growing tense and angry seemed idyllic compared with the next three days. We spent every possible moment towing *Argo* over the *Titanic* site, learning as much as we could about the wreckage—trying to fly over a jagged hulk, that is, as close as we dared, and then closer still, without getting tangled or crashing.

Although we had read a great deal about the *Titanic* and studied many pictures, none of that research prepared us for the impact of the actual sighting. The scale of the ship seemed inhuman, as if a mountain had crashed to the seafloor—now a broken mountain of rusting iron, which had once displaced 46,000 tons of water. We saw that the hull, originally 885 feet long, had indeed split in two before sinking. The front section stood upright now on the bottom in surprisingly good condition, its prow buried sixty feet deep in gray muck. The middle part of the ship was missing. The stern had spun around 180 degrees and now stood alone, facing backward. It looked collapsed, all but destroyed from internal implosions.

On September 3, as a storm front approached with gale-force winds, we raised *Argo* out of the water and secured it on deck. ANGUS, our indestructible old standby, took its place underwater. Throughout that afternoon and evening, the Dope-on-a-Rope exposed thousands of frames of 35 mm color film. As those long rolls came out of the developing room, we pored over the images as eagerly as we had marveled at pictures of white clams living in the deep-sea oases ANGUS had photographed years earlier. All around the *Titanic,* we saw thousands and thousands of objects that had drifted to the bottom or somehow spilled out intact, including a porcelain teacup sitting upright, a silver serving tray, many bottles of wine, and—saddest of all—pairs of empty shoes with old-fashioned buttons, splayed out on the seafloor at an angle and distance suggesting that their owners had worn them all the way to the bottom.

Storm winds howled around *Knorr* on the morning of September 4, and the seas began building to twenty feet. It was time to go; our research ship had other commitments. We dragged ourselves away from the site and headed back to port.

Before long, the wind and waves seemed insignificant. Radio messages had been flying between ship and shore since the day of discovery, and soon a full-blown media storm erupted. Everyone wanted exclusive news, exclusive pictures. On the morning we found the *Titanic,* after daylight broke, a long-range helicopter beat its way out from Canada across the Atlantic. A television crew inside requested an exclusive interview. We handed over, instead, three copies of the first videotape ever made of *Titanic*'s wreckage, one addressed to Woods Hole, one to IFREMER, and one to the world press, which we defined as follows: every news organization. Less than two hours later, that stunning footage was broadcast as a "CBS exclusive." Immediately, John Steele, the new director at Woods Hole, began fielding howls of protest from other networks, as well as threats of a lawsuit.

Two days later, we loaded the first color pictures from ANGUS, high-quality still photos, onto another chopper. To prevent any more hanky-panky, we also sent along two of our team members as couriers—one American, one French. It didn't work. My own boss, John Steele—thoroughly intimidated by the media sharks—whisked his packet to American broadcasters before the other packet could arrive in France aboard the Concorde, the world's fastest passenger jet. That violated our agreement with IFREMER to release photos in both countries simultaneously. As a result, Americans began seeing these images on television while they were still dawdling toward Paris at merely supersonic speed. Producers for the French networks, not wanting to be scooped on this French co-discovery, were forced to buy footage via satellite from U.S. broadcasters for immediate airing. They were furious. We had insulted them, IFREMER, and the glory of France. It looked as though our partnership would dissolve amid an ugly international squabble.

On the morning of September 6, as *Knorr* approached the calmer waters of Vineyard Sound, the media frenzy was still intensifying. Yet a touching welcome awaited Michel and me, and our teams. Behind the Woods Hole officials, the navy brass, and security personnel, and well behind the television cameras and microphones, a good part of the local population had gathered. Bands were playing, horns honking, and flags waving. Someone shot off a cannon to salute us. I had never imagined how strongly our search for a great ocean liner would grip and stir the public.

The race to find the *Titanic* was over. We had plumbed the eternal darkness with our tethered eyeball, and we had found the unmarked grave. For science, a new era in deep-sea exploration could finally begin.

CHAPTER

9

RECOVERING OUR PAST BY REMOTE CONTROL

With the right controls, you can make an ROV [remotely operated vehicle] dance on the head of a pin.

—*Dana Yoerger*

Early in May 1972 the *Fred T. Berry*, an obsolete 390-foot destroyer built during World War II, went down in a burst of explosions: the U.S. Navy used it for target practice, then scuttled it off Key West, Florida. The *Berry* rested on the bottom in 360 feet of water for more than a year, its rigging and cables waving peacefully in the current. Then its victims arrived. On June 17, 1973, a twenty-three-foot submersible named *Johnson Sea-Link* motored slowly toward the old gray hulk. The pilot knew enough to be wary; he knew that large wrecks can snare intruding subs. But the lure was set. The destroyer had become an artificial reef, and observers inside the *Sea-Link* wanted to see what kinds of creatures had moved in and set up shop. As the submersible edged closer, a surprisingly strong current caught it and quickly swept it, like struggling prey, into a mass of tentacle-like cables. Soon the sub was helplessly entangled.

The *Sea-Link* had two separate passenger chambers with no connecting passageway. Pilot Jock Menzies and scientist Robert Meek occupied the forward compartment, a transparent viewing sphere made of five-inch-thick acrylic plastic. Two more observers occupied the rear compartment, an aluminum lock-out chamber that divers could enter and exit once it had been pressurized to equal the outside water pressure. The *Sea-Link* had not been seriously damaged when the *Berry* snagged it, and its four occupants were in no immediate danger. But a horrifying thought gripped their minds nonetheless. All lives now depended on the performance of two air scrubbers, one in each chamber, filled with eight pounds of Baralyme. As the ocean chilled their submersible, the men knew that the lime would become less reactive and would absorb less carbon dioxide. Their own inhaling and exhaling would then slowly suffocate them.

An urgent call for help brought a navy submarine-rescue vessel to the site within six hours, but the strong current made maneuvering difficult. This ship, it became clear, could do nothing to free the submersible. A diving bell, rushed all the way from San Diego, raised hopes for a time, but it too got trapped in the rigging. Twenty-four hours had passed before a second rescue ship, a well-equipped research vessel, positioned itself above *Sea-Link.* Its closed-circuit television and grappling hook improved the chances of rescue, but the situation was now desperate. Menzies, the trapped pilot, talked on the underwater acoustic telephone to guide a surface crew lowering the hook. Finally it got a firm grip, overpowered the ensnaring cables, and raised the submersible.

Menzies and Meek survived. The acrylic wall of their pressure sphere, a fair insulator, had kept the chamber from chilling too fast. In addition they had put their air scrubber near a small motor that gave off a little heat. But the other chamber, with its heat-sucking aluminum wall, had chilled rapidly. Its two occupants could do nothing to stem the loss of heat or the rise of carbon dioxide. Both died before the *Sea-Link* returned to the surface. One of them, Clayton Link, was the son of the submersible's designer.

Edwin Link, Clayton's father, agonized over the accident. Yet aside from deploying more rescue ships (at a cost the navy was not likely to support), he could think of few practical ideas for improving safety. Exploring anything in deep water, he knew, has always been risky. The strong current and loose cables that trapped *Sea-Link* could just as easily have snared any other submersible. He did not blame the pilot; it can be hard to tell where strong currents are flowing until you are caught in them. Furthermore, the *Sea-Link* design was basically sound, and there are severe limits to how much life support a submersible can carry. The risks, in short, were inherent to exploration. Link said as much in testimony to the Coast Guard Board of Inquiry. "It is inevitable," he wrote, "that as man continues to invade the ocean depths, similar accidents are bound to occur in spite of every precaution."

Afterward, of course, manned submersibles continued to invade far greater depths, not always with utmost caution. And just as Link predicted, life-threatening accidents continued to occur. In 1975, for instance, *Alvin* became stuck in a fissure on the Mid-Atlantic Ridge. Fortunately, pilot Jack Donnelly managed to thrash his way out. On another occasion, in 1983, *Alvin* drifted into a maze of engulfing debris while sampling mud near a large fragment of the hull of *Thresher* (ironically, the same sunken vessel that had prompted so many calls for better rescue equipment). At that moment, any twisted piece of *Thresher* might have snagged the little submersible. Pilot Ralph Hollis suddenly realized the danger when he started bumping into obstacles on all sides. Blinded by suspended sediments, and with his magnetic compass not to be trusted because the craft was surrounded by huge, overarching chunks of metal, he managed nevertheless to inch his way out rather than deeper into the tomb.

Three years after that near-disaster, I was descending nervously in *Alvin* toward the bottom of the North Atlantic, two and a half miles below. Hollis was again the pilot, and another potential trap lurked in the dark abyss. We were plunging toward the *Titanic*, one of the most enticing wrecks of all time. During the next two weeks, according to my plan, we would poke all around the behemoth, both outside and inside if possible. I had long dreamed, in fact, of somehow descending the luxury liner's grand staircase. Yet I knew that trying such a maneuver in *Alvin* would be suicidal. However, during the year that had elapsed since the last *Titanic* expedition by my Deep Submergence Laboratory (DSL), an elaborate scheme had begun to take shape.

— *Jason Junior*

Ever since we had begun making plans to look for the *Titanic,* we had intended the effort to unfold in two phases. Simply finding the lost ship would be more than enough of a challenge for the first expedition. If any time remained, we would document as much as we could with *Argo* and ANGUS. The second phase would take place the following summer, when we planned to investigate the wreckage far more thoroughly. That mission had support from the navy, which wanted to continue advancing our *Argo/Jason* program. On that second trip we planned to dive to the bottom in *Alvin* and visit *Titanic* in person.

My return in a manned submersible provoked merciless ribbing. I had already stated, after all, that vehicles like *Alvin* were doomed. The whole point of finding the *Titanic* had been to demonstrate a new paradigm for deep-sea exploration. So why did I want to return in *Alvin?* The reason was straightforward. *Jason,* the robotic companion we were building for *Argo,* was not ready. Yet even though diving in *Alvin,* we would still be advancing the new paradigm. In addition to its usual three-man crew, *Alvin* would be taking along a fourth passenger, riding outside in the science basket. *Alvin* was about to serve as garage and home base for the remotely operated vehicle *Jason Junior,* or *JJ,* our prototype for the bigger *Jason* vehicles to come.

Alvin was essentially acting as a midwife for the technology that would eventually replace it. The humans inside,

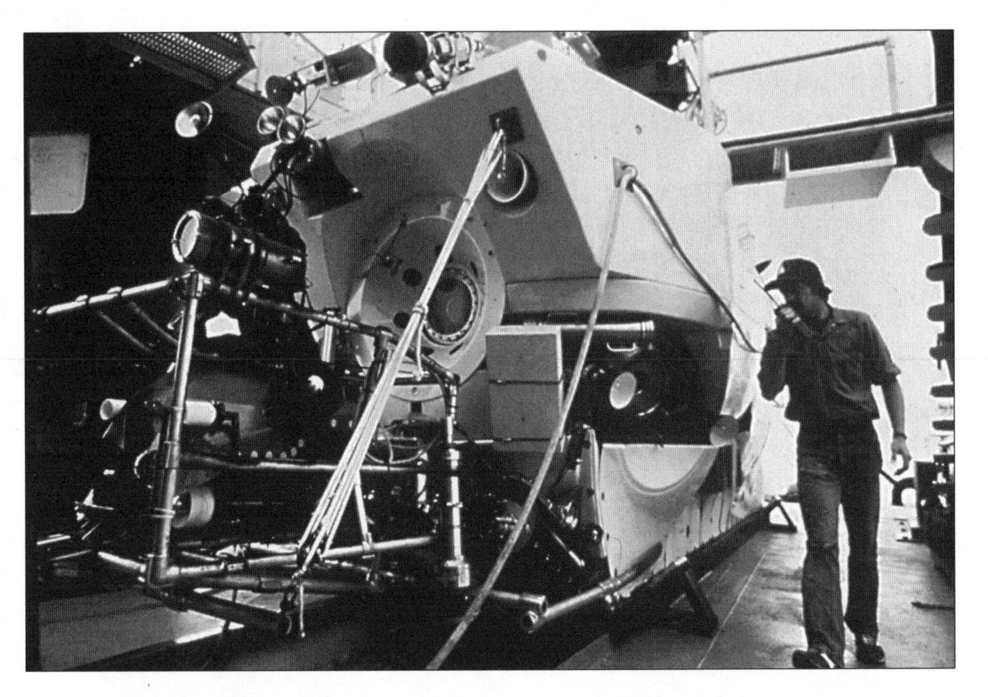

The remotely operated vehicle *Jason Junior* rests securely inside its tiny garage on the front of *Alvin* as the submersible undergoes its final checks prior to diving to the wreck site of the *Titanic* in July 1986. (Source: Perry Thorsvik/National Geographic Society Image Collection.)

crouched at *Alvin*'s tiny viewports, would get a firsthand look at *Titanic*'s exterior. But their view would not be confined to the portholes of a submersible. *JJ* would go where we dared not take *Alvin*—including, we hoped, deep inside the sunken liner's ghostly hull. Observers inside *Alvin* would then, in effect, go there as well by means of telepresence.

Our stand-in for the real *Jason*, a two-and-a-half-foot blue cube, weighed 200 pounds in air; we packed it with syntactic foam to make it neutrally buoyant underwater. *JJ* could not be controlled from the surface, but its short, 250-foot neutrally buoyant tether allowed an operator aboard *Alvin* to "fly" it. With no mechanical gripping arm or storage baskets, *JJ* could not place instruments or gather samples, as its successor would be able to do. But it could certainly roam and explore. After leaving *Alvin,* the little robot would send real-time images through its tether to a small monitor we had placed in the science rack at the back of *Alvin*'s sphere. *JJ* had a color television camera, a 35 mm still camera, powerful lights, and motors to propel it, all of which made it seem like a free-roving version of *Argo.* We called it our "swimming eyeball."

Unfortunately, my engineering team had not yet been able to test *JJ* on a deep dive. We wanted to hurry it out to *Titanic*'s graveyard because others were also preparing to explore the site; we suspected that the wreckage would not remain undisturbed. Before the pillaging started, we wanted to document everything, inside and out, like careful forensic scientists. So our engineers were still slaving over *JJ* as we left, working in nonstop shifts. They had to finish building it as we cruised across the Atlantic.

The second *Titanic* expedition began in early July 1986, when we departed Woods Hole aboard the research vessel *Atlantis II,* a modern tender that had finally replaced old *Lulu.* Right from the start, this expedition had a vastly different feel to it than the initial search. The first expedition had run high on extremes—anxiety, despair, elation. This time we simply aimed our ship at a satellite fix 900 miles east of Cape Cod and drove to work. We found *Titanic* by floating over its known position and watching our ship's echo sounder print an outline of the wreck, exactly where we expected it to be. (Unfortunately, a group of U.S. investors would follow precisely the same procedures the very next summer. Diving in *Nautile,* a new French submersible they had leased from our French partners, they made quite a haul in salvaged artifacts.)

The morning of July 13, 1986, dawned calm and gorgeous. Under the bright summer sun, we rolled *Alvin* out of its hangar and aft along the deck of *Atlantis II,* toward the A-frame crane that would lower the submersible into the water. On this first dive, pilot Ralph Hollis and copilot Dudley Foster, both old *Alvin* hands, would accompany me to the sunken ship. Hollis and I had often descended together, but this time I sensed an awkward tension between us. Although equally committed to deep submergence activities at Woods Hole, we were now involved with competing programs: his allegiance remained with the older manned-submersible technology, while I was now championing the new ROV systems. This mission would soon reveal some of the best and worst qualities of both technologies.

At 8:35 we began our descent. It would take us nearly two and a half hours, sinking at 100 feet per minute, to reach the depth where *Titanic* lay. Usually, nothing much happens on these long descents, and observers on board have been known to nap. But this was not to be one of those days. Halfway down, an alarm went off. Seawater was leaking into our battery bank, it kept reminding us. We switched to auxiliary batteries, located inside the pressure sphere, but our time on the bottom would have to be short.

When we approached the seafloor, Hollis dropped our descent weights. Finally, we could drive *Alvin* in a lateral direction—but which one? Our navigator on *Atlantis II* was not getting a good signal from one of the transponders. Since he could not track our position, he could not guide us toward the *Titanic.* We might have located it ourselves had it been within a few hundred yards of us, but *Alvin*'s search sonar had stopped working. Now *Alvin*'s lights illuminated only the immediate foreground: a featureless terrain of fine mud. Seeing no sign of the *Titanic* within that small area, we started moving. Since we surmised that a current flowing from the south had pushed us north of our goal, we turned south. Soon, against the shrill, insistent buzz of the battery alarm, *Atlantis II* confirmed our guess via the acoustic telephone. "*Alvin,*" we heard, "this is *A-II.* Tracking is now working. *Titanic* should bear fifty yards to the west of your present location."

Hollis steered *Alvin* west, as directed. Almost at that moment, I realized that a change had come over the mucky bottom. Suddenly it began to slope upward—unnaturally so.

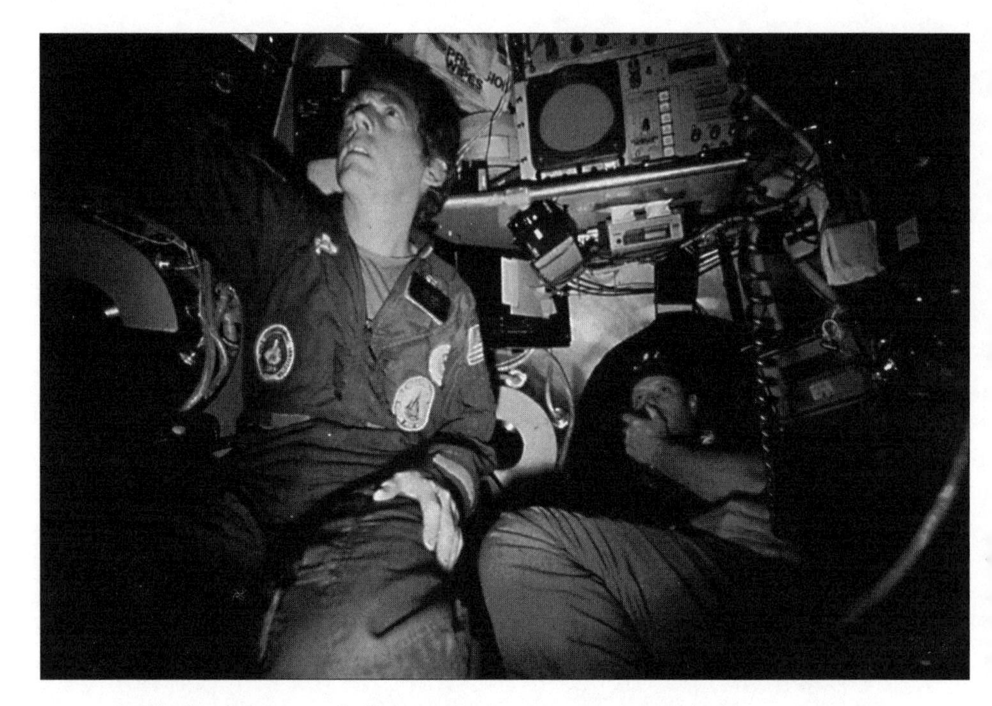

Inside *Alvin*'s pressure hull, the author uses an acoustic telephone to communicate with the surface while veteran pilot Dudley Foster (left) checks the submersible's altitude above the bottom. (Source: Martin Bowen/Woods Hole Oceanographic Institution.)

Surely something huge had plowed this slope of mud and small boulders into such a steep contour. *Alvin* stopped. And there, in front of us, a black wall of steel plates rose up and up, seemingly forever. Riveted together in Belfast long before any of us were born, it stretched upward farther than we could see into the inky blackness. It was as if we had discovered the ancient walls of Troy in the middle of the night. The walls of *Titanic* stood before us, frozen in time.

No sooner had we gotten our single breathtaking glimpse than Hollis dropped *Alvin*'s iron ballast and began our ascent, fearful of the damage being done to our batteries as the seawater leak continued. According to safety rules his own *Alvin* group had established, he should have aborted the dive before we even reached bottom. Fortunately, technicians were able to fix the leak after we surfaced.

On the second dive it was *Jason Junior*'s turn to start leaking. Ten minutes into our descent, Martin Bowen, *JJ*'s operator, saw a warning on his control panel: seawater was flooding one of the ROV's motors. If we turned back quickly to seal the leak, Hollis could not take us down again today. (To do so would mean that we would be returning to the surface after suppertime, and that would be against the rules.) Bowen had to shut down power to the whole robot to save the electronics, which meant that the rest of the dive would belong entirely to *Alvin*. We descended for another two hours—one stubborn pilot, one angry and frustrated team leader, and one disappointed *JJ* operator.

As we neared bottom, our surface navigator again guided us toward *Titanic*'s bow. This time we approached it head on, a sight to make even veteran deep-sea explorers cringe. All of a sudden an enormous prow emerged from the darkness, not thirty feet away, as if still plowing ahead under full power—and on the brink of running us over. Once again we gawked in amazement.

As we ascended toward the decks high above, the wash from *Alvin*'s propellers sent gobs of rust showering down in great cascades. We were going up because we needed to find a landing place. This would be a particularly difficult task, since we had no way of knowing whether the steel plating beneath *Titanic*'s vanished wooden decks was still strong enough to support even *Alvin*'s light underwater weight. Yet we needed to find secure footing on the wreck itself if we were to send *Jason Junior* inside on future dives. Hollis maneuvered with exquisite control, landing us like a feather, and the rusting ship's horizontal surfaces held firm. After leaving the bow *Alvin* traversed to the stern, where we encountered a surprisingly strong current that swept us toward mangled wreckage near *Titanic*'s shorn-away midsection. A forest of jagged metal rushed past our viewports. Any one of those sharp points seemed capable of spearing *Alvin* through its fiberglass sail and pinning us to the wreck, but again we escaped unscathed.

When we finally returned to the surface, stormy waves knocked *Jason Junior* from its basket, then tossed the little robot up and down as it dangled from its tether. To rescue *JJ*, our support divers had to cut the cable. My DSL engineering team worked all night on the traumatized rover, meticulously resealing every spot that might leak and repairing the vital tether. By now our stand-in for the senior *Jason* had failed to work in shallow water before we left Woods Hole, failed to work in the deep ocean, and jumped ship at the first opportunity afterward. Although frustrating, none of this seemed unusual. Every complex new technology goes through similar shakedowns.

The first real test of *Jason Junior* came during the third dive. We delayed launching *Alvin* for several hours as one problem after another cropped up in *JJ*, but finally we descended. Dudley Foster took us directly to *Titanic*'s deck, and then Martin Bowen took us down the main staircase. With *Alvin* parked at the lip of a wide opening, he edged our squat little box out of its garage and into the gaping gloom of what had been three flights of magnificent stairs. The steps had cascaded from one landing to the next toward the ship's first-class public rooms deep inside the hull; now only their framework remained. Down the little robot went, its tether slowly paying out along the steps trod by John Jacob Astor and Benjamin Guggenheim during the last hours of their lives—and then farther down, past the landing above the A-Deck foyer where an ornate clock depicted Honor and Glory crowning Time.

Near the bottom, Bowen steered *Jason Junior* toward a room off the staircase. He was gaining confidence by the minute, easing into his role as telepresent pilot. Hunching over the console in his lap, eyes locked onto the little monitor screen, he pointed out a large fixture up ahead that was glittering in *JJ*'s floodlights. "Look," he whispered. "Look at that chandelier."

Actually, it was almost a chandelier. What we had found was an ornate light fixture, beaded with crystals and dangling from the ceiling on its electrical cord.

Over the course of ten more dives, we used *Alvin* and *JJ* to make as thorough a survey of *Titanic* as any of us in either group—the *Alvin* gang or my DSL team—could have wished. On our fourth dive, *JJ* did its best work of the expedition, ranging over such varied wonders as the remains of Captain Edward J. Smith's cabin and the crow's nest from which Fred Fleet, *Titanic*'s lookout, had first spotted the fateful iceberg. We even saw the old telephone handset he had used to warn the bridge. Our sixth dive, with *JJ* left behind for repairs, took us to the debris field near *Titanic*'s severed stern. Roaming there with *Alvin*'s lights and cameras, we added to the macabre inventory recorded by ANGUS the year before, photographing dinnerware, bedposts, space heaters, sinks and toilets and bathtubs, faucets and spittoons, doorknobs, champagne bottles (with corks still in place), and shiny copper pots—even ornate windows with their panes still intact. We also happened across a purser's safe, with its door rusted shut. That door did not keep us from peeking inside, because the back of the safe had rotted away. If it had ever contained any treasures, they had long since disappeared.

We never intended to remove any contents from the safe or from any other part of *Titanic*, and we never did. But before leaving, we made two contributions: a pair of small plaques. One of them, donated by the Explorers' Club, commemorated our expedition. We dropped it gently from *Alvin*'s mechanical arm onto a bow capstan. The other, a belated epitaph, honored the 1,522 men, women, and children who perished in the disaster. We placed it on the stern. Looking at *Titanic*'s shattered stern as we pulled away, I could think only of the passengers who had stayed on board, crowding closer and closer together on that deck as it tilted higher and higher—and then snapped off from the rest of the ship.

A New Archeology Beckons

Back at the Deep Submergence Laboratory in Woods Hole, our engineering team continued its work on the full-size *Jason.* At the same time, we turned our attention to new research possibilities. Our finding of many well-preserved artifacts at the *Titanic* site had encouraged us to look for older shipwrecks. We knew that some ancient wrecks had already been found, especially in the Mediterranean Sea. A few were remarkably old—dating back more than 3,000 years—and remarkably well preserved. Parts of their wooden hulls had often sunk deep into mud, which protected the planks from shipworms and wood-boring mollusks. Even where hulls had not survived, archeologists had been able to learn a great deal about ancient civilizations from bits of pottery, tools, cargo containers, and other items carried aboard the ships. These artifacts yielded clues about patterns of trade, for instance, which further suggested cultural exchanges.

All the knowledge derived from marine archeology—a small discipline with few practitioners—had accumulated from painstaking underwater excavation carried out over years and decades. That work had often bumped up against an impenetrable barrier—depth. Every ancient ship that had ever been excavated had gone down in shallow coastal waters. Many more undoubtedly sank in deeper waters, but archeologists had no easy way to locate them, and no suitable tools to excavate deep sites if they could be found. For the most part, fieldwork was still being done by scuba divers who rarely went deeper than 100 or 200 feet. *Argo* and *Jason,* on the other hand, could plumb the abyss down to 20,000 feet. Once we finished developing *Jason,* we believed that we could dramatically push back this important archeological frontier. How much ancient history lost in the deep sea might then be recovered?

From 1986 to 1988, we researched ancient trade routes with archeologists at Harvard University. We concentrated on the Mediterranean Sea, looking especially at routes that had crossed deep water. This research confirmed a hunch that had long excited me. The deep ocean, I had always suspected, is the world's largest museum. In the Aegean Sea alone, ships have been traveling from village to village for at least 9,000 years. Even before recorded history, humans were fetching obsidian from islands there for Stone Age tools, in boats that can only be imagined—rafts, dugouts, perhaps hide-covered frames. Some of those boats probably sank, and in later years many boats certainly sank. For example, more than 600 ancient Persian ships perished in the eastern Mediterranean in one year alone, during two storms. How many more went down all around the world, over many thousands of years? Some of those ships no doubt broke apart, and their debris floated away. They will be lost forever. But others surely survive more intact on the bottom. Sitting there like time capsules, they will reveal more than just nautical history. Virtually everything that humans have ever made has probably been carried across water on some sort of ship at one time or another.

As our *Argo/Jason* system matured, we gained confidence that we could use it to explore the seafloor beneath ancient trade routes—and, for the first time, unlock this deep museum. But it was not a new dream. We were following the lead of a remarkable pioneer, extending his vision and methods with newer, more powerful technology. This pioneer, it turned out, would join us years later on one of our expeditions.

George Bass and *Asherah*

In the early 1960s, when scientists like Robert Dietz and Allyn Vine were agitating for deep-diving craft they could use to explore the abyss, visions of manned submersibles also entered the mind of a young archeologist at the University of Pennsylvania. While still a graduate student, George Bass had become a leader in the new field of underwater archeology. In 1960, he directed a team of seasoned scuba divers who excavated a Bronze Age vessel in 100 feet of water off Turkey, the oldest ship yet discovered. The following year he conducted "the most scientific underwater excavation ever," he boasted, on a seventh-century Byzantine wreck. "We wanted plans that would enable a shipwright to build a Byzantine hull plank by plank and nail by nail and then be able to lade it."

Such a detailed reconstruction of the past required painstaking excavation. Bass's team worked for weeks before moving a single object—labeling each significant fragment of every artifact on the seafloor and then mapping, photographing, and drawing everything. To ensure precision, they mounted surveyors' instruments on tables underwater and set up an elaborate metal framework as a reference grid. And they repeated this labeling, mapping, photographing, and drawing inch by inch, layer by layer, as they dug into the site. All this effort required incredible patience. Each diver had less than forty-five minutes a day to work on the bottom. Four hours, by contrast, expired every day on the surface as divers commuted to the site. More hours passed by unproductively as the divers ascended slowly to allow decompression—and one of them nevertheless got the bends.

Despite such thoroughness, Bass's attempt to extend traditional land-based methods underwater provoked many doubts. Skeptical archeologists passed harsh judgments on his work without bothering to become familiar with it. "There is no time underwater," one said. "Time is money, and you must work fast to get the results as quickly as possible." Saying that an excavator worked "fast" was the ultimate insult. It suggested sloppy salvage or looting for treasure—not true archeology.

Late in 1963, after another summer of difficult work, Bass was desperate to improve efficiency. Anything he could do to make excavation easier (short of skimping on accuracy) would greatly improve conditions in the field. His funding had been so tight, with so much of it devoted to diving equipment and other seafaring essentials, that divers typically lost thirty pounds of weight on an expedition owing to hard work, long hours, and poor diet. Bass was also determined to expand his research into deeper water. Since the 1950s, bottom trawlers and then sponge divers, who used metal helmets and air hoses connected to the surface, had been recovering classical bronze statues from depths around 300 feet—well beyond the reach of his scuba diving team. One sponge diver brought up an exquisite bronze bust not far from the area where Bass was excavating, and he said that it came from a shipwreck.

Bass imagined whole cargoes of precious artworks lying in cold storage down below. He had visions of searching the seafloor to find them, staying down hour after hour—much longer than scuba divers could stay down and much deeper. He had visions, in other words, of finding and exploring these wrecks by using a deep submersible.

The question was whose submersible to use. This was in 1963, when popular enthusiasm for deep-sea exploration was beginning to build from slow simmer to full boil. *Alvin* was under construction, but Bass knew that such a state-of-the-art submersible would dive far deeper than he needed to go and, more important, would be far too expensive. He saw a photograph of another kind of sub—a two-man "sport submarine"—in *Newsweek,* and a scuba-diving colleague volunteered to travel the country via Greyhound bus looking at plans for such bargain-basement subs. But then Electric Boat—the chief builder of U.S. military submarines—heard of Bass's search, and three salesmen paid him a visit in his cramped cubicle at the university. Bass was convinced at first that he could not afford the diving craft they had in mind, given the meager funding he had from the National Science Foundation and the National Geographic Society. But after a lot of hard bargaining, they struck a deal. Electric Boat eliminated many features to lower the price, and Bass committed himself to raising more money.

The ship would be named *Asherah,* after the Phoenician goddess of the sea, and it would dive to 600 feet, far deeper than Bass had ever gone using scuba tanks, although not as deep as *Alvin* would go or Cousteau's *Soucoupe* had already gone. Yet on its technical merits *Asherah* would become a worthy addition to that pioneering fleet. Sixteen feet long, packed with syntactic foam, and weighing only four and a half tons, it would carry two crew members in extremely high-tech quarters. "When the hull was almost completed," Bass wrote in a book about his early explorations, "I scrambled up a ladder and lowered myself through the hatch. An entire telephone company seemed jammed into the five-foot sphere." All in all, it presented quite a contrast to the Mediterranean fishing scows Bass would be using to haul his new sub to its work sites.

Construction started in January 1964, and work continued around the clock to complete *Asherah* by May for a summer expedition. When launching day arrived, Bass's wife approached the sub with a bottle to christen it—but she couldn't find it! Although it had been painted bright orange, *Asherah* was so little that it had almost disappeared under festive blue bunting. When it finally was launched in traditional military fashion, with salutes and flags and a brass band playing, *Asherah* reminded Bass of "a great orange ice cream cone floating on its side." From somewhere in the crowd, amid the solemn fanfare, a female voice squealed, "How cute!"

Asherah was not a toy, of course, any more than the equally cute *Alvin,* launched about a month later. *Asherah* had six viewports, a twenty-four-hour air supply, a ton of batteries, and motors that would push it at speeds of up to 4 knots—much faster than the *Soucoupe.* And yet it proved extremely maneuverable, with electric propellers on its sides that could be rotated to go up or down, forward or backward, or to hover. "I was impressed by the ease with which we could inch over the bottom," Bass wrote, "as delicately as a diver." *Asherah* would justify its huge expense, he believed, by cutting excavation work time and costs. Bass had developed stereoscopic cameras like the ones used in airplanes for contour mapping; he intended to "fly" them in *Asherah* over sites he needed to survey and photograph. He would also use the sub's mechanical arm to excavate, and he envisioned keeping a supervisor inside it all day underwater, directing scuba divers over an acoustic speaker.

Bass planned to use *Asherah* first to assist an ongoing excavation of a late Roman wreck in 140 feet of water. After that, he planned to search for deeper shipwrecks. Right away, at the Roman wreck site, *Asherah* began paying him back. In one half-hour mapping pass with the new cameras, his two-man submersible crew got as much accomplished as a dozen scuba-diving archeologists working weeks underwater could have done. However, Bass never took *Asherah* into deeper water that summer to search for the statue-bearing ships that so intrigued him. His base lay near an unofficial (and disputed) boundary between Greek and Turkish waters, and Greeks were fighting Turks on Cyprus. Full-scale war seemed imminent. Bass's camp—with its neat rows of tents, navy ship, bright orange submersible, and radio equipment—looked like any other military target. Instead of dodging bombs and bullets, Bass decided to call it a season and shipped *Asherah* back to the United States.

From 1965 to 1967, *Asherah* earned its keep as a leased submersible. One group used it to conduct biological studies off the East Coast; another evaluated commercial fisheries in the Pacific; another inspected submerged cables off Washington State. Also during that time, scientists from the University of Rhode Island used *Asherah* to explore a submerged barrier spit and lagoon complex off Block Island. Like the work of K. O. Emery in *Alvin* at about the same time, their findings provided further evidence that ancient shorelines had formed—and had probably been inhabited—during the most recent glacial period.

It was not until 1967 that Bass could take advantage of *Asherah* once again for his own research. He still wanted to find the ancient ships that had carried bronze statues. By this time, he had learned a valuable lesson about deep submersibles: they are not the best tool to use for searching the ocean floor.

Bass had already tried looking for his bronze-bearing shipwrecks by other visual searching techniques, such as towing a television camera. His team even tried a contraption called the Towvane—something like a tethered bathysphere with vanes to force it deep underwater as a fearless observer peered out from inside. (Bass's surface crew chief, meanwhile, said he had been "scared to death," fearing that the cable would snag or the Towvane would crash into rocks.) At the end of all this, Bass conceded, "we had simply rediscovered a basic oceanographic principle: the sea is very large."

Bass now knew that the most effective strategy involved a broad sonar search to pinpoint likely targets, followed by dives in a submersible to inspect each target. To assist him in the deep-water search he was planning for 1967, Bass called on Fred Spiess of the Scripps Institution of Oceanography. Spiess, a mastermind of deep-towed sonar technology, agreed to help. He sent one of his newest side-scanning sonar "fish" to Turkey along with his best technician and also a young graduate student.

With the help of a sponge diver who had found one of the ancient bronze statues, Bass delineated two search areas of one mile and two miles on a side. This time the on-again, off-again war between Greece and Turkey did not interfere. Instead of hostile planes or ships, the main enemy became the utter tedium of a methodical sonar search. Bass's first impressions best capture the true nature of this work. "*A search* is a mixture of excitement and boredom," he wrote in his book. "Eager anticipation rises each morning with the sun, but slowly turns to disappointment with the monotony of passing hours."

The sonar team on loan from Scripps found a dozen possible targets at the larger site. At the smaller site they found one shipwreck candidate. It was nothing special, some sort of bump—the sonar trace showed no details. Bad weather set in right after these searches, preventing *Asherah* from making follow-up dives. After the Scripps team left, a second new side-scan sonar device, along with a team under the direction of

Martin Klein from EG&G International, an electronics company, joined the expedition. Summer was nearly over, and Bass had already returned to his teaching duties, leaving associates in charge of this last operation. The new sonar team made another pass over the smaller search area. They found the same bump the Scripps team had found, but to them it looked rather special. "Good God," Klein shouted when he saw it, "we've really got something big."

This time the weather cooperated, and *Asherah* was ready to dive. The moment Bass's group had been waiting for all summer had finally arrived. What would *Asherah*'s crew find on the bottom? Bass's published reconstruction conveys the same excitement every submersible group on the verge of discovery has felt:

> Yukel [Egdemir, *Asherah*'s pilot] and Don [Rosencrantz, the engineer in charge of *Asherah*] lowered themselves into the submarine from the dinghy. The sound of steel against steel rang across the water as the hatch dropped in place. . . .
>
> The bubbling sound of flooding ballast tanks melded with the rising whine of twin electric motors as Yukel threw on full power to force the submarine under. . . .
>
> Matt [Kaplan], in the rowboat, holding steady near the buoy, lowered a hydrophone for underwater communications.
>
> "We're at two hundred fifty feet and still can't see bottom . . . we're at two hundred seventy feet. . . ." Don relayed a stream of unessential reports to assure the topside crew of the submarine's safe descent.
>
> Suddenly sounds of shouting nearly drowned out crashing and whistling noises over the underwater radio. Was the *Asherah* collapsing?
>
> Only Don and Yukel were unconcerned. At 285 feet the *Asherah* had reached the bottom. . . . Visibility was limited to only a foot and a half, but the shapes of amphoras—amphoras everywhere—could not be mistaken. [In ancient times, merchants used amphoras to transport a variety of products, including wine, olive

oil, and fish sauce. These terra-cotta jars resist corrosion and commonly mark the presence of an ancient shipwreck.]

"It's the biggest wreck I've ever seen!" Don shouted, as Yukel whistled and cheered. "It's a wreck, it's a wreck. We landed right on it."

The crashing noise heard through the hydrophone at the moment of discovery remained a mystery until *Asherah* resurfaced. It turned out to have been wild banging from a tambourine that Don Rosencrantz had inexplicably taken on board.

Asherah's jubilant crew had finally found one of the bronze-statue-bearing wrecks that George Bass had sought for so long. It was an ancient merchant galley. The two men made a quick pass over the site and managed to take a few photographs in cloudy water before resurfacing. It was the last day of the expedition, and the last dive *Asherah* would ever make for archeology.

What killed the sub was basically fear—or prudent common sense, if you take an administrator's point of view. The insurance premium alone for the few dives *Asherah* had made during the summer of 1967 had cost as much as a small archeological excavation on land. The University of Pennsylvania then decided that more insurance would be needed. Any further use of *Asherah* would require a $5 million liability policy—at a cost that would have swamped Bass's meager operating budget. "We were forced to sell the *Asherah* just as we were beginning to realize her full potential," he wrote sadly.

By the mid-1970s, *Asherah* was listed in a large U.S. Navy survey of manned submersibles as "inactive." It had become one of the many small diving craft that customers stopped using and owners retired as the deep-sea exuberance of the 1960s gave way to the financial crunch of the 1970s. "I believe the *Asherah* was the first commercially built and sold submersible ever in America," Bass recently recalled, "so I am quite proud of what we did back then when I was still a graduate student." He last heard of *Asherah* in the mid-1990s, when someone called him from Florida to say he had seen the little orange submersible on a trailer in a parking lot. It had been sitting there for several years and no one had paid the parking bill, so apparently it was abandoned. Bass never got the caller's name or telephone number, which he regrets, because *Asherah* itself would now qualify as a wonderful museum piece.

Bass continued working in shallower waters, and he went on to found the Institute of Nautical Archaeology, a leading research center in the growing new discipline he had helped establish. But it would be many long years—roughly two decades—before marine archeology could blossom anew in very deep waters. By then my group at Woods Hole was pushing back the old frontier using ROVs, after first testing them on a modern shipwreck.

— *Argo* Looks into Amphora Alley

After our second *Titanic* expedition, we needed expert guidance on where to look for ancient shipwrecks. Our studies with the Harvard archeologists concentrated on the Tyrrhenian Sea, a triangular body of water bounded on its western side by the islands of Corsica and Sardinia, on its northeastern side by the Italian peninsula, and on its southern edge by the island of Sicily and the Tunisian coast of North Africa, where the ancient city of Carthage once stood. The vast majority of the Tyrrhenian Sea is deep: its basin drops off to depths in excess of 10,000 feet. During the rise and fall of the Roman Empire (about 300 B.C. to A.D. 400), thousands of commercial and military ships went down in these waters.

We decided to conduct a preliminary expedition in the Tyrrhenian Sea using *Argo*. For several weeks, starting in May 1988, we flew the tethered eyeball along reconnaissance lines that crossed the dominant trade routes. We were hoping that debris from ancient wrecks would show up on our monitor screens in much the same way that debris from the *Titanic* had finally revealed itself. This time, however, we could not count

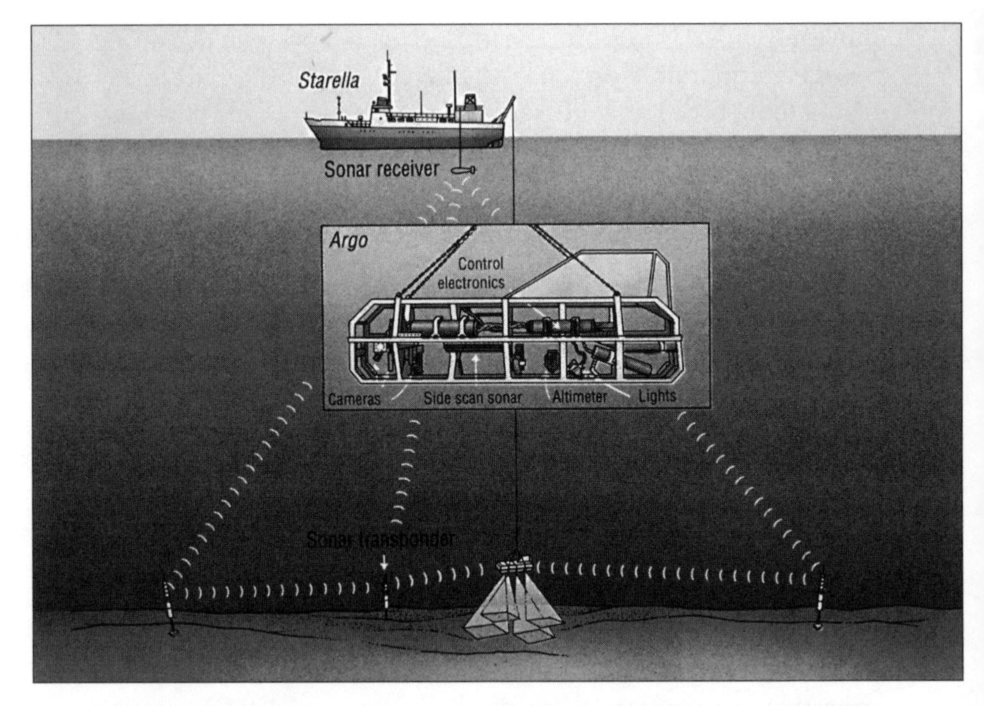

Artist's rendering of the early *Argo* system as deployed in the mid- to late 1980s, in this case the 1988 use of the system in the Mediterranean Sea from the research vessel *Starella*. (Source: Jack McMaster/Margo Stahl.)

on seeing large pieces of wreckage—especially a hull. In truth, we could not be sure whether ancient ships would have left any signature as distinctive as the debris trails from modern ships. We speculated, however, that they might have left signs of a panicky passage—that crews trying to run before a storm, for instance, would have thrown cargo overboard. In deep seas well away from sediment-depositing rivers, we thought that such cargoes might still be sitting on the bottom, even if the ships that dumped them escaped sinking, or if wood-boring mollusks had long ago digested the main wreckage.

The Strait of Sicily looked promising to us. One of the safest routes from Carthage to Rome crossed this stretch of open water (which separates Tunisia from Sicily) before hugging the coastline the rest of the way. Unfortunately, a few days before our arrival a large storm passed through the region. Its waves had stirred up the bottom sediments, which in this area can hang suspended in the water for many days, hovering like a cloud just above the seafloor. When we lowered *Argo,* we found the visibility to be less than three feet; it was impossible to search this important trade route.

One by one, other sites proved equally disappointing. When we traversed a north-south route near Sardinia, we could see the bottom fairly well on one attempt, not so well on another. But not a single trace of the several thousand ships we knew had gone down in this basin appeared on our screens. When we ran similar search lines along the northern coast of Sicily, we saw devastation. Intense bottom trawling had resurfaced the seafloor: long linear gouges marked the paths of countless nets dragged across the muck with their mouths held open. Farther north, off the Italian seaport of Naples, *Argo* had to turn back because hundreds of miles of fishing nets deployed in the area made tethered towing impossible.

We continued expanding our search just as we had with *Titanic,* crossing as many trade routes in as many places as possible. As our anxiety and exhaustion mounted, this expedition began to resemble the search for *Titanic* in another unpleasant way: the level of bickering increased, and confidence in our search strategy declined. That was understandable. After more than 200 miles of towing, *Argo* had found a grand total of one possible debris field. It lay north of Skerki Bank off the coast of Tunisia, along a route we considered improbable—across 350 miles of open water, the most direct but least sheltered route between ancient Carthage and Rome's port city of Ostia. We had discovered in this area a number of amphora in approximately 2,500 feet of water. One long concentration of them, which might have been dumped or gone down inside ships, we called "Amphora Alley." Yet we had seen *only* amphoras there—no shipwrecks.

Electronic still image of Roman amphora photographed by *Argo* in 1988 just northwest of the *Isis* wreck site.

With a growing sense of futility as our time at sea ticked away, we returned to Skerki Bank and Amphora Alley to survey it more thoroughly. The alternative would have been worse: making wilder and wilder stabs at less and less probable sites. Within a day or so, we had run a gridlike series of towing tracks less than a mile apart. They revealed an enormous but spotty debris field, covering some twenty square nautical miles. We saw hundreds of amphoras, yet little to recommend one spot where a few lay half-buried in muck over another spot somewhere else. But then, as *Argo* began scanning this vast area more closely, looking first at the largest concentration of amphoras, we finally saw hints of a shipwreck slide across our monitors. Tiny pieces they were, in 2,500 feet of water—not of the ship itself, the hull or planking. We saw instead only a series of depressions containing more durable items: an anchor, grinding stones, and small pieces of pottery.

These remnants turned out to belong to a merchant ship that had probably been traveling from Africa to Rome. We dubbed it *Isis* in memory of the Egyptian goddess and patron of sailors. It was this shipwreck that gave us the first opportunity to test the deep-ocean mapping and sampling abilities of our new, full-size *Jason* one year later.

The Long Gestation of *Jason*

Our work on *Jason* had begun much earlier, as soon as the Deep Submergence Laboratory was established in 1982. Our choice of an archeological target for *Jason*'s initial mission, which occurred much later, did not determine its design, because we had always wanted to use *Jason* for more than probing shipwrecks. In fact, we designed *Jason* primarily for mapping the Midocean Ridge, an enormous undertaking that had begun years earlier with Project FAMOUS. Along the way, we intended to adapt the techniques geologists would use to other branches of deep-sea exploration, including marine archeology.

The Project FAMOUS expeditions had started with broad sonar surveys and other towed-instrument sweeps, and they had ended with humans diving miles beneath the surface in submersibles to make detailed visual inspections. We knew that deep-sea archeology would follow the same pattern: broad surveys to search for wrecks, then stationary work above interesting sites. Our new paradigm called for *Argo* to make the preliminary surveys and *Jason* to do the close inspecting. That would relieve humans of their deep-diving chores, but we wanted one further improvement: a seamless fit between our mapping and imaging instruments on both vehicles. Every detail that *Jason* zoomed in to inspect had to match perfectly the broader surveys that *Argo* could make.

The quality of mapping and imaging we wanted to achieve required more sensitive sonar devices and cameras, which became available in the 1980s. It also required precise navigation, so that we could carry out properly spaced surveying runs and accurately combine the results. And finally it required precise mechanical control, because even the best instruments in the world, kept on track by the best navigation, would produce poor results if mounted on a rolling, pitching platform. This meant that our vehicles would have to be controlled with a level of precision not possible at the time we had begun working on *Jason.* As a result, Dr. Dana Yoerger, an expert on precision control and advanced robotics, recognized that he would have to come up with a sophisticated new control system.

Yoerger needed better control on three levels: the surface ship, the free-roving *Jason* below it, and the cable connecting them. First of all, he knew that a research ship on the surface would drift and bob unpredictably in varying wind and sea conditions. Second, the cable would transmit this random motion all the way down to *Hugo, Jason*'s suspended garage, thousands of feet underwater. So *Hugo* would be swaying, wiggling, and yo-yoing at the end of its long cable while *Jason* remained tethered to *Hugo* by a shorter cable. And third, *Jason* itself would be moving independently from those surface and cable forces, in response to commands from its operator.

In the early 1980s, the oil and gas industry developed a technique called "dynamic positioning" to hold a ship close to one spot on the surface (often near a drilling platform). Dynamic sensors and powerful thrusters corrected for the motion of wind and waves. If the wind began pushing the ship north, say, an engine would kick in to compensate, pushing it south with equal force. Yoerger's engineering team adapted this technique to the Woods Hole research vessel *Knorr* and carried out an initial test in moderate seas (winds of 8–15 knots) in September 1987. The results were excellent. Without the use of anchors, *Knorr* remained within eighty feet of its assigned position when using an acoustic transponder system installed on the bottom, and within forty feet when using a global satellite tracking system. That would be good enough to keep *Jason* well within the larger area it would be inspecting far below, roughly 300 feet on a side.

Meanwhile Yoerger's team carefully studied cable behavior. To do this, they instrumented a 0.68-inch tow cable to track its high-frequency motions, primarily those caused by vibrations during towed and stationary maneuvers. This experiment also proved highly successful. It confirmed the previous modeling done by computer, and it led to new insights regarding cable vibration, which also turned out to be manageable.

The next control problem involved *Jason* itself. The key to precision here was inserting a relay vehicle between the surface ship and *Jason*. Eventually *Hugo*, our Huge *Argo*, would perform this function. The full connection between *Jason* and the surface would then work like this: First, the surface ship's dynamic positioning system would keep it within a circle with a forty- to eighty-foot radius. Its unpredictable motion within that circle would continually tug and drag *Hugo*, hanging like a passive weight on the end of a long, taut cable. *Jason*, meanwhile, would be roving away from *Hugo* on a second tether, slightly buoyant and slack. *Hugo* would be wiggling one end of that tether, but being slack, this second tether would not push or yank the free-floating *Jason* at its other end. *Jason* would therefore remain free of surface action and subject to precise control by its operator.

Before we built *Hugo*, we developed a smaller unnamed relay vehicle to stand in for it—basically just a platform with instruments. It acted as a weight on the end of the long cable stretching back to the surface ship, and it allowed us to watch what happened when we tested the system. Little did we know the vital role that this simple platform would soon play in the survival of our project.

Even with a relay vehicle damping out motions from the surface, we still needed to maneuver *Jason*. The precise control we envisioned far exceeded anything yet attained with

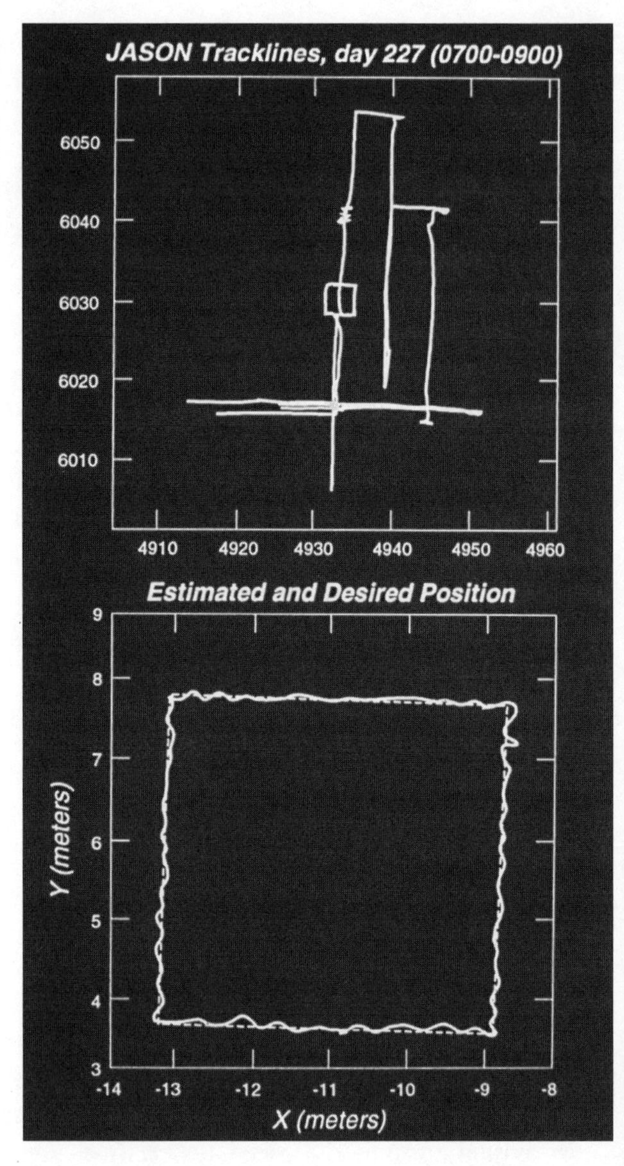

Jason test tank results demonstrating the vehicle's ability to follow a desired trackline while under automated control without the assistance of a pilot. (Source: Dana Yoerger.)

either manned or unmanned vehicles. Yoerger's search for perfection led him to study in great detail the dynamic behavior of existing ROVs. Characteristically, he did this not by reading about them in the literature but by experimenting with the actual vehicles in our lab. One by one he ran them over set courses repeatedly, first in our test tank indoors and then off the dock. As a result, the development of *Jason* took more than five years. By 1988, our DSL engineers had nearly eliminated the rocking and yawing off course that had plagued other deep-sea craft. To achieve this, they mounted many control sensors on *Jason.* A compass and gyroscope, for instance, determined its heading and the rate of drift from that heading, which automatic controls could then correct. Accelerometers detected changes in speed; they provided instantaneous, three-dimensional data on how the vehicle was responding to thrust from its motors, buffeting currents, and other factors. An inclinometer, meanwhile, registered the amount of tilting in two axes, keeping track of pitch and roll.

None of these exquisite controls did a thing to enhance *Jason*'s appearance (although as proud parents we were extremely fond of it). Out of water, *Jason* had all the gracefulness of the proverbial beached whale. You saw an inert, 2,600-pound, nine-foot-long, vaguely beetle-shaped cube. This blue-sided, red-backed beetle had about the same size and heft as a small car. When submerged, however, it became quite agile. For one thing it became weightless, although it was certainly still massive. But that mass was exquisitely balanced, and its seven well-placed electric motors could push it precisely in any direction without the slightest uncontrolled twist, spin, or tumble.

When Yoerger was satisfied he could make this bulky tuna "dance on the head of a pin," as he put it, he addressed tracking precision. Existing bottom-mounted transponder systems would not be adequate. We wanted to track *Jason*'s three-dimensional position several times a second to a precision of one or two inches. To meet those needs, Yoerger's team developed a high-frequency acoustic navigation system called the Sonic High Accuracy Ranging and Positioning System (SHARPS), which could resolve distances to a precision of less than one inch and make as many as ten position readings per second. In the open ocean, a three-transponder SHARPS array would cover an area approximately 100 yards on a side. That would allow us to stake out a site for *Jason*'s explorations large enough to include most shipwrecks, small geological features, or the interesting environment around hydrothermal vents.

With *Knorr* upgraded to hold its position on the surface, and with *Jason* approaching the deftness needed down below, we needed only two custom-made cables to complete our system. The first one, 30,000 feet long and 0.68 inch thick, connected our temporary relay vehicle (a stand-in for *Hugo*) to the surface. Inside this cable's steel armor, three copper conductors and three single-mode optical fibers carried electrical power, commands, and data. We designed the second, much shorter cable to connect *Hugo* and *Jason*. Slightly positive in buoyancy, it was approximately 100 yards long. Like the tow cable, it also had three copper conductors and three single-mode optical fibers, but it used synthetic Spectra fibers instead of steel armor to provide strength at a reduced size and weight. Eventually, we would outfit both *Jason* and its relay vehicle to take full advantage of all three optical fibers. Observers in the control van would have access to four continuous video channels; they could switch them to view the scene through any of eight cameras on either vehicle.

By the early spring of 1989, we had finished building and testing the system and were preparing the much larger *Hugo* to replace its temporary stand-in platform. Our second expedition to the Tyrrhenian Sea was scheduled to begin in April. At last, we could launch *Jason* on a real deep-sea mission. Unlike our previous search trip with *Argo*, we would now be able to include far more observers in our virtual crew than the control van itself could accommodate. Many thousands more would be watching from satellite transmissions relayed to stations on shore.

Telepresence Ups the Ante

We had no trouble whatsoever recruiting a larger observing crew. Soon after discovering the *Titanic,* we noticed one fascinating consequence: the amazing impact the discovery had on young students. After both *Titanic* expeditions, we received thousands of letters from students asking how they might become involved in deep-sea exploration. We saw their interest as an opportunity to motivate larger numbers of them to continue their education in science and technology. We could do this, we thought, by including them as observers in future expeditions.

One of the original goals we had set for the *Argo/Jason* system was to allow more scientists to participate, not only by working inside *Jason*'s control van at sea but also by viewing satellite transmissions to their laboratories on shore. If such a telepresence could be established with scientists, why not with young students? They could assemble at downlink sites around the world and watch the same views, at the same time, as researchers aboard our surface ship. So we decided to inaugurate a series of live broadcasts, called the Jason Project, when we began using *Jason* on its first deep-sea mission. Such a program, in addition to enriching education, might also help promote this newly emerging technology to a scientific community still skeptical of *Jason*'s usefulness.

For the first Jason Project, we arranged to broadcast programs to twelve museums in the United States and Canada. To enhance the effect of telepresence, the museums built replicas of our control van so that students could see the views from *Jason* just as we did, on a bank of computer consoles. For mass viewing, the museums added three large television projector screens, placed in auditoriums that seated 300–400 students and teachers. Two-way audio links between all the sites

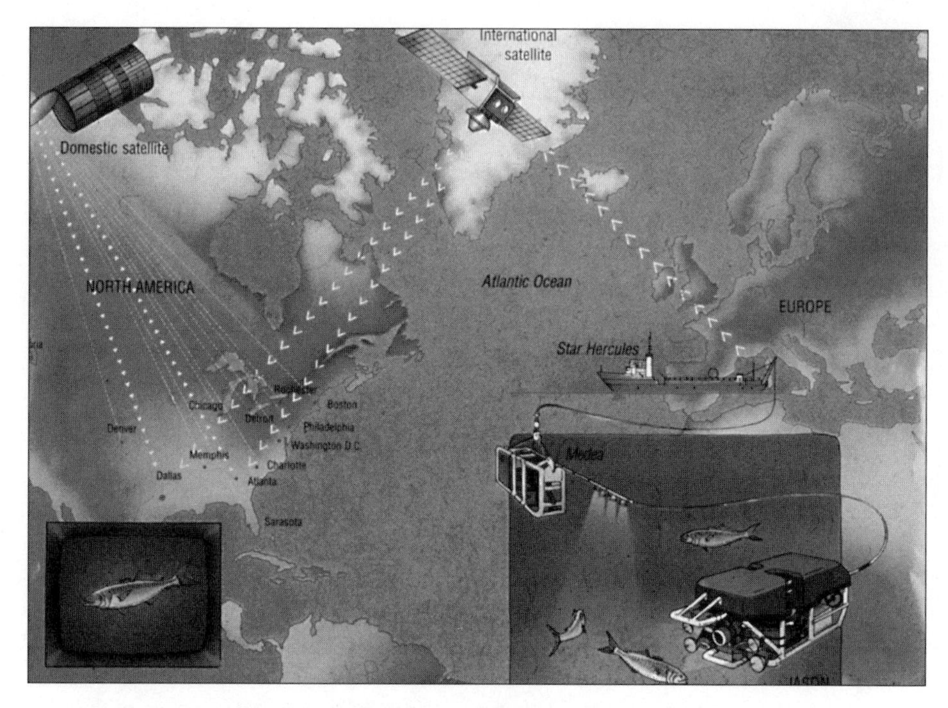

Telecommunications hookup used to provide "live" access for students and teachers in North America during 1989 Jason Project expedition to the Mediterranean Sea. (Source: Jack McMaster/Margo Stahl.)

allowed students to ask questions during the eighty-four live programs we arranged to broadcast over a two-week period in May 1989. Each program would reach thousands of students live at the downlink sites, and all together some 250,000 over the two-week period.

With the Jason Project up and running, the new era I had begun to envision while at Stanford seemed almost within reach. But the sword of instant communication would cut two ways. It would promote our successes with the *Argo / Jason* system but also broadcast embarrassing failures. That seemed fair enough—a dose of real-world exploring rather than a fake story with impossibly cool heroes and a guaranteed happy ending.

One of many downlink sites used during the Jason Project to provide hundreds of thousands of students and teachers the opportunity to participate "live" in field research.

We quickly finished building *Hugo, Jason*'s huge underwater garage and relay vehicle; *Argo/Jason* became the *Hugo/Jason* system. We then returned to the Tyrrhenian Sea in April 1989 with *Jason* and *Hugo* loaded aboard the *Star Hercules,* an oil-field support ship with bow and stern thrusters set up for dynamic positioning. Our transit from England had been harrowing. In the Bay of Biscay, a monstrous rogue wave had struck and washed over the *Star Hercules,* damaging both *Hugo* and *Jason.* It had also flooded our vans and ruined hundreds of thousands of dollars worth of electronic gear inside them. We had just enough remaining equipment, spare parts, and time for repairs (with technicians working around the clock) to keep us on schedule for the live broadcasts. But when we arrived at Skerki Bank, sea conditions there were also far from ideal. A major storm had just passed through. On April 26, despite high swells, we prepared for a test broadcast. For the first time, we were going to launch *Hugo* in deep water with *Jason* securely fastened inside its large aluminum frame.

Hugo, four times larger than *Argo,* our original search vehicle, weighed 10,000 pounds in air. A huge box made of large aluminum I-beams, it was relatively light for its size. But the problem was not its weight in air; it was the weight in water. *Jason, Hugo*'s passenger, weighed nothing at all in water. *Hugo,* by contrast, weighed some 7,000 pounds when submerged. That heavy load made me nervous, even though our Jason Project engineer assured me that the cable could handle far more weight. By now, with years of experience at sea, I knew that cables could find ways to upset calculations—like the ones that had rusted from the inside out, dropping *Alvin* to the bottom of the ocean, or like *Argo*'s long tether, which had jumped into the winch axle during our search for *Titanic,* nearly ending that expedition. (I did not at that time know of still another precedent: Jacques Cousteau's cable had broken during a test of his new *Soucoupe* in 1957, when his support ship pitched in a wave and snapped it. That first complete *Soucoupe* hull ended up on the bottom of the Mediterranean, never to be recovered.)

"Good morning from the Mediterranean," I said into my headset microphone, while looking toward the lens of an active television camera. I was broadcasting live to the twelve museum sites when *Hugo* and *Jason* dropped into the water. Out of the corner of my eye, I could see the launch progressing; I forced myself to smile and sound cheerful. As I talked, the stern of the *Star Hercules* rose up on the crest of a large passing wave. As the stern descended into the following trough, *Hugo* sank underwater at a much slower speed. I could see the cable go slack. The *Star Hercules* quickly rose up the crest of the next large swell. With the surface ship rising and *Hugo* still sinking, the snap loading on the cable skyrocketed. Suddenly I saw the monitor showing views from *Jason* go blank.

The optical fiber cable had snapped. I shouted, ripped off my headset, and bolted from the control van. So ended our first broadcast.

Thankfully, that broadcast was only a test. Producers for the Turner network explained what had happened to the twelve museums and to hundreds of television technicians. Meanwhile, on *Star Hercules,* a real disaster was unfolding. Crew members had taken shelter when they saw the cable go slack; its severed, whiplashing end had not injured anyone. But now the brutish *Hugo,* with our irreplaceable *Jason* inside, was free-falling to the ocean floor some 2,500 feet below.

—— The Hasty Birth of *Medea*

Inside the control van, Dana Yoerger and our chief navigator noticed that the emergency transponder on *Hugo* had activated. That allowed them to see its remarkably slow fall, at a speed of only 4 knots. While most of us on deck were knotting up inside, thinking that years of work had just been lost, *Hugo*'s unstreamlined shape and large surface area were saving the day—causing enough drag in the water to keep it from falling fast and smashing *Jason* to bits on the seafloor. Yoerger tracked the huge garage all the way to the bottom as it came to rest on soft sediment.

Long before it settled, we had already begun piecing together a recovery sled to bring it back. There was only one candidate on board: the temporary relay vehicle we had used as a stand-in for *Hugo* the year before, while testing our new cable. It was little more than a weight with instruments attached, but now that we were rushing it into active duty, we felt that it deserved a name. We dubbed it *Medea* in honor of the mythical Jason's wife.

Technicians worked all night to install cameras, floodlights, a sonar transponder, and control circuitry on *Medea*'s metal frame. Beneath this frame, suspended on a long steel chain, they attached a large grappling hook. The next morning, *Medea* was ready. The weather, however, was not; we had to flee a vicious storm.

On April 29, two days before our first scheduled broadcast, we returned to Skerki Bank. The seas were still rough. Using the dynamic positioning system on *Star Hercules* to minimize cable motion, we carefully lowered our new grappling rig to the bottom and positioned it above *Hugo.* Martin Bowen, who had flown *Jason Junior* so deftly through the *Titanic,* took the flyer's console, from which he could raise and lower the cable. Yoerger, meanwhile, attempted to maneuver *Medea* by jiggling thrusters on *Star Hercules.* As the long cable shifted and wiggled, Bowen made his move. He dropped the hook—right on top of *Hugo*'s upright frame. But instead of grabbing *Hugo*'s chain bridle as planned, it fell down to one side, latching onto one of *Hugo*'s main aluminum beams. It was not a firm grip. We lifted anyway.

Hugo fell off twice on the way up as the cable went slack from rising and falling waves. But each time that *Hugo* fell off it also, amazingly, fell back onto the hook, which snagged it again in midwater as the stern of *Star Hercules* rose up on the next approaching wave. When *Hugo* broke through the surface, swaying wildly, *Jason* was crashing around inside it. Before we could secure *Hugo, Jason* escaped and fell back into the water. Fortunately its neutral buoyancy kept it near the surface. We went after it through the heavy seas in a Zodiac, paddling with oars when the motor stalled, and we finally managed to secure it with lines and haul it back to *Star Hercules.*

Clearly, *Hugo*'s design had been a mistake. We never used it again. Instead, we hastily devised a new launching strategy for *Jason,* with the newly christened *Medea* as its relay platform.

We now had to launch two vehicles separately. First to go over the side was *Jason.* Its pilot then drove the robot, tethered the whole time to *Medea,* away from the *Star Hercules.* Once *Jason* reached the end of its tether, crew members hoisted *Medea* into the water with the crane. Weighing some 600 pounds instead of *Hugo*'s 10,000 pounds, it put a smaller and more acceptable strain on the cable. Both vehicles then proceeded together toward the seafloor, falling at the same rate. Underwater, the vehicles took turns eyeing each other. *Jason* watched *Medea* on its stern video camera as *Medea* descended passively on its long, sturdy cable. Once the bottom came into view on *Medea*'s downward-looking camera, our winch operator stopped unreeling the cable. *Jason* then continued down, motoring under its own power, and eventually entered *Medea*'s field of view. The unplanned eye-in-the-sky perspective we gained from *Medea* later proved invaluable, especially when obstacles or suspended sediments obscured *Jason*'s vision. And surprisingly enough, the new launch procedure worked flawlessly.

—— *Jason* Completes the Mission

During the early part of May, hundreds of thousands of students and teachers around the world watched a team of oceanographers, engineers, archeologists, and conservationists explore the debris field littered with amphoras that we had discovered the previous summer. They also watched as our group recovered artifacts from the exposed upper section of the fourth-century Roman trading ship we had named *Isis*.

As we expected, the most delicate aspect of this project turned out to be the recovery of more than fifty-four ancient artifacts. Our engineering team at DSL had to develop a unique manipulator system to handle this kind of object. Scientific missions tend to be unpredictable and commonly demand fine control, so we had designed *Jason*'s manipulator not for particularly heavy lifting but with an ability to exert very low forces and torques. That made *Jason* a careful worker on the bottom, but not much of a pack mule. For this Skerki Bank mission, we modified *Jason*'s manipulator to hold a large set of tongs. *Jason* used them to encircle and lift amphoras without applying pressure. Once it had securely grasped an amphora in the tongs, the robot thrusted up off the bottom and deliv-

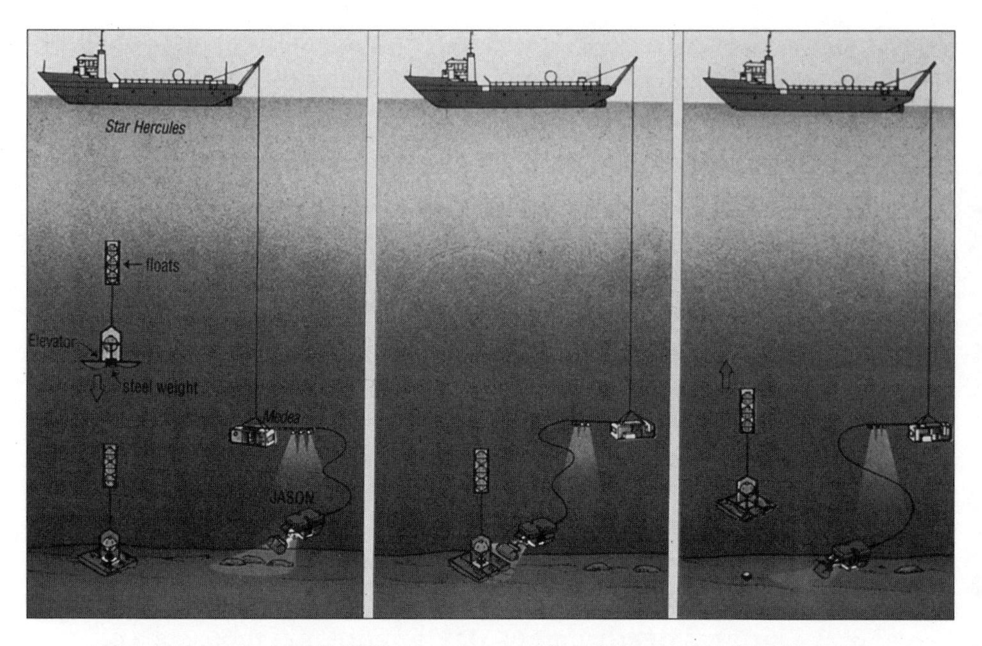

Elevator system used in 1989 to recover artifacts from the *Isis* shipwreck site. After lifting an artifact from the bottom, *Jason* places it into a soft mesh compartment in the elevator. An acoustic command sent from the support ship causes the elevator to drop weight and return to the surface. (Source: Jack McMaster/Margo Stahl.)

ered its cargo to an elevator—a simple frame with netting for its floor and walls—which rested on the bottom a short distance away. When the elevator's various compartments were full, we sent an acoustic signal from the surface that caused weights to drop, and floats attached to the elevator then lifted it back to the surface. Conservators aboard *Star Hercules* emptied the elevator's compartments, and our crew dropped it back to the bottom again.

In all, *Jason* made fifteen dives, and the elevator made thirteen round trips without causing any damage to the artifacts it held. Aside from the amphoras, these included a millstone from the galley, a cup and pitcher, and a small terra-cotta lamp. The lamp was still blackened with soot, possibly from the day

Isis went down in a gale like the one that had chased us away only days earlier.

Our observers on shore watched all this from many perspectives: *Jason*'s cameras, *Medea*, and cameras aboard *Star Hercules* showing the crew at work. Expedition members paused to explain what we were doing as we fielded questions. We had not expected any traces of the vessel's wooden structure to remain, so we all became excited when our survey photographs revealed the outline of a classic Roman merchant vessel in the mud, roughly 100 feet long. The wash from one of *Jason*'s thrusters even revealed the ends of a few planks. The most exciting moment, though, must surely have been when the amphora debris field first slid into sight. Our crew in the van, not knowing if or when we would find it again, broke out in spontaneous cheering, and we learned later that students had cheered too—as if standing right there with us, finding the deepest ancient shipwreck site ever discovered.

After the expedition, project archeologist Anna McCann of Trinity College was able to determine—from the artifacts we had collected and from close inspection of other debris through *Jason*'s cameras—that our site was a composite: other ships had also gone down there. In addition to the trading ship *Isis*, now dated to A.D. 356, at least six others, McCann predicted, should be found in the same area. A few years later, a new series of expeditions proved her right.

Observers on Shore Begin Driving

In contrast to the wreck of *Isis*, we knew quite well why ships had sunk at a site we explored the following summer. On August 7, 1813, two American warships, *Hamilton* and *Scourge*, drifted lazily on Lake Ontario. The evening was calm, the water still. The War of 1812 was, for the moment, uneventful. The American crews had been preparing to fight British ships from the city of Hamilton, but now they slept at their guns. Suddenly a violent squall caught them napping, with

Science team in the *Jason* control van during the detailed 1990 investigation of the American warships *Hamilton* and *Scourge*, lost in a squall on Lake Ontario during the War of 1812. (Source: John Earle/Odyssey Corp.)

sails rigged. As often happens in summer, the lake changed its character in a matter of minutes. And within that short time, both schooners flipped over and rapidly sank. They were not seen again until 1975, when a search using side-scan sonar located both ships resting upright on the bottom in some 310 feet of water. Subsequent expeditions documented their remarkable state of preservation, including the skeletal remains of their crews.

In 1990, we conducted the second Jason Project on Lake Ontario. This *Medea / Jason* expedition had two primary goals. First, we wanted to map the two ships precisely as part of an ongoing effort to determine whether they should be left on the bottom or raised and placed in a museum. Second, we wanted to broadcast our activities live in sixty programs similar to the ones we had transmitted in 1989. Instead of having our dis-

Jason, outfitted with a wide variety of acoustic and visual sensors, is lowered over the side of its support barge during the investigation of the *Hamilton* and the *Scourge.* (Source: Robert D. Ballard/Odyssey Corp.)

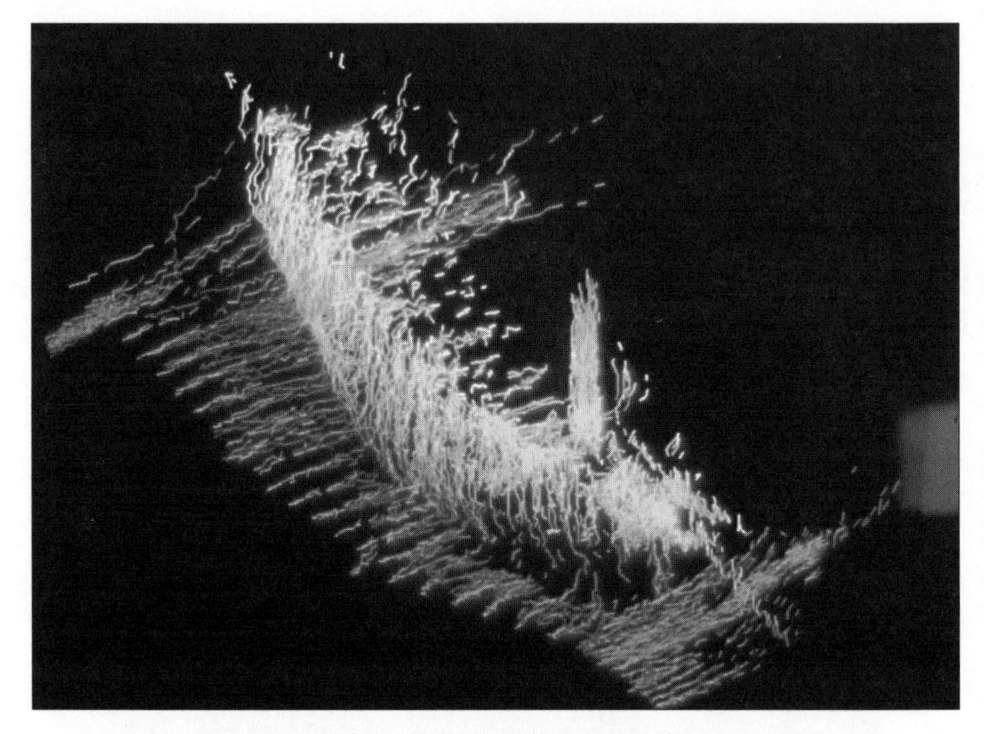

Three-dimensional characterization of the warship *Hamilton* based on a series of vertical scans made by a digital sonar while *Jason* was precisely tracked using the SHARPS system. (Source: Deep Submergence Laboratory.)

tant observers watch us recover artifacts, however, we would have them watch how we used high-precision tracking and sensors to document the two ships in great detail. Also, if everything worked as planned, our observers this time would do more than just watch. A few would actually participate in a long-distance control experiment by maneuvering *Jason* around the shipwreck site.

During the first phase of our work, *Jason* carried several visual and acoustic sensors, including a digital scanning sonar and a narrow-beam sonar. Using the SHARPS system to track *Jason* accurately, we were able to scan the hull of the *Hamilton* carefully and build a three-dimensional model of its current shape.

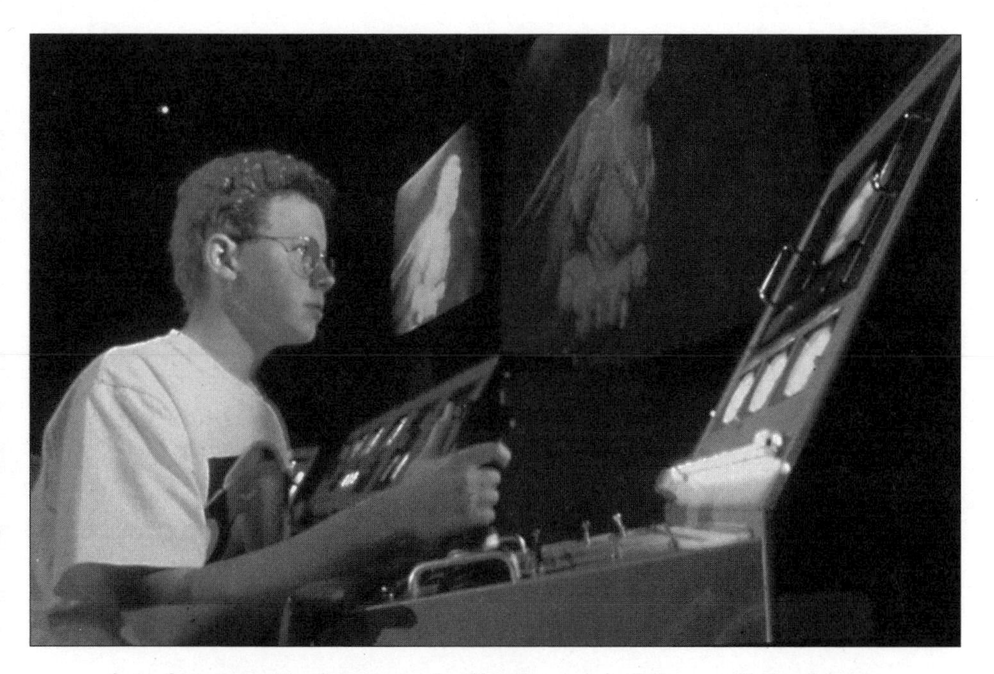

A student operating *Jason* near the *Hamilton* on the bottom of Lake Ontario from a Jason Project downlink site hundreds of miles away. (Source: Mark D. Thiessen/National Geographic Society Image Collection.)

On a different survey of each ship, we mounted a digital still camera on *Jason,* which then made a series of automated runs along and over the ships. After collecting the data set, an image processor on our surface ship removed particle backscatter and compensated for uneven lighting. Since we had installed a Pixar workstation on board, our imaging technician was then able to construct a mosaic of the *Hamilton* while still in the field. Flicking back and forth between processed images brought up on the display, he panned, zoomed, and blended them together until the mosaic was complete.

At the end of each broadcast, a student at one of the North American downlink sites got into position at a control console. Our *Jason* pilot at sea could then transfer as much control of the vehicle to the student as he felt the student could handle.

Then, using a joystick and watching a computer graphics display, the student flew *Jason* for several minutes. This experiment worked beautifully, and it clearly demonstrated that scientists on shore would be able to pilot a remote-controlled vehicle during future deep-sea expeditions.

And so, by the summer of 1990—within a decade of my first sketches of the concept at Stanford—we had demonstrated all the components of a new paradigm for deep-sea exploration. Along the way we also demonstrated other capabilities—not always the ones we had planned. For instance, we rescued a sunken robot. In some respects, our rescue efforts resembled the attempt to raise the *Johnson Sea-Link* in 1973. Yet our task was not nearly as grim and urgent. Our sunken robots needed no air. And if rescue had been impossible? We would have measured the cost in dollars and delays, rather than lost lives and broken hearts.

CHAPTER

10

SHOULD HUMANS CONTINUE TO DIVE?

Two Paradigms

Il faut aller voir. [We must go and see.]

—*Jacques Cousteau's motto*

You'll never see as much in *Alvin* as you can see with *Jason* and *Medea.*

—*Robert Ballard*

Just as the marine science community needed more than twenty years of experience with manned submersibles before it granted them wide acceptance, researchers today will take an equally long time to accept their replacement by remotely operated vehicles (ROVs). Resistance to this inevitable shift has been strong. Some of it is typical of the early stages of any paradigm shift, when the majority is still trying to catch up with the previous trend. By the early 1980s, the majority of scientists in the oceanographic community had never set foot in a submersible. Only a few had actually seen the incredible riot of life forms surrounding an active hydrothermal vent. But after the discovery of these ecosystems, everyone wanted to get in on the action and dive themselves. Using unmanned ROVs just didn't seem the same.

Nothing can match one's first mind-boggling, gut-wrenching immersion in the deep ocean—as everyone who has tried it from Beebe onward can attest. The vehicle one used to get there basks for a while in the reflected glow of this excitement. Eventually, however, the excitement and novelty of any new technology wear off. One sees its drawbacks and limitations more clearly and becomes more open to new approaches—sometimes radically new ones. Such paradigm shifts typically require at least a generation. Afterward, the mania for the older technology seems quaint. What was all that excitement about? Oh yes. Dirigibles. Bathyscaphs. Small submersibles.

Such shifts have occurred countless times in the past. It is difficult to imagine today, for example, the excitement mixed with sheer terror that passengers in the early 1800s reported on taking their first train ride. Lurching and swaying and clickety-clacking behind a fiery locomotive belching smoke and screeching steam, at sustained speeds exceeding ten or even fifteen miles per hour, some actually screamed in fright. Others quickly sensed that the vast distances isolating one town from the next had suddenly shrunk to insignificance, and that their world had changed forever (which was true). Some of the more farsighted among them realized that railroads would become a dominant form of transportation, a gleaming steel network on land replacing rivers and canals, to which all human geography would inevitably adjust (which was true for a time).

Nowadays, at the forefront of oceanographic research, manned submersibles have probably just passed the peak of their paradigm reign, like railroads many decades ago. Deep-sea exploration by means of telepresence has been gaining momentum, like virtual travel on the Internet. There is still a long way to go, and the technology for ROVs still has much room (and holds much promise) for improvement.

Submersible Use Expands

By the end of 1982, U.S. and French manned submersible programs, with the investigation of hydrothermal vents as their primary activity, had settled into a familiar routine. Program after program emerged that combined the strengths of three technologies. In a typical expedition or series of expeditions, multi-narrow-beam sonars arrived first to map additional segments of the Midocean Ridge. Next came follow-up surveys by towed vehicles such as ANGUS, a similar French system called RAI, or the Scripps Deep-Tow system to pinpoint active vent sites. Finally, research groups deployed manned submersibles such as *Alvin*, *Sea Cliff*, *Sea Turtle*, *Cyana*, and *Nautile* to investigate specific vents or clusters of vents. As these programs expanded, *Alvin* easily became the most productive research submersible of all time. By 1997 Allyn Vine's little dream-come-true had taken scientists on more than three thousand dives and spent well over three years undersea—and it was still being booked to capacity.

Although the Midocean Ridge dominated many research agendas, scientists continued to use manned submersibles for more traditional studies. Marine geologists roved everywhere from seamounts to submarine canyons, and biologists examined life at all levels, from midwater fishes and jellies and marine snow to microbes that live in the benthic muck (see the references in Further Reading). The French deep submergence programs became by far the most global in scope. The submersibles *Cyana* and *Nautile* visited the Japan and Kuril Trenches, the Puerto Rico Trench, many sites in the Mediterranean and Red Seas, many Atlantic and Pacific sites on the Midocean Ridge, and coastal rock formations off France, Portugal, Spain, and parts of Africa. But as French research projects expanded worldwide, other countries launched manned submersibles that they often used closer to home.

The scope of submersible-based research in Japan increased dramatically in the 1980s thanks to the effective use of *Shinkai 2000*. It gradually gained acceptance for traditional work in marine biology—a subject that Japan, a major fishing nation, has pursued with abiding interest. It also gave a quick boost to geological research on the ocean floor. Given Japan's long history of destructive earthquakes, Japanese scientists had hopes of providing the public with improved seismic forecasting. (One such effort, called the Kaiko Program, included a joint series of dives with the French and their *Nautile* submersible.) Researchers worked for many years to establish long-term monitoring stations on the seafloor, and they have used *Shinkai 2000* to install and service instruments such as tiltmeters and seismographs. (In 1990 a newer version, *Shinkai 6500,* plunged four miles deep in the Pacific, a record for small submersibles, and it will significantly extend the range of this work on the bottom.)

Canadian and Russian deep-diving programs during the 1980s centered on the use of the *Pisces IV* submersibles, followed in the 1990s by Russia's two *Mir* submersibles. Russian investigators carried out geological studies in the rift zone of the Reykjanes Ridge in the North Atlantic, while Canadian programs included a number of geological investigations in the northern parts of both the Atlantic and Pacific Oceans. Like Japan, Canada has also carried out midwater biological studies primarily in the fishing areas off British Columbia. The Canadian work with submersibles has also included studies of iceberg scour on the Atlantic continental shelf and research on mass wasting processes around the continental margins of the Grand Banks, the Baffin Shelf, the Labrador Shelf, and the Laurentian Fan (beyond the mouth of the St. Lawrence River). An earthquake in 1929 triggered massive landslides and turbidity currents all around the Laurentian Fan, and these studies traced long-term effects.

— Submersibles Meet Their Match

By the 1990s, however, the tide was turning. ROV systems such as *Medea/Jason* were being used more often to explore the Midocean Ridge, although the use of manned submersibles remained strong. This trend began off the coast of Washington State, in the rift valley of the Juan de Fuca Ridge. During July and August 1991, *Jason* carried out its first major, comprehensive science program there. On this expedition, for the first time, we used *Jason*'s full deep-ocean capabilities to operate the vehicle twenty-four hours a day in a complex and rugged volcanic terrain. It was precisely the sort of setting we had designed the *Argo/Jason* (now *Medea/Jason*) system to map and explore. In particular, we expected our vehicles to eliminate a blind spot that had plagued previous mapping and imaging expeditions.

An interpretive gap had long existed between oceanographers' large-scale and small-scale imaging technologies—that is, between sonar and visual systems. Scientists diving in *Alvin* often saw that depressions on the bottom were much deeper than sonar maps had indicated, and volcanic terrain looked more rugged than sonar mapping showed it to be. Narrow ravines, for instance, often separated different peaks that sonar images had contoured as a single volcanic hump. These discrepancies greatly complicated the detailed mapping efforts of the 1970s. It was hard to place the actual terrain being visually explored, with its important geological details, correctly on the broad survey maps.

Argo, with its 100-kHz side-scan sonar, might have narrowed the gap if it had worked as intended. But although its visual imagery met all expectations, it proved to be an unreliable acoustic platform, commonly fishtailing as it was towed through the water and blurring the resolution of the resulting sonar images.

To remedy this problem we developed a new side-scan sonar system, called DSL-120 to denote its 120-kHz frequency. Towed behind a depressor weight that, like *Medea,* removed surface motions, the DSL-120 has two transducers on each side of the vehicle mounted at a spacing of a half wavelength, in this case a few inches. This mounting configuration makes it possible to determine the phase of the returning signals and from that information to calculate the micro-topography to either side of the vehicle's path.

In essence, the DSL-120 combines the strengths of a side-scan sonar system with those of a multi-narrow-beam sonar. The observer gets not only a "shadow-graph" of the bottom terrain, indicating its varying degrees of roughness, but a bathymetric map as well. Since the DSL-120 is towed 100 yards above the bottom, it can produce a high-resolution topographic map having a contour interval of one meter.

With this family of vehicles—the DSL-120, *Argo,* and the *Medea / Jason* system—scientists can bridge the gap between large- and small-scale mapping systems. They can begin by looking at the big picture and then zoom in on specific targets in greater detail without getting lost along the way.

Before our 1991 cruise, geologists had carried out extensive studies in the Juan de Fuca area using a wide variety of mapping systems, including swath-mapping sonar, high-resolution side-scan sonar, deep-towed still and video cameras, and *Alvin* with human observers. Although these studies had all advanced scientific understanding of the Midocean Ridge, the data sets could not be merged to form a coherent, unified picture. We now had the tools to provide such a picture. To test their usefulness, John Delaney of the University of Washington, as co–chief scientist, planned and executed the July–August 1991 *Jason* expedition.

During the first phase of the expedition, the U.S. Navy support ship *Laney Chouest* towed our new DSL-120 over a twelve-mile-long segment of the rift valley axis, guided by a network of bottom-mounted acoustic transponders. This gave us a detailed contour map with a higher resolution than the previous sonar surveys. Next we lowered our *Medea / Jason* system to map a smaller zone within this area at still higher resolution. For twenty-nine hours, *Jason* ran survey lines about fifty yards apart at an altitude of between thirty and fifty feet above the bottom. Its 200-kHz side-scan sonar system recorded overlapping swaths of data, while an electronic still camera snapped an image every twenty seconds.

After this intermediate-scale mapping, *Jason* began a series of small-scale studies within an area measuring 500 by 350 yards. This required us to lay down a finer-scale network of 300-kHz EXACT transponders, similar to the SHARPS system but wireless. Like SHARPS, this system could resolve distances to a precision of less than one inch. (Actually, *Jason* set up the transponders. We put them on an elevator and dropped them to the bottom, and then *Jason* deployed them to cover the areas we wanted to investigate.) We then prepared *Jason* to measure a few physical and chemical properties of the seawater by mounting conductivity, temperature, and manganese sensors on the front of the robot. *Jason* took closely spaced readings along precisely navigated tracks at altitudes of ten, fifty, and eighty yards above one of the hydrothermal vents. With this information, we mapped a hydrothermal plume in three dimensions. We could then literally see how warm water from the vent was welling up and spreading nutrients through the surrounding seawater.

In all, *Jason* spent more than 190 hours on the bottom, collected nearly 700 hours of video imagery (the total from several cameras), obtained more than 18,000 electronic still photos, made approximately forty EXACT-navigated runs in hydrothermal plumes, and produced digitized maps of the bottom terrain on several scales. This unique data set would make it possible for a scientist to zoom in on specific targets when using *Jason* in any part of the area we had mapped. In this case the area included a network of gjars or fissures, small blocks of seafloor crust shifted up or down by tectonic faulting, and numerous hydrothermal sites with tall sulfide deposits and "black smoker" chimneys. While investigating these sites, *Jason*'s pilot could operate the robot either manually or by automatic control, using the EXACT tracking system to run precise data-gathering lines. During this final phase of our investigation, *Jason*'s two high-quality color cameras obtained spectacular video and electronic still images.

By the time our expedition ended, we had demonstrated *Medea/Jason*'s superiority to many previous combinations of deep-sea research vehicles. But we had also learned when to quit. The ocean is rarely a passive subject on any cruise, and this time it dished out a new lesson on the vulnerability of our vehicles.

Since ROVs can be launched and recovered more easily than manned submersibles, and since they do not place human lives at risk underwater, there is a tendency to deploy them in worse weather or to keep them down longer as the weather deteriorates. We decided to keep *Jason* down on August 6. For several days, we had been watching a low-pressure area south of our position. It was moving east. Both *Alvin* and *Nautile*, which were working south of us, had headed into port to avoid the approaching storm. But we decided to stay and continue working, hoping that the storm would keep to its current track and pass us to the south. It did not. During the night of August 6, it suddenly turned north and began moving faster, and its winds intensified. Before we could react, we found ourselves in a building sea with wind gusts of 31 knots. It was too late to recover *Medea/Jason*. We had no choice but to ride out the storm.

By 8:00 the following morning, the wind was blowing in excess of 60 knots, and seas were running at fifteen to twenty feet. That was when our winch suddenly failed and the take-up drum began to unspool cable, creating a giant knot. To make matters worse, the winch was mounted on a deck area that was flooded, as big waves of green water broke repeatedly over the bow. Fifteen of our crew literally poured out onto that deck and worked desperately to respool the cable back onto the drum. While we were struggling at this task, we had to do something about our vehicles down below. Since we could neither winch in nor winch out, we tied off the cable to prevent it from unwinding any farther and ran with the sea, hoping that the drag on the cable would cause *Medea/Jason* to swing back and upward behind us. If it did not rise above the rugged volcanic terrain we had been mapping, it would probably crash into steep fault scarps.

It took us three hours to respool the cable. For many hours afterward the weather remained too rough to effect a recovery. The storm had pushed us well away from the Juan de Fuca Ridge; *Jason* and *Medea* were at least in deeper water. All we could do now to revive the expedition was reverse course and slowly cruise back over the twenty-four miles we had lost during the winch crisis. Finally, with the wind and seas dropping, we brought the two vehicles back aboard the *Laney Chouest* after sixty-two hours underwater. To our great surprise, we found them in working condition and sent them back to the bottom a few hours later.

Aside from our failure to dodge the storm, our expedition to the Juan de Fuca Ridge proved highly successful. Far more than the previous archeology cruises, which did not demonstrate the full capabilities of *Jason* and its family of vehicles, the mapping and measuring we did in 1991 produced valuable information—data that other researchers could use—on a subject of major scientific importance: the Midocean Ridge. This work moved us much closer to convincing the marine science community that ROVs were now coming of age.

Real Scientists Virtually Come Aboard

Two years later, during an expedition to the Sea of Cortez, we realized *Medea*/*Jason*'s ultimate application.

When I had first presented the concept to our navy sponsors in 1982, I had said that our new scheme for deep-sea exploration would ultimately allow scientists to participate from their laboratories ashore. My Deep Submergence Laboratory team demonstrated how this would work during the first two Jason Projects but did not realize the goal completely. During our expedition to the Mediterranean Sea in 1989, scientists as well as hundreds of thousands of students and teachers watched our broadcasts live as *Jason* recovered artifacts from

Isis. During the same expedition, many of the same viewers watched another series of broadcasts, when *Jason* investigated an active hydrothermal vent, the first one discovered in the Mediterranean. In both instances, however, their participation was passive. During the second Jason Project, broadcast from Lake Ontario, students ashore actually took control of *Jason* via satellite. But that was largely a test of the concept; it did not involve shore-based scientists.

It was not until 1993 that scientists joined our crew while remaining on shore, fully participating by means of telepresence through long-distance data links. The cruise took place in February and March aboard the *Laney Chouest,* the same ship we had used two years earlier on the Juan de Fuca Ridge. This time we were investigating the Guaymas Basin, in an area known to have active hydrothermal vents. We first surveyed the site with a SEABEAM sonar mounted on the *Laney Chouest* and then deployed *Medea / Jason* near the bottom to make more detailed maps of smaller areas, collecting a set of data on varying scales similar to the data set we had collected at the Juan de Fuca site. Here, in the Guaymas Basin, the area we selected for detailed mapping included an active high-temperature vent and a deposit of polymetallic sulfides.

Once we had gathered and processed this nested set of maps and data, we distributed the information via a satellite Internet link to participating research institutions, such as Woods Hole, the University of Washington, and the University of Rhode Island. This made it possible for scientists on shore to observe two- and three-dimensional displays of *Jason* moving over the bottom. At the same time, they had access to the data being collected, as well as the data gathered earlier, through either a real-time network feed or files they could download from the network.

Researchers then used *Jason* to gather data for their own investigations. Some made real-time observations using real-time video images and digital data streams and then directed our crew at sea to collect biological samples in a certain manner or make high-temperature water measurements. One participant, biologist Cindy Van Dover of Woods Hole, was able to operate *Jason* remotely to conduct a close-up visual inspection of an active vent community complete with clams and tube worms. Another scientist, geophysicist Russ McDuff of the University of Washington, supervised a series of precisely spaced track lines one meter apart, run at various altitudes above a high-temperature vent. Operating *Jason* in automatic mode using real-time displays onshore, he collected conductivity and temperature measurements within a three-dimensional grid. While our expedition was still under way, this scientist left his laboratory and traveled across the country to attend a meeting. Our crew at sea continued making data runs as he had directed, and he continued monitoring the experiment by telephone using a modem and a laptop computer from an airport terminal! As a result, he was able to detect flaws in the instrumentation and data transmittal and direct our seagoing team to make appropriate corrections.

Clearly, this expedition surpassed all the previous *Medea / Jason* demonstrations. It showed that many researchers who could not otherwise join a deep-sea expedition can now virtually "come aboard" anytime, from almost anywhere on shore, to observe, make measurements, collect data, and conduct real-time experiments. The Guaymas Basin cruise in 1993 gave us a brief, imperfect glimpse of a new era in deep-sea exploration. Since then, more and more scientists have begun using *Medea / Jason* to explore the Midocean Ridge—during recent expeditions to the Mid-Atlantic Ridge and the Juan de Fuca Ridge, and a return trip in 1998 to the Guaymas Basin.

Meanwhile, at the Institute for Exploration—my new home base in Mystic, Connecticut—we have been busy planning and designing a new trio of remotely operated vehicles: *Echo,* a side-scan sonar sled for broad surveys and searches; *Argus,* a tethered relay vehicle with optical cameras; and *Hercules,* a neutrally buoyant robot similar to *Jason.* We will inaugurate *Argus* on an expedition to the Black Sea in the summer of 2000, where George Bass—the archeologist who dove in *Asherah* in the 1960s—and others from his group will join us. We will be searching in 7,000-foot-deep waters there for shipwrecks and evidence of very old submerged settlements—flooded, some researchers believe, by a catastrophic deluge 7,600 years ago.

— Leaving the Body Behind

Although we have learned a lot over the past two decades, all expeditions put together have probably investigated less than 1 percent of the seafloor. The exploration of the world's oceans has just begun. Thanks to the emerging technologies of robotics and telecommunications, the next generation of deep-sea explorers may actually see, for the first time, more of the earth's solid surface than all previous generations combined. They will then at last put our knowledge of this ocean planet on a par with that of Mars and of the far side of the moon.

Until recently, humans have been able to enter the realm of eternal darkness only in very small numbers, encased by expensive machines. Now we don't need those diving machines. As a result, exploration should become far more democratic. Although only one person at a time can actually lay hands on the pilot's controls, almost everyone will be able at least to observe the unfolding adventure. Yet right now, in the deep sea, two eras overlap. Robots are sending views from the bottom with laser-light pulses through fibers of glass while humans are still descending in hard little spheres, surrounded by syntactic foam. In the long run those submersibles may be doomed, but I see no reason to rush them into early retirement.

Tethers, however, remain a problem. They snap, they tangle, they restrict. Sometimes they even get robots into trouble. In the early days, deep-ocean explorers could hardly wait to get rid of the bathysphere's tether. Piccard's new bathyscaph achieved that feat. It could take a researcher all the way to the bottom with enough freedom of motion to do some exploring.

Small submersibles vastly expanded that freedom. Now we can cut the ultimate tether—the one that binds our questioning intellect to vulnerable human flesh. Through telepresence, a mind detaches itself from the body's restrictions and enters the abyss with ease, and with lightning-quick fiber optic nerves. As Jacques Cousteau used to say, the ideal means of deep-sea transport would allow us to move "like an angel." Our minds can now go it alone, leaving the body behind. What could be more angelic than that?

A NOTE ON SOURCES

Some passages in this book describe events and scenes in great detail. They contain dialogue, actions, descriptions, and even reconstructions of a person's thoughts or feelings. The information in these passages came from the participants themselves. Often I was one of those participants, and therefore the source. In some instances—on an expedition, say—I did not witness all the action firsthand, but I got the details from others who did.

One scene I describe in some detail—the last minutes inside *Thresher*—obviously had no surviving witnesses. I saw the debris field; I figured out how *Thresher* broke apart and the pieces sank. Other careful investigators, whose knowledge I trust, figured out how the accident occurred. The rest comes from experience—service in the navy, years of diving in submersibles, and knowledge of equipment and procedures. That's how I know, for example, that the hull would have creaked and groaned.

For detailed passages describing events that occurred before my time, participants were still the source. In the early days, virtually everyone who made a notable descent in a deep-diving craft became at least a minor celebrity, and therefore wrote a book. Those books contain the detailed information. (They often elaborate on articles that appeared in the *National Geographic*.) For William Beebe, the main source is *Half Mile Down* (New York: Harcourt Brace, 1934). For Otis Barton, *The World Beneath the Sea* (New York: Crowell, 1953). For

Auguste Piccard, *Earth, Sky, and Sea* (New York: Oxford University Press, 1956). For Jacques Piccard, *Seven Miles Down: The Story of the Bathyscaph* Trieste (New York: Putnam, 1961). For Georges Houot and Pierre Willm, *2000 Fathoms Down* (New York: Dutton, 1955). For George Bass, *Archaeology Beneath the Sea* (New York: Walker and Company, 1975), plus a few memories he shared later through e-mail. For Jacques Cousteau, there are too many sources to name. He published prolifically; the contents of his books, articles, and films overlap. In *The Living Sea* (New York: Harper & Row, 1963) he has a lot to say about the *Soucoupe*.

In addition, Richard Ellis, who has written several books about the ocean, including *Deep Atlantic: Life, Death, and Exploration in the Abyss* (New York: Knopf, 1996), barely missed interviewing Otis Barton (who died in 1992). He did interview members of Barton's family, however, and was kind enough to share what he had learned about the first bathyscaph's designer.

Finally, a couple of books on submersibles stand out as important sources of descriptive detail. Needless to say, I knew a lot about *Alvin* firsthand. Nevertheless, Victoria Kaharl's *Water Baby: The Story of* Alvin (New York: Oxford University Press, 1990), refreshed my memory on many occasions. Over the years, I had heard many of the same stories she heard, sometimes in slightly different versions. The other book is a real treasure. In 1975, R. Frank Busby compiled information on every known submersible constructed by that date in *Manned Submersibles* (Washington, D.C.: Office of the Oceanographer of the Navy, 1976). There is a page on each sub, with its history, description, technical information, and often photographs. If you find it hard to believe that a hundred deep-diving vessels could have been built by the early 1970s, you can look them all up here.

FURTHER READING

People ask me what's the most fabulous discovery I've ever made. And I always say it's been in the library.

—*George Bass*

1931

Beebe, W. A round trip to Davy Jones's Locker. *Natl. Geogr. Mag.* 59:653–778.

———. Deep sea diving (2nd Bermuda oceanographic expedition). *Monaco International Hydrographic Review* 8:245.

1933

Beebe, W. Preliminary account of deep sea dives in the bathysphere with especial reference to one of 2,200 feet. *Proc. Natl. Acad. Sci. USA* 19:178–188.

1934

Beebe, W. A half mile down. *Natl. Geogr. Mag.* 66:661–704.

1952

Sasaki, T. The bathysphere *Kuroshio. J. Sci. Res. Inst. Jpn.* 46(1284):147–161.

1953

Ishida, M. Record of sounds of the squid under the sea. Rept. Div. Surv. Fish. Ground in 1952, Sci. Res. Inst.

Motoda, S. Observation of the motion of plankton under the sea surface. Rept. Div. Surv. Fish. Ground in 1952, Sci. Res. Inst.

Saito, J. On the operation of *Kuroshio,* the undersea observation chamber. Rept. Div. Surv. Fish. Ground in 1952, Sci. Res. Inst.

Suzuki, N., and K. Kato. Studies on suspended materials or marine snow in the sea. *Bull. Fac. Fish. Hokkaido Univ.* 4:132.

Tsujita, T. Studies on naturally occurring suspended organic matter in waters adjacent to Japan. *Rec. Oceanogr. Works Jpn.* 1:94–100.

1954

Cousteau, J. To the depths of the sea by bathyscaphe. *Natl. Geogr. Mag.* 56:67.

Houot, G. S. Two and a half miles down. *Natl. Geogr. Mag.* 56:80.

Houot, G. S., and P. Willm. Le bathyscaphe à 4,050 m au fond de l'océan. *Éditions de Paris* 1.

Nishizawa, S., S. Fukuda, and N. Inoue. Photographic study of suspended matter and plankton in the sea. *Bull. Fac. Fish. Hokkaido Univ.* 5:36–40.

Piccard, A. *Au fond des mers en bathyscaphe.* Paris: Arthaud.

Tamura, T., and N. Inone. Observation of the amount of starfish by *Kuroshio,* the undersea chamber. *Mon. Rept. Hokkaido Region Fish Lab.* 11(9).

1955

Cousteau, J. Diving through an undersea avalanche. *Natl. Geogr. Mag.* 108:538.

Furnestin, J. Une plongée en bathyscaphe. *Rev. Trav. Inst. Peches Mar.* 19:435–442.

Houot, G. S. Le bathyscaphe *F.N.R.S. III* au service de l'exploration des grandes profondeurs. *Deep-Sea Res.* 2:247–249.

Ito, K., and Aoyama. Observation of the seine net operated on the bottom. Rept. Div. Surv. Fish. Ground in 1952, Sci. Res. Inst.

Lagunov, I. I. Experience in underwater observations from a hydrostat. *Rybn. Khoz.* (Moscow) 8:54–57.

Sasa, Y. *Kuroshio,* undersea observation chamber, a new weapon for submarine geological work. *J. Japn. Assoc. Petrol. Technol.* 20(4):1–10.

Sasaki, T., N. Okami, S. Watanabe, and G. Oshiba. Measurements of the angular distribution of submarine daylight. *J. Sci. Res. Inst. Jpn.* 49:103.

Tregouboff, G. Sur l'emploi de la tourelle submersible Galeazzi pour des observations biologiques sous-marines à faibles profondeurs. *Bull. Inst. Oceanogr. Monaco* 1070.

1956

Pérès, J. M., and J. Picard. Nouvelles observations biologiques effectuées avec la bathyscaphe *F.N.R.S. III* et considérations sur le système aphotique de la Méditerranée. *Bull. Inst. Oceanogr. Monaco* 1075.

Tregouboff, G. Prospection biologique sous-marine dans la région de Villefranche-sur-Mer en Juin 1956. *Bull. Inst. Oceanogr. Monaco* 1085:1–24.

1957

Bernard, F. Plancton observé durant trois plongées en bathyscaphe au large de Toulon. *C. R. Acad. Sci. C* 245:1968–1971.

Houot, G. S. Promenades dans les canyons sous-marins. *Rev. Geogr. Phys. et Geol. Deux. Ser.* 1:56–57.

Pérès, J. M., J. Piccard, and M. Ruivo. Résultats de la campagne de recherches du bathyscaphe *F.N.R.S. III,* Organisée par le Centre National de la Recherche Scientifique sur les Côtes du Portugal. *Bull. Inst. Oceanogr. Monaco* 1092.

Piccard, J., and R. S. Dietz. Oceanographic observations *Trieste* (1953–1956). *Deep-Sea Res.* 4:221–229.

1958

Bernard, F. Plancton et benthos observés durant trois plongées en bathyscaphe au large de Toulon: Résultats scientifiques des Campagnes du bathyscaphe *F.N.R.S. III. Ann. Inst. Oceanogr.* 35(235):287–326.

Botteron, G. Étude de sédiment récolté au cours de plongées avec le bathyscaphe *Trieste* au large de Capri. *Bull. Univ. Lausanne* 124.

Brouardel, J. Appareils de prélèvements: Résultats scientifiques des Campagnes du bathyscaphe *F.N.R.S. III,* 1954–1957. *Ann. Inst. Oceanogr. Paris* 235:255–258.

Dietz, R. S. Deep sea research in bathyscaphe *Trieste. New Sci.* 3(74):30–32.

Dietz, R. S., R. V. Lewis, and A. B. Rechnitzer. The bathyscaph. *Sci. Am.* 198(4):27–33.

Fage, L. Les campagnes scientifiques du bathyscaphe *F.N.R.S. III,* 1954–1957: En résultats scientifiques des campagnes du bathyscaphe *F.N.R.S. III,* 1954–1957. *Ann. Inst. Oceanogr. Paris* 35(235):237–243.

Houot, G. S. Le bathyscaphe *F.N.R.S. III* et la recherche scientifique: En résultats scientifiques des campagnes du bathyscaphe *F.N.R.S. III,* 1954–1957. *Ann. Inst. Oceanogr.* 35(235):243–244.

———. Le bathyscaphe et l'exploration des grandes profondeurs. *Colloq. Int. CNRS* 7.

Lewis, R. V. Preliminary considerations of acoustic measurements made by the Underwater Sound Lab. aboard the bathyscaph, *Trieste,* during the summer of 1957. *U.S. Nav. J. Underwater Acoust.* 8:155–157.

Maxwell, A. E. The bathyscaph—A deep-water oceanographic vessel, Part 1: A report on the 1957 scientific investigations with the bathyscaph *Trieste. U.S. Nav. J. Underwater Acoust.* 8:149–154.

Pérès, J. M. Remarques générales sur un ensemble de quinze plongées effectuées avec le bathyscaphe *F.N.R.S. III:* En résultats scientifiques des campagnes du bathyscaphe *F.N.R.S. III,* 1954–1957. *Ann. Inst. Oceanogr. Paris* 35(235):259–285.

———. Trois plongées dans le canyon du Cap Sicie, effectuées avec le bathyscaphe *F.N.R.S. III* de la Marine Nationale. *Bull. Inst. Oceanogr. Monaco* 1115.

Piccard, J. *Le bathyscaphe et les plongées du* Trieste *1953–1957.* Comité pour la Recherche Océanographique au Moyen du Bathyscaphe *Trieste.* Lausanne: CRO.

———. 11,000 Meter unter dem Meeresspiegel. Die Tauchfahrt des Bathyskaphs *Trieste. Dtsch. Hydrogr. Z.* 15:38.

Tregouboff, G. Le bathyscaphe au service de la planctonologie: En résultats scientifiques des campagnes du bathyscaphe *F.N.R.S. III,* 1954–1957. *Ann. Inst. Oceanogr. Paris* 3S(235):327–341.

———. Prospection biologique sous-marine dans la région de Villefranche-sur-Mer au cours de l'année 1957. I. Plongées en bathyscaphe. *Bull. Inst. Oceanogr. Monaco* 1117.

Willm, P. *Le bathyscaphe,* pp. 187–192. Brussels: Centre Belge d'Océanographie et de Recherches Sous-Marines.

———. Noted techniques, par les ingénieurs de genie maritime: En résultats scientifiques des campagnes du bathyscaphe *F.N.R.S. III,* 1954–1957. *Ann. Inst. Oceanogr. Paris* 35(235):245–254.

1959

Dietz, R. S. 1,100 meter dive in the bathyscaphe *Trieste. Limnol. Oceanogr.* 4:94–101.

Disteche, A. pH measurements with a glass electrode withstanding 1,500 kg/cm^2 hydrostatic pressure. *Rev. Sci. Instrum.* 30:474–478.

Jerlov, N. G. Maxima in the vertical distribution of particles in the sea. *Deep-Sea Res.* 5:173–184.

Jerlov, N. G., and J. Piccard. Bathyscaph measurements of daylight penetration into the Mediterranean. *Deep-Sea Res.* 5:201–204.

Orkin, P. A. Deep-sea diving by bathyscaph. *Times Sci. Rev.* 2:10–12.

Pérès, J. M. Deux plongées au large du Japon avec le bathyscaphe français *F.N.R.S. III. Bull. Inst. Oceanogr. Monaco* 1134.

———. Observation en bathyscaphe de l'lnstabilité des Vases bathyales méditerranéennes. *Bull. Rec. Trav. Sta. Mar. d'Endoume* 29(17):3.

Tregouboff, G. Prospection biologique sous-marine dans la région de Villefranche-sur-Mer en mars 1959. *Bull. Inst. Oceanogr. Monaco* 1156.

1960

Dangeard, L. Glissements de vase sous-marine et phénomènes de compaction: Observations faites en bathyscaphe. *C. R. Acad. Sci. C* 251: 2224–2225.

Disteche, A., and M. DuDuisson. Mesures directes de pH aux grandes profondeurs sous-marines. *Bull. Inst. Oceanogr. Monaco* 1174.

Lomask, M., and R. Frasseto. Acoustic measurements in deep water using the bathyscaph. *J. Acoust. Soc. Am.* 32:1028-1033.

Mackenzie, K. V. Formulas for the computation of sound-speed in sea water. *J. Acoust. Soc. Am.* 32:100.

Pérès, J. M. Le bathyscaphe, instrument d'investigation biologique des mers profondes. *Bull. Rec. Trav. Sta. Mar. d'Endoume* 33(20):1725.

———. Observations sur les sédiments à partir du bathyscaphe *F.N.R.S. III* où par photographies profondes. *Bull. Rec. Trav. Sta. Mar. d'Endoume.* 33(20):25–27.

Rechnitzer, A. B. The bathyscaph *Trieste* project. A source of operational experience and insights for planning future deep submergence development. In: Proceedings of the National Academy of Science Conference on Deep Diving Submarines, 12–13 September.

1961

Dangeard, L. À propos des phénomènes sous-marins profonds de glissement et de resédimentation. *Cah. Oceanogr.* 13:68–72.

Mackenzie, K. V. Sound-speed measurements utilizing the bathyscaph *Trieste. J. Acoust. Soc. Am.* 33:1113–1119.

Piccard, J., and R. S. Dietz. *Seven Miles Down.* New York: G. P. Putnam's Sons.

Rechnitzer, A. B., and D. Walsh. The U.S. Navy bathyscaph *Trieste.* In: Proceedings of the 10th Pacific Science Congress.

Tregouboff, G. Prospection biologique sous-marine dans la région de Villefranche-sur-Mer en juillet–août 1960. *Bull. Inst. Oceanogr. Monaco* 1220.

———. Rapport sur les travaux interessants la planctonologie méditerranéenne publiés entre juillet 1958 et octobre 1960. *Rapp. P. V. Reun. Comm. Int. Explor. Sci. Mer Mediterr. Monaco* 16(2):55–65.

———. Rapport sur les travaux relatifs à la planctonologie méditerranéenne publiés entre juillet 1956 et juin 1958. *Rapp. P. V. Reun. Comm. Int. Explor. Sci. Mer Mediterr. Monaco* 15:191–225.

Walsh, D. The future use of deep submersible vehicles. *Bull. Houston Geol. Soc.* 4:12–17.

1962

Bernard, F. Contribution du bathyscaphe à l'étude du plancton: Avantages et incovénients. Contributions to the Symposium on Zooplankton Production. *Rapp. P. V. Reun. Comm. Int. Explor. Sci. Mer Mediterr. Monaco* 158.

Dietz, R. S. The sea's deep scattering layer. *Sci. Am.* 207:44–50.

Disteche, A. Electrochemical measurements at high pressure. *J. Electrochem. Soc.* 109:1084–1092.

LaFond, E. C. Bathyscaph dive 84 (abstract). *J. Geophys. Res.* 67:3573.

———. Deep current measurements with the bathyscaphe *Trieste. Deep-Sea Res.* 9:115–116.

Mackenzie, K. V. *In situ* sound-speed measurements aboard the French bathyscaph *Archimède* at Japan (abstract). *J. Acoust. Soc. Am.* 34:1974.

———. Bathyscaph aids sound-speed measurements. *Bur. Ships J.* 12(May): 33–35.

———. Further remarks on sound-speed measurements aboard the *Trieste. J. Acoust. Soc. Am.* 34:1148–1149.

Magnien, C. Mesures de la célérité des ondes ultrasonorées dans l'eau de mer faites à bord du bathyscaphe *F.N.R.S. III* en Méditerranée en Janvier 1961. *Ann. Geophys.* 18:300–302.

Oulianoff, N. Ripple marks cruissees (rhomboides) el le problème general de fossilisation des rides. *C. R. Acad. Sci. C* 254:148–150.

Tregouboff, G. Prospection biologique sous-marine dans la region de Villefranche-sur-Mer en janvier 1961. *Bull. Inst. Oceanogr. Monaco* 1226.

Walsh, D. The bathyscaph as an acoustic vehicle. *Nav. Res. Rev.,* April, pp. 14–18.

1963

Barham, E. G. Siphonophores and the deep scattering layer. *Science* 140:826–828.

———. The deep scattering layer as observed from bathyscaph *Trieste.* In: Proceedings of the XVI International Zoological Congress, 20–27 August, Vol. 4, pp. 298–300.

Batzler, W. E., and E. G. Barham. Acoustic scattering from a layer of siphonophores (abstract). *J. Acoust. Soc. Am.* 35:792–793.

Delauze, H. Les installations scientifiques du bathyscaphe *Archimède. Bull. Assoc. Fr. Etude Grandes Profond. Ocean.* 2:6–8.

Delauze, H., and J. M. Pérès. Aperçu sur les résultats de la campagne au Japon du bathyscaphe *Archimède. Bull. Rec. Trav. Sta. Mar. d'Endoume* 30(45):3–8.

Dill, R. F. Submarine canyons investigated by bathyscaph and SCUBA diving. In: *Submarine Geology,* 2nd ed., ed. F. P. Shepard, pp. 314, 321, 348. New York: Harper and Row.

Hamilton, E. L. Sediment sound velocity measurements made *in situ* from bathyscaph *Trieste. J. Geophys. Res.* 68:5991–5997.

Moore, D. G. Geological observations from the bathyscaph *Trieste* near the edge of the continental shelf off San Diego, Calif. *Bull. Geol. Soc. Am.* 74:1057–1062.

Okubo, I. Étude chimique de l'eau relevée de la fosse des Kouriles par le bathyscaphe *Archimède. La Mer* 1(1):3–6. (Also *Bull. Hakodate Mar. Obs.* 10:3–6.)

Sasaki, T., K. Ozama, and I. Okubo. On the chemical elements of sea water in the Kuril Trench taken by the bathyscaphe *Archimède. Bull. Hakodate Mar. Obs.* 10:3–6. (Also *La Mer* 1(1):3–6.)

1964

Buffington, E. C. Structural control and precision bathymetry of La Jolla submarine canyon, California. *Mar. Geol.* 1:44–58.

Delauze, H. Généralités sur la fosse de Porto Rico. *Bull. Rec. Trav. Sta. Mar. d'Endoume.* 31(47):139–148.

Inoue, N. An undersea observation vessel, *Kuroshio,* and its photographic apparatus. In: Proceedings of the 10th Pacific Science Congress: Physical Aspects of Light in the Sea, ed. T. E. Tyler, pp. 7–10.

Keach, D. C. Down to *Thresher* by bathyscaph. *Natl. Geogr. Mag.* 125:765–777.

Martin, G. W. *Trieste:* The first ten years. *U.S. Nav. Inst. Proc.* 90(Aug.):52–54.

Pickwell, G. V., E. G. Barham, and J. W. Wilson. Carbon monoxide production by a bathypelagic siphonophore. *Science* 144:860–862.

Reyss, D. Observation faites en Soucoupe plongeante dans deux vallées sous-marines de la mer Catalane: Le rech du Cap et le rech Lacaze-Duthiers. *Bull. Inst. Oceanogr. Monaco* 1308.

Shepard, F. P., J. R. Curray, D. L. Inman, E. A. Murray, E. L. Winterer, and R. F. Dill. Submarine geology by diving saucer. *Science* 145:1042–1046.

Vaissiere, R., and C. Carpine. Compte rendu de plongées en Soucoupe plongeante SP-300. *Bull. Inst. Oceanogr. Monaco* 1314.

1965

Bolgarov, N. SEVER-2 on the ocean bottom. *Teknika-Molodezhi* 1:5.

Dean, J. R. Deep submersibles used in oceanography. *Geogr. J.* 131:70-72.

Dill, R. F. Bathyscaph observations in La Jolla Fan Valley. In: Proceedings of the 39th Annual Meeting of the Society for Economic Paleontologists and Mineralogists, New Orleans.

Guille, A. Éxploration en Soucoupe plongeante Cousteau de l'entrée nord et de la baie de Rosas (Espagne). *Bull. Inst. Oceanogr. Monaco* 1357.

———. Obsérvations faites en Soucoupe plongeante à la limite inferieure d'un fond à *Ophiothrox quinquemalulate* D. Ch. au large de la côte du Roussellon. *Rapp. P. V. Reun. Comm. Inst. Explor. Sci. Mer Mediterr. Monaco* 18(2):115–118.

Karius, R., P. M. Merifield, and D. M. Rosencrantz. Stereo-mapping of underwater terrain from a submarine. In: Proceedings of the Marine Technology Society Symposium on Ocean Science and Engineering, June, pp. 1167–1177.

Kiselev, O. N. Underwater observations from the SEVER-1 bathystat. In: *Razvitiye morskikh podvodnykh issledovaniy.* Okeanogr. Komissii, Akad. Nauk. SSSR. Moscow: Izd-vo Nauka, pp. 21–27.

Moore, D. G. Erosional channel wall in La Jolla Sea-Fan Valley seen from bathyscaph *Trieste II. Bull. Geol. Soc. Am.* 76:385–392.

Pérès, J. M. Aperçu sur les résultats de deux plongées effectuées dans le ravin de Puerto-Rico par le bathyscaphe *Archimède. Deep-Sea Res.* 12:883–891.

Rebeyrol, Y. Les derniers exploits d'*Archimède. Science et Vie* 107(569):104–108.

Reyss, D., and J. Soyer. Étude de deux vallées sous-marines de la mer Catalane (Compte rendu de plongées en Soucoupe plongeante SP-300). *Bull. Inst. Oceanogr. Monaco* 1356. (Also *Rapp. P. V. Reun. Comm. Int. Explor. Sci. Mer Mediterr. Monaco* 18:75–81.)

Sakhalov, I. N. Undersea research vehicles. *Sudostroyenie,* August, pp. 18–19.

Shepard, F. P. Diving saucer descents into submarine canyons. *Trans. N.Y. Acad. Sci.* 2 27(2):292–297.

———. Submarine canyons explored by Cousteau's diving saucer. In: Proceedings of the 17th Symposium of the Colston Research Society, University of Bristol, 5–9 April, pp. 303–311.London: Butterworths.

Theodor, J. Aperçu sur l'histoire de la plongée et de la photographie sous-marine au service de la biologie. *Bull. Lab. Arago Univ. Paris Vie et Milieu* 19:347–350.

Tregouboff, G. Rapport sur les travaux concernant la planctonologie méditerranéenne publiés entre octobre 1960 et septembre 1962, bathyscaphe *F.N.R.S. III. Rapp. P. V. Reun. Comm. Int. Explor. Sci. Mer Mediterr. Monaco* 17:409–411.

1966

Azhazha, V. G., and O. A. Sokolov. *Podvodnaya lodka v nauchnom poiske.* Moscow: Izd-vo Nauka.

Barham, E. G. An unusual pelagic flatfish observed and photographed from a Diving Saucer. *Copeia* 1966(4):865–867.

———. Deep scattering layer migration and composition: Observations from a Diving Saucer. *Science* 151:1399–1403.

Bellaiche, G., and G. Pautot. Quelques observations morphologiques et sédimentologiques effectuées à bord du bathyscaphe *Archimède* au large des Maures et de l'Esterel. *Bull. Soc. Geol. France* 8:769–772.

Borikou, P. A., B. P. Brovko, and P. A. Kapin. Investigation of the ocean employing new submarine vehicles. *Okeanol. Akad. Nauk. SSSR* 6:1113–1120.

Boulouard, C., and H. Delauze. Analyse palynoplantologique de sédiments prélevés par le bathyscaphe *Archimède* dans la fosse de Japon. *Mar. Geol.* 4:461–466.

Church, R. Oceanographic exploration by small manned submersibles. In: Proceedings of the Symposium on Modern Developments in Marine Science, Los Angeles, 21 April, American Institute of Aeronautics and Astronautics, pp. 127–131.

Giermann, G. Phénomènes géologiques dans la bale d'Aspra Spitia (Golfe de Corinthe, Grèce) étudiés à l'aide de la soucoupe plongeante et de la troika. *Bull. Inst. Oceanogr. Monaco* 1364.

———. Tauchkugel Soucoupe plongeante und Photoschlitten Troika: Zwei neue Werkzeuge für die geologishe Unterwasserkartierung. *Dtsch. Hydrogr. Z.* 19:170–177.

Hall, J. B. Operational accomplishments with *Asherah.* In: Proceedings of the Symposium on Modern Developments in Marine Science, Los Angeles, 21 April, American Institute of Astronautics and Aeronautics, pp. 127–131.

Kiselev, O. N. Nekotorye dannye o povedenii treski. In: *Biologicheskiye i Okeanograficheskiye Usloviya Obrazovaniya Promslovykh Skopleniy Ryb.* Vsesoyuznyy Nauchno-Issledovatel'skiy Institut Morskogo Rybnogo Khozyaystvo i Okeanografii. Moscow: Izd-vo Pischevaya promyshlennost', *Trudy,* no. 60, pp. 169–172.

Lagunov, I. I. Nekotorye resul'taty podvodnykh nablyudenii v Barentsevom more. In: *Biologicheskiye i Okeanograficheskiye Usloviya Obrazuvaniya Promslovykh Skopleniy Ryb.* Vsesoyuznyy Nauchno-Issledatel'skiy Institut Morskogo Rybnogo Khozyaystvo i Okeanografii. Moscow: Izd-vo Pischevaya promyshlennost', pp. 161–167.

Neumann, A. C., and E. W. Hull. Half mile down. *Geo-Mar. Technol.* 2(1): 16–27.

Pérès, J. M. Le rôle de la prospection sous-marine autonome dans les recherches de biologie marine et d'océanographie biologique. *Experientia* 22:417–488.

Pinder, A. C. The *Yomiuri. Geo-Mar. Technol.* 2(9):24–25.

Shepard, F. P., and R. F. Dill. *Submarine Canyons and Other Sea Valleys.* Chicago: Rand McNally.

Terry, R. D. *The Deep Submersible.* North Hollywood, Calif.: Western Periodicals.

Zaferman, M. L., and V. I. Znamenskll. Underwater observations from the ship *Tunets. Rybn. Khoz.* (Moscow) 42(5):38–40.

1967

Arnold, H. A. Manned submersibles for research. *Science* 158:84–91.

Barham, E. G., N. J. Ayer, and R. E. Boyce. Macrobenthos of the San Diego Trough: Photographic census and observations from bathyscaph *Trieste. Deep-Sea Res.* 14: 773–784.

Bellaiche, G. Contribution à la connaissance géologique de la fosse du Japon à la suite de trois plongées en bathyscaphe *Archimède. Seances Soc. Geol. Fr.* 7:290.

Buffington, E. C., E. L. Hamilton, and D. E. Moore. Direct measurements of bottom slope, sediment sound velocity and attenuation, and sheer strength from *Deepstar 4000.* In: Proceedings of the 4th U.S. Navy Symposium on Military Oceanography, Vol. 1, pp. 81–90.

Busby, R. F. Undersea penetration by ambient light and visibility. *Science* 158:1178–1179.

Dangeard, L. Observations faites en "soucoupe" plongeante dans le canyon de Planier au large de Marseille. *Bull. Soc. Linnol. Normandie* 10:227–230.

Dill, R. F. Military significance of deeply submerged sea cliffs and rocky terraces on the continental slope. In: Proceedings of the 4th U.S. Navy Symposium on Military Oceanography, Vol. 1, pp. 106–120.

Dmitriyev, A. N. The underwater laboratory "Benthos-300." *Sudostroyeniye* 2.

Gibson, T. G., and J. Schlee. Sediments and fossiliferous rocks from the eastern side of the Tongue of the Ocean, Bahamas. *Deep-Sea Res.* 14:691–702.

High, W. L. The submarine *Pisces* as a fisheries tool. *Comm. Fish. Rev.* 29(4):21–24.

LaFond, E. C. *Deepstar* studies the sea floor interface. *U.S. Nav. Ships Tech. News,* December, pp. 34–37.

———. Movements of benthonic organisms and bottom currents as measured from the bathyscaphe *Trieste.* In: *Deep-Sea Photography,* ed. J. B. Hersey, pp. 295–302. Johns Hopkins Oceanographic Studies No. 3. Baltimore: Johns Hopkins University Press.

Lenz, J., and T. Hjalmar. Tauchbeobachtungen an Plankton und an Echostreuschichten. *Helgol. Wiss. Meeresunters.* 14:534–546.

McMaster, R. L., and L. E. Garrison. A submerged Holocene shoreline near Block Island, Rhode Island. *J. Geol.* 75:335–340.

Manheim, F. T. Evidence for submarine discharge of water on the Atlantic continental slope of the southern U.S., and suggestions for further research. *Trans. N.Y. Acad. Sci.* 2 29:839–853.

Riedl, R. Die Tauchmethode, ihre Aufgaben und Leitungen bei der Erforschung des Litorals; eine kritische Untersuchung. *Helgol. Wiss. Meeresunters.* 15:294–352.

Rucker, J. B., N. T. Stiles, and R. F. Busby. Sea-floor strength observations from the DSRV *Alvin* in the Tongue of the Ocean, Bahamas. *Southeast. Geol.* 8(1):1–7.

Schlee, J. Geology from a deep-diving submersible. *Geotimes* 12(4):10–13.
Shepard, F. P. Submarine canyon origin based on deep-diving vehicle and surface ship operations. *Rev. Geogr. Phys. Geol. Dynam.* 2 9(5):347–356.
Strasburg, D. W., E. C. Jones, and R. T. B. Iversen. Use of small submarine for biological and oceanographic research. *J. Cons. Perm. Int. Explor. Met.* 31:418–426.
Trumbull, J. V. A., and M. J. McCamis. Geological exploration in an east coast submarine canyon from a research submersible. *Science* 158:370–372.
Zarudski, E. F. K. Swordfish rams the *Alvin*. *Oceanus* 13(4):14–18.

1968

Anderson, V. C., and D. D. Lowenstein. Improvements in side-looking sonar for deep vehicles. In: Proceedings of the Society of American Marine Science Institutes Symposium, Miami, Florida, 16–19 January.
Backus, R. H., J. E. Craddock, R. L. Haedrich, D. L. Shores, J. M. Teal, A. S. Wing, G. W. Mead, and W.D. Clarke. *Ceratoscopelus maderensis:* Peculiar sound-scattering layer identified with this myctophid fish. *Science* 160:991–993.
Ball, M. M., P. R. Supko, and A. C. Neumann. Gravity measurements on the floor of the Florida Straits (abstract). *Trans. Am. Geophys. Union* 49(1):196.
Barham, E. G. A window in the sea. *Oceans Mag.* 1:55–60.
Bass, G. F. New tools for undersea archeology. *Natl. Geogr. Mag.* 134:403–423.
Bellaiche, G. Précisions apportées à la connaissance de la pente continentale et de l'abyssale à la suite de trois plongées en bathyscaphe *Archimède*. *Rev. Geogr. Phys. Geol. Dynam.* 10:137–145.
Brock, V. E., and T. C. Chamberlain. A geological and ecological reconnaissance off Western Oahu, Hawaii, principally by means of the research submarine *Asherah*. *Pac. Sci.* 221:373–394.
Church, R., and E. C. Buffington. California black coral. *Oceans Mag.* 1(2):41–44.
Dangeard, L., M. Rioult, J. Blanc, and L. Blanc-Vernet. Résultats de la plongée en soucoupe no. 421 dans la vallée sous-marine de Planier, au large de Marseille (note de géologie marine). *Bull. Inst. Oceanogr. Monaco* 67(1384).
Edwards, R. L., and K. O. Emery. The view from a storied sub: The *Alvin* off Norfolk, Va. *Comm. Fish. Rev.* 30(8–9):48–55.
Emery, K. O. Positions of empty pelecypod valves on the continental shelf. *J. Sed. Petrol.* 38:1264–1269.
Emery, K. O., and D. A. Ross. Topography and sediments of a small area off the continental slope south of Martha's Vineyard. *Deep-Sea Res.* 15:415–422.
Franceschetti, A. P. Deep research vehicle use in oceanography. *U.S. Nav. Ship Syst. Comm. Tech. News* 17(1):2–15.

Gaul, R. D., and W. K. Clarke. Gulfview diving log May 27–June 12, 1967. *Gulf Univ. Res. Corp. Pub.* 106.

Gibson, T. G., J. E. Hazel, and J. F. Mello. Fossiliferous rocks from submarine canyons off northeastern United States. *U.S. Geol. Surv. Prof. Pap.* 600D:222–230.

Good, D. E. Instrumentation for the measurement of physical and chemical properties at the sea floor interface with a deep submersible. In: Proceedings of the 4th Society of American Marine Science Institutes Symposium, Florida, 16–19 January, pp. 252–259.

Kreitner, F. J. Instrumentation required for oceanographic missions using deep submersibles. In: Proceedings of the 4th Society of American Marine Science Institutes Symposium, Florida, 22–26 January.

Mackenzie, K. V. Accurate depth determinations during *Deepstar 4000* dives. *J. Ocean Technol.* 2:61–67.

MacLeish, K. A taxi for the deep frontier. *Natl. Geogr. Mag.* 133:139–150.

Markel, A. L. What has the manned submersible accomplished? *Oceanol. Int.*, July–August, pp. 35–38.

Milliman, J. D., and F. T. Manheim. Observations in deep-scattering layers off Cape Hatteras, U.S.A. *Deep-Sea Res.* 15:505–507.

Monod, T. Un précurseur du bathyscaphe au XVIII siècle: la (lanterne aquatique) de Benoist de Meillet. Premier Congrès International d'Histoire de l'Océanographie, Monaco 1966. *Bull. Inst. Oceanogr. Monaco Spec. Pap.* 2.

Neumann, A. C. The submersible as a scientific instrument. *Oceanol. Int.*, July–August, pp. 39–42.

Neumann, A. C., and M. M. Ball. Submersible observation of sediment movement, bottom currents, and bedrock over the Bahamian and Floridian escarpments of the Straits of Florida (abstract). *Trans. Am. Geophys. Union* 49:196.

Niino, H. Underwater investigation of submarine valleys in Toyama Bay. *J. Mar. Geol.* 4:1–9.

Pelletier, B. R. The submersible *Pisces* feasibility study in the Canadian Arctic. *Mar. Sed.* 4:69–72.

Rainnie, W. O. Adventures of *Alvin*. *Ocean Ind.* 3(5):22–28.

———. The use of deep submersible *Alvin* in oceanography. In: *Selected Papers from the Governor's Conference on Oceanography*, ed. S. Freedgood and M. A. Griffin, pp. 129–143. New York: Rockefeller University Press.

Rebikoff, D. Mosaic photogrammetry survey of the continental shelf off Port Everglades: In: Proceedings of the Marine Technology Society Symposium on Ocean Science and Engineering on the Atlantic Shelf, 19–20 March, Philadelphia, pp. 249–256.

Ross, D. A. Current action in a submarine canyon. *Nature* 218:1242–1244.

Rucker, J. B., N. T. Stiles, and R. F. Busby. Bottom sediment strength observations from the submersible *Alvin*. *Geol. Soc. Am. Spec. Pap.* 115:497–498.

Saila, S. B. A quarter of a mile below the sea in *Deepstar 4000. Maritimes* 12:1–4.

Sasaki, T. *Shinkai,* a deep-diving research vehicle. *Jpn. Electron. Eng.,* October, pp. 52–56.

Schlee, J. S., W. O. Rainnie, M. J. McCamis, V. P. Wilson, and D. M. Owen. Geological observations from DSRV *Alvin. Geol. Soc. Am. Spec. Pap.* 101:452–453.

Thompson, L. G. D. Gravity and magnetic instrumentation systems on deep oceanographic survey vehicles. Geophysical Corporation of America Technical Division Publication TR 68-7G.

Thompson, L. G. D. Gravity meter tests on the deep research vehicle *Aluminaut,* Vieques, Puerto Rico. Geophysical Corporation of America Technical Division Publication TR 68-13G.

University of Michigan. Project SUBMICH: A Report and Appraisal of the Use of a Research Submarine in the Great Lakes. Special Report No. 33. Great Lakes Research Division, University of Michigan.

Waller, R. A., and R. I. Wicklund. Observations from a research submersible: Mating and spawning of the squid, *Dorytheuthis plei. BioScience* 18: 110–111.

Wicklund, R. I. A naturalist's log. *Underwater Nat.* 5:24–27.

———. Observations on the feeding behavior of the false Albacore. *Bull. Am. Litt. Soc.* 5:30–31.

Wigley, R. L. Can submersible vehicles be used effectively in studies of cold water shelf fisheries? *Fish. News Int.* 7:32–34.

Wigley, R. L., and K. O. Emery. Submarine photos of commercial shellfish off northeastern U.S. *Comm. Fish. Rev.* 50(5):24–27.

1969

Boylan, L. Soviet-Bloc submersible development. *Mar. Technol. Soc. J.* 3(2):21–38.

Bullis, H. R., Jr. Calico scallop submarine survey. U.S. Bureau of Commercial Fisheries, Pascagoula, Miss., Cruise Report 9/69.

Drake, L. L., and H. Delauze. Gravity measurements near Greece from the bathyscaphe *Archimède. Ann. Inst. Oceanogr.* 46(1):71–77.

Gonet, O. Étude gravimetrique du Lac Leman à bord de la mesoscaphe *Auguste Piccard. Mat. Geol. Suisse Geophys. Comm. Geotech. Suisse Geophys.* 8.

Hathaway, J. C., and E. T. Degens. Methane-derived marine carbonates of Pleistocene age. *Science* 165:690–692.

Hawkins, L. Visual observations of manganese deposits on the Blake Plateau. U.S. Naval Oceanographic Office Informal Report No. 68-99. (Also *J. Geophys. Res.* 74:7009–7017.)

Milne, A. R. A small research submarine in the arctic. *Arctic* 22:69–70.

Pollio, J. Applications of underwater photography. *Mar. Technol. Soc. J.* 3:65–78.

Pruna, A., and R. L. Hairs. Panoramic reconstruction of the sea floor environment using underwater photography, direct observations, and field sketches. *Mar. Technol. Soc. J.* 5:73–78.

Ross, D. A. Geologic observations in two east coast canyons from DSRV *Alvin*. *Geol. Soc. Am. Spec. Pap.* 121:256.

Ross, D. A., J. V. A. Trumbull, and C. D. Hollister. Geologic observations in Corsair Canyon from DSRV *Alvin*. *Geol. Soc. Am. Spec. Pap.* 121:372.

Sanders, J. E., K. O. Emery, and E. Uchupi. Microtopography of continental shelf by side-scanning sonar. *Bull. Geol. Soc. Am.* 80:561–572.

Tyler, A. V. Observations Made by the St. Andrews Staff during Submersible Operation, 1968. Technical Report, Fisheries Research Board, Canada, No. 102.

Webb, M. *Lamellibrachia barhami*, gen. nov., sp. nov. (Pogonophora), from the northeast Pacific. *Bull. Mar. Sci.* 19:18–47.

1970

Andrews, J. E., F. P. Shepard, and R. J. Hurley. Great Bahama Canyon. *Geol. Soc. Am. Bull.* 81:1061–1078.

Ballard, R. D., and K. O. Emery. *Research Submersibles in Oceanography*. Washington, D.C.: Marine Technology Society.

Emery, K. O., R. D. Ballard, and R. L. Wigley. A dive aboard *Ben Franklin* off West Palm Beach, Florida. *Mar. Technol. Soc. J.* 4(2):7–16.

Franqueville, C. Étude comparative du macroplancton en Méditerranée nord-occidentale par plongées en soucoupe SP 350, et pêches au chalut pélagique. *Mar. Biol.* 5:172–179.

Grice, G. D., and K. Hulsemann. New species of bottom-living calanoid copepods collected in deep water by DSRV *Alvin*. *Bull. Mus. Comp. Zool.* 139(4):185–230.

1971

Ballard, R. D., and Uchupi, E. Geological observations of the Miami Terrace from the submersible *Ben Franklin*. *Mar. Technol. Soc. J.* 5(2):43–48.

Farmanfarmanian, A. Microbial degradation of organic matter in the deep sea. *Science* 171(3972):672–675.

Houot, G. *20 ans de bathyscaphe*. Paris: Arthaud.

1972

Ballard, R. D., and E. Uchupi. Carboniferous and Triassic rifting: A preliminary outline of the tectonic history of the Gulf of Maine. *Geol. Soc. Am. Bull.* 83:2285–2302.

Jannasch, H. W., and C. O. Wirsen. *Alvin* and the sandwich. *Oceanus* 16:20–22.

Richards, A. F., and M. Perlow, Jr. Variability of geotechnical properties of lutite in Wilkinson Basin, Gulf of Maine, as measured in place from submersible *Alvin* (abstract). *Am. Assoc. Pet. Geol. Bull.* 56(3): 647–648.

1973

Perlow, M., Jr., and A. F. Richards. Geotechnical variability measured in place from a small submersible. *Mar. Technol. Soc. J.* 7(4):27–32.

Polloni, P. T., G. T. Rowe, and J. M. Teal. "Biremis blandi" (Polychaeta: Terebellidae), new genus, new species, caught by D.S.R.V. *Alvin* in the Tongue of the Ocean, New Providence, Bahamas (technical report). *Mar. Biol.* 20:170–175.

Turner, R. D. Wood-boring bivalves: Opportunistic species in the deep sea. *Science* 180(4093):1377–1379.

1974

Ballard, R. D., and Uchupi, E. Geology of the Gulf of Maine. *Am. Assoc. Petrol. Geol. Bull.* 58(6):1156–1158.

Bellaiche, G., J. L. Cheminee, J. Francheteau, R. Hekinian, X. Le Pichon, H. D. Needham, and R. D. Ballard. Inner floor of the Rift Valley: First submersible study. *Nature* 250:558–560.

Neumann, A. C. Cementation, sedimentation structure on the flanks of carbonate platform, NW Bahamas. In: *Recent Advances in Carbonate Studies,* abstract volume, ed. L. C. Gerhard and H. G. Multer, pp. 26–30. Special Publication No. 6. Rutherford, N.J.: Fairleigh Dickinson University, West Indies Laboratory.

1975

ARCYANA. Transform fault and rift valley from bathyscaphe and diving saucer. *Science* 190:108–117.

———. Un premier bilan du project FAMOUS. *La Recherche* 53:167–170.

Ballard, R. D. Improving the usefulness of deep-sea photographs with precision tracking. *Oceanus* 18(3):40–43.

———. Photography from a submersible during project FAMOUS. *Oceanus* 18(3):31–39.

Ballard, R. D., W. B. Bryan, J. R. Heirtzler, G. Keller, J. G. Moore, and T. H. van Andel. Manned submersible observations in the FAMOUS area: Mid-Atlantic Ridge. *Science* 190:103–108.

Ballard, R.D., and Uchupi, E. Triassic rift structure in the Gulf of Maine. *Am. Assoc. Petrol. Geol. Bull.* 59(7):1041–1072.

Bellaiche, G., and J. L. Cheminee. l'Opération FAMOUS: Étude detaillée d'une fraction de la dorsale medio-atlantique. *Rev. Geogr. Phys. Geol. Dyn.* 17:209–218.

———. Les formes volcaniques observées pendant l'opération FAMOUS dans le plancher interne du rift medio-atlantique par 36 degrees 50 minutes north. *C. R. Acad. Sci. D* 282:519–522.

Heirtzler, J. R., and W. B. Bryan. The floor of the Mid-Atlantic Rift. *Sci. Am.* 233:79–90.

Rowe, G. T., P. T. Polloni, and R. L. Haedrich. Quantitative biological assessment of the benthic fauna in deep basins of the Gulf of Maine. *J. Fish. Res. Board Can.* 32(10):1805–1812.

1976

Heirtzler, J. R., and J. F. Grassle. Deep-sea research by manned submersible. *Science* 194:294–299.

Hekinian, R., J. G. Moore, and W. B. Bryan. Volcanic rocks and processes of the Mid-Atlantic Ridge rift valley near 36 deg. 49 min. N. *Cont. Miner. Pet.* 58:83–110.

Mullins, H. T., G. W. Lynts, M. M. Ball, and A. C. Neumann. Hydrocarbon potential of northwestern Bahama Platform. *Am. Assoc. Pet. Geol. Bull.* 60(4):700–701.

Phillips, J. D., K. R. Peal, and W. M. Marquet. An integrated approach to seafloor geologic mapping on the Mid-Atlantic Ridge: ANGUS, *Alvin* and sonarray. In: *Oceans '76,* pp. 13–18. Publication 76CH11189. New York: IEEE.

Schlager, W., R. L. Hooke, and N. P. James. Episodic erosion and deposition in the Tongue of the Ocean (Bahamas). *Geol. Soc. Am. Bull.* 87(8):1115–1118.

Wiebe, P. H., S. H. Boyd, and C. Winget. Particulate matter sinking to the deep-sea floor at 2000 meters in the Tongue of the Ocean, Bahamas, with a description of a new sedimentation trap. *J. Mar. Res.* 34(3):341–354.

Wirsen, C. O., and H. W. Jannasch. Decomposition of solid organic materials in the deep sea. *Environ. Sci. Technol.* 10(9):880–886.

1977

ARCYANA. Rocks collected by bathyscaph and diving saucer in the FAMOUS area of the Mid-Atlantic Rift valley: Petrological diversity and structural setting. *Deep-Sea Res.* 24:565–589.

Ballard, R. D. Notes on a major oceanographic find. *Oceanus* 20(3):35–44.

Ballard, R. D., and Moore, J. G. *Photographic Atlas of the Mid-Atlantic Ridge Rift Valley.* New York: Springer-Verlag.

Ballard, R. D., and T. H. van Andel. Morphology and tectonics of the inner rift valley at latitude 36°50′N on the Mid-Atlantic Ridge, *Geol. Soc. Am. Bull.* 88:507–530.

———. Project FAMOUS: Operational techniques and American submersible operations. *Geol. Soc. Am. Bull.* 88:495–506.

Bellaiche, G. Généralités sur les fonds sédimentaires observés par submersibles dans le rift medio-atlantique et la faille transformante. *Bull. Bur. Rech. Geol. Min. (BRGM) Sect. IV* 4:311–317.

———. The French research submersibles and the diving campaigns in the Mediterranean. *Rapp. P. V. Reun. Comm. Int. Explor. Sci. Mer Mediterr. Monaco* 24(6):147–148.

Bryan, W. B., and J. G. Moore. Compositional variations of young basalts in the Mid-Atlantic ridge rift near lat. 36 deg. 49 min. N. *Geol. Soc. Am. Bull.* 88(4):556–570.

Bryan, W. B., and G. Thompson. Basalts from DSDP leg 37 and the FAMOUS area: Compositional and petrogenic comparisons. *Can. J. Earth Sci.* 14:875–885.

Cohen, D. M. Swimming performance of the gadoid fish *Antimora rostrata* at 2400 meters. *Deep-Sea Res.* 24(3):275–277.

Cola, T., and K. D. Turner. Genetic relations of deep sea wood borers. *Bull. Am. Malacol. Union* 43:1-9-25.

Corliss, J. B., and R. D. Ballard. Oasis of life in the cold abyss. *Natl. Geogr. Mag.* 152(4):441–453.

Fehn, U., M. D. Siegel, G. R. Robinson, H. D. Holland, D. L. Williams, A. J. Erickson, and K. E. Green. Deep-water temperatures in the FAMOUS area. *Geol. Soc. Am. Bull.* 88(4):488–494.

Grassle, J. F. Slow recolonisation of deep-sea sediment. *Nature* 265(5595): 618–619.

Grassle, J. F., and J. P. Grassle. Temporal adaptations in sibling species of Capitella. In: *Ecology of Marine Benthos,* ed. B. C. Coull, pp. 177–189. Columbia: University of South Carolina Press.

Haedrich, K. L., and G. T. Rowe. Megafaunal biomass in the deep sea. *Nature* 269(5624):141–142.

Heirtzler, J. R. Detailed structure of Mid-Atlantic rift valley floor. *Tectonophysics* 38:7–10.

Heirtzler, J. R., R. D. Ballard, F. Aumento, W. G. Melson, J. M. Hall, H. Bougault, L. Dmitriev, J. F. Fischer, T. L. Wright, G. A. Miles, R. D. Hyndman, and M. F. J. Flower. Submersible observations at the Hole 332B area: Initial reports of the Deep Sea Drilling Project, covering Leg 37 of the cruises of the drilling vessel *Glomar Challenger,* Rio de Janeiro, Brazil to Dublin, Ireland, May–July 1974. *Init. Rep. Deep Sea Drilling Project* 37:363–365.

Heirtzler, J. R., P. T. Taylor, R. D. Ballard, and R. L. Houghton. A visit to the New England Seamounts. *Am. Sci.* 65:446–472.

Heirtzler, J. R., and T. van Andel. Project FAMOUS: Its origin, programs, and setting. *Geol. Soc. Am. Bull.* 88:481–487.

Houghton, R. L., J. R. Heirtzler, R. D. Ballard, and P. T. Taylor. Submersible observations of the New England seamounts. *Naturwissenchaften* 64:348–355.

Jannasch, H. W., and C. O. Wirsen. Microbial life in the deep sea. *Sci. Am.* 236(6):42–52.

Keller, G. H. The submersible: A unique tool for marine geology. In: *Submersibles and Their Use in Oceanography and Ocean Engineering,* ed. R. A. Geyer, pp. 213–234. New York: Elsevier.

Neumann, A. C., J. W. Kofoed, and G. H. Keller. Lithotherms in the Straits of Florida. *Geology* 5(1):4–10.

Nozaki, Y., J. K. Cochran, K. K. Turekian, and G. Keller. Radiocarbon and ^{210}Pb distribution in submersible-taken deep-sea cores from Project FAMOUS. *Earth Planet. Sci. Lett.* 34(2):167–173.

Ramberg, I. B., and T. H. van Andel. Morphology and tectonic evolution of the rift valley at lat. 36 deg. 30 min. N. Mid-Atlantic Ridge. *Geol. Soc. Am. Bull.* 88(4):577–586.

Turner, R. D. Wood mollusks and deep sea food chains. *Bull. Am. Malacol. Union* 43:13–19.

Uchupi, E., R. D. Ballard, and J. P. Ellis. Continental slope and upper rise off western Nova Scotia and Georges Bank. *Am. Assoc. Petrol. Geol. Bull.* 61(9):1483–1492.

White, W. R., and W. B. Bryan. ^{87}Sr/^{86}Sr, K, Rb, Cs, Sr, Ba, and rare earth geochemistry of basalts from the FAMOUS area. *Geol. Soc. Am. Bull.* 88:571–576.

Wilber, R. J., and A. C. Neumann. Porosity controls in subsea cemented rocks from deep-flank environment of Little Bahama Bank. *Am. Assoc. Pet. Geol. Bull.* 61(5):841.

1978

Aybulatov, N. A., V. P. Brovko, N. N. Grebtsov, Y. Y. Pavlyuchenko. The first oceanographic studies from the submersible *Argus* in the Black Sea. *Oceanol. Acad. Sci. USSR* 18(5):606–609.

Corliss, J. D., R. D. Ballard, K. Crane, J. Dymond, J. D. Edmond, T. H. van Andel, R. P. Von Herzen, and D. L. Williams. Hydrothermal warm springs on the Galapagos Rift. *Science* 203:1073–1083.

Crane, K. Structure and tectonics of the Galapagos inner rift, 86 degrees 10′W. *J. Geol.* 86:715–730.

CYAMEX Scientific Team. Découverte par submersible de sulfures polymétalliques massifs sur la dorsale du Pacifique oriental, par 21 deg. N (Projet RITA). *C. R. Acad. Sci. D* 287:1365–1368.

Grassle, J. F. Diversity and population dynamics of benthic organisms. *Oceanus* 21(1):42–49.

Grassle, J. F., and J. P. Grassle. Life histories and genetic variation in marine invertebrates. In: *Marine Organisms: Genetics, Ecology, and Evolution,* ed. S. Battaglia and J. A. Beardore, pp. 347–364. NATO Conference Series IV, Marine Sciences, Vol. 2. New York: Plenum Press.

Hermes, O. D., R. D. Ballard, and P. O. Banks. Upper Ordovician peralkalic granites from the Gulf of Maine. *Bull. Geol. Soc. Am.* 89(12): 1761–1774.

Jannasch, H. W. Experiments in deep-sea microbiology. *Oceanus* 21(1):50–57.

Jenkins, W. J., J. M. Edmond, and J. B. Corliss. Excess ^{3}He and ^{4}He in Galapagos submarine hydrothermal waters. *Nature* 272(5649):156–158.

Keller, G. H., and F. P. Shepard. Currents and sedimentary processes in submarine canyons off the Northeast United States. In: *Sedimentation in Submarine Canyons, Fans, and Trenches,* ed. D. J. Stanley and G. Kelling, pp. 15–31. Stroudsburg, Pa.: Dowden, Hutchinson & Ross.

Le Pichon, X. Exploration of the ocean beds. *Recherche* 9(88):314–322.

Mullins, H. T. Deep Carbonate Bank Margin Structure and Sedimentation in the Northern Bahamas. Ph.D. thesis, University of North Carolina.

Mullins, H. T., G. W. Lynts, A. C. Neumann, and M. M. Ball. Characteristics of deep Bahama channels in relation to hydrocarbon potential. *Am. Assoc. Pet. Geol. Bull.* 62(4):693–704.

Neumann, A. C. Carbonate slopes. *Am. Assoc. Pet. Geol. Bull.* 62(3):549.

———. Lithotherms in the Straits of Florida (reply). *Geology* 6:7–8.

Ryan, W. B. F., M. B. Cita, E. L. Miller, D. Hanselman, W. D. Nesteroff, B. Hacker, and M. Nibbelink. Bedrock geology in New England submarine canyons. *Oceanol. Acta* 1(2):233–254.

Schlager, W., and N. P. James. Low-magnesium calcite limestones forming at the deep-sea floor, Tongue of the Ocean, Bahamas. *Sedimentology* 25(5):675–702.

Smith, K. L., Jr. Benthic community respiration in the N.W. Atlantic Ocean: In situ measurements from 40 to 5200 m. *Mar. Biol.* 47:337–347.

———. Metabolism of the abyssopelagic rattail *Conyphaenoides armatus* measured in situ. *Nature* 274:362–364.

Smith, K. L., Jr., G. A. White, M. B. Laver, and J. A. Haugsness. Nutrient exchange and oxygen consumption by deep-sea benthic communities: Preliminary in situ measurements. *Limnol. Oceanogr.* 23(5):997–1005.

Turekian, K. K., J. K. Cochran, and D. J. DeMaster. Bioturbation in deep-sea deposits: Rates and consequences. *Oceanus* 21(1):34–41.

1979

Ballard, R. D., W. B. Bryan, K. Davis, J. de Boer, S. DeLong, H. Dick, K. O. Emery, P. J. Fox, M. Hempton, F. Malcolm, W. G. Melson, R. Spydell, J. Stroup, G. Thompson, R. Wright, and E. Uchupi. Geological and geophysical investigation of the Mid-Cayman Rise spreading center: Initial results and observations. In: *Deep Drilling Results in the Atlantic Ocean: Ocean Crust,* ed. M. Talwani, C. B. Harrison, and D. E. Hayes, pp. 66–93. Maurice Ewing Series 2. Washington, D.C.: American Geophysical Union.

Ballard, R. D., and J. F. Grassle. Return to oases of the deep. *Natl. Geogr. Mag.* 156(5):680–705.

Ballard, R. D., R. T. Holcomb, and T. H. van Andel. The Galapagos Rift at 86°W: 3. Sheet flows, collapse pits, and lava lakes of the rift valley, *J. Geophys. Res.* 84(B10):5407–5422.

Bellaiche, G., F. Coumes, F. Irr, F. Roure, and J. R. Vanney. Structure of the French Riviera submarine canyons: Evidence of a polygenetic history from a submersible study ("Cyaligure" campaign). *Mar. Geol.* 31(1–2):M5–M12.

Bryan, W. B. Regional variation and petrogenesis of basalts from the FAMOUS area, Mid-Atlantic Ridge. *J. Pet.* 20:293–325.

Bryan, W. B., G. Thompson, and P. J. Michael. Compositional variation in a steady-state zoned magma chamber: Mid-Atlantic Ridge at 36 degrees 50′N. *Tectonophysics* 55(1–2):63–85.

Corliss, J. B., J. Dymond, L. I. Gordon, J. M. Edmond, R. P. von Herzen, R. D. Ballard, K. Green, D. Williams, A. Bainbridge, K. Crane, and T. H. van Andel. Submarine thermal springs on the Galapagos Rift. *Science* 203(4385):1073–1083.

Corliss, J. B., L. I. Gordon, and J. M. Edmond. Some implications of heat/mass ratios in Galapagos Rift hydrothermal fluids for models of seawater-rock interaction and the formation of oceanic crust. In: *Deep Drilling Results in the Atlantic Ocean: Ocean Crust,* ed. M. Talwani, C. G. Harrison, and D. E. Hayes, pp. 391–402. Maurice Ewing Series 2. Washington, D.C.: American Geophysical Union.

CYAMEX Scientific Team. Massive deep-sea sulfide ore deposits discovered on the East Pacific Rise. *Nature* 277:523–528.

Dayal, R., A. Okubo, I. W. Duedall, and A. Ramamoorthy. Radionuclide redistribution mechanisms at the 2800-m Atlantic nuclear waste disposal site. *Deep-Sea Res.* 26(12A):1329–1345.

Edmond, J. M., J. B. Corliss, and L. I. Gordon. Ridge crest hydrothermal metamorphism at the Galapagos spreading center and reverse weathering. In: *Deep Drilling Results in the Atlantic Ocean: Ocean Crust,* ed. M. Talwani, C. G. Harrison, and D. E. Hayes, pp. 383–390. Maurice Ewing Series 2. Washington, D.C.: American Geophysical Union.

Edmond, J. M., C. Measures, R. E. McDuff, L. H. Chan, R. Collier, B. Grant, L. I. Gordon, and J. B. Corless. Ridge crest hydrothermal activity and the balances of the major and minor elements in the ocean: The Galapagos data. *Earth Planet. Sci. Lett.* 46(1):1–18.

Edmond, J. M., C. Measures, B. Mangum, B. Grant, F. R. Sclater, R. Collier, and A. Hudson. On the formation of metal-rich deposits at ridge crests. *Earth Planet. Sci. Lett.* 46(1):19–30.

Fornari, D. J., D. W. Peterson, J. P. Lockwood, A. Malahoff, and B. C. Heezen. Submarine extension of the southwest rift zone of Mauna Loa Volcano, Hawaii: Visual observations from US Navy Deep Submergence Vehicle DSV *Sea Cliff. Bull. Geol. Soc. Am. Part 1* 90(5):435–443.

Grassle, J. F., C. J. Berg, J. J. Childress, J. P. Grassle, R. R. Hessler, H. J. Jannasch, D. M. Karl, R. A. Lutz, T. J. Mickel, D. C. Rhoads, H. L. Sanders, K. L. Smith, G. N. Somero, R. D. Turner, J. H. Tuttle, P. J. Walsh, and A. J. Williams. Galapagos '79: Initial findings of a deep-sea biological quest. *Oceanus* 22(2):2–10.

Harbison, G. R., and R. B. Campenot. Effects of temperature on the swimming of salps (*Tunicata,* Thaliacea): Implications for vertical migration. *Limnol. Oceanogr.* 24(6):1081–1091.

Jannasch, H. W. Chemosynthetic production of biomass: An idea from a recent oceanographic discovery. *Oceanus* 22(4):59–63.

———. Microbial turnover of organic matter in the deep sea. *BioScience* 29(4):228–232.

Jannasch, H. W., and C. O. Wirsen. Chemosynthetic primary production at East Pacific sea floor spreading centers. *BioScience* 29(10):592–598.

Keller, G. H., D. N. Lambert, and R. H. Bennett. Geotechnical properties of continental slope deposits—Cape Hatteras to Hydrographer Canyon. In: *Geology of Continental Slopes,* ed. L. J. Doyle and O. H. Pilkey, pp. 131–151. Society of Economic Paleontologists and Mineralogists Special Publication 27. Tulsa: Okla.: Society of Economic Paleontologists and Mineralogists.

Le Pichon, X., J. Angelier, J. Boulin, D. Bureau, J. P. Cadet, J. Dercourt, G. Glacon, H. Got, D. Karig, N. Lyberis, J. Mascle, L. E. Ricou, and F. Ghiebault. A study of the Hellenic Trench by submersible observation of active tectonics. *C. R. Acad. Sci. D* 289(16):1225–1228.

Lonsdale, P. Submersible exploration of the Gulf of California plate boundary. In: *Summaries, 5th Meeting of Investigation Centers of Baja California and Scripps Institution of Oceanography.* La Paz, Mexico: Centro de Investigaciones Biologicas de Baja California.

MacIlvaine, J. C., and D. A. Ross. Sedimentary processes on the continental slope of New England. *J. Sediment. Pet.* 49(2):563–574.

Mart, Y., G. A. Auffret, J. M. Auzende, and L. Pastouret. Geological observations from a submersible dive on the western continental slope of the Armorican Massif. *Mar. Geol.* 31(3-4):M61–M68.

Mironov, A. N., and A. M. Podrazhanskiy. Bottom fauna studies during the 21st cruise of the R/V *Dmitriy Mendeleyev* using the *Pisces-7* deep-water vehicle. *Oceanol. Acad. Sci. USSR* 19(5):625–626.

Mullins, H. T., and A. C. Neumann. Deep carbonate bank margin structure and sedimentation in the northern Bahamas. In: *Geology of Continental Slopes*, ed. L. J. Doyle and O. H. Pilkey, pp. 165–192. Society of Economic Paleontologists and Mineralogists Special Publication 27. Tulsa, Okla.: Society of Economic Paleontologists and Mineralogists.

———. Geology of the Miami Terrace and its paleo-oceanographic implications. *Mar. Geol.* 30(3-4):205–232.

———. Seismic facies and depositional processes of modern off-platform carbonate rocks in northern Bahamas. *Am. Assoc. Pet. Geol. Bull.* 63(3): 500.

Polloni, P., R. Haedrich, G. Rowe, and C. H. Clifford. The size-depth relationship in deep ocean animals. *Int. Rev. Ges. Hydrobiol.* 64(1):39–46.

Robins, C. R., D. M. Cohen, and C. H. Robins. The eels, *Anguilla* and *Histiobranchus*, photographed on the floor of the deep Atlantic in the Bahamas. *Bull. Mar. Sci.* 29(3):401–405.

Rowe, G. T. Monitoring with deep submersibles. In: *Monitoring the Marine Environment*, ed. D. Nichols, pp. 75–85. New York: Praeger.

Rowe, G. T., and W. D. Gardner. Sedimentation rates in the slope water of the northwest Atlantic Ocean measured directly with sediment traps. *J. Mar. Res.* 37(3):581–600.

Rowe, G. T., and R. L. Haedrich. The biota and biological processes of the continental slope. In: *Geology of Continental Slopes*, ed. L. J. Doyle and O. H. Pilkey, pp. 49–59. Society of Economic Paleontologists and Mineralogists Special Publication 27. Tulsa: Okla.: Society of Economic Paleontologists and Mineralogists.

Schlager, W., and A. Chermak. Sediment facies of platform-basin transaction, Tongue of the Ocean, Bahamas. In: *Geology of Continental Slopes*, ed. L. J. Doyle and O. H. Pilkey, pp. 193–208. Society of Economic Paleontologists and Mineralogists Special Publication 27. Tulsa, Okla.: Society of Economic Paleontologists and Mineralogists.

Smith, W., J. F. Grassle, and D. Kravitz. Measures of diversity with unbiased estimates. In: *Ecological Diversity in Theory and Practice*, ed. J. F. Grassle et al., pp. 177–191. Fairland, Md.: International Cooperative Publishing.

Stehling, K. R. A Submersible Physics Laboratory Experiment. U.S. NOAA Technical Report OOE1.

Turekian, K. K., J. K. Cochran, and Y. Nozaki. Growth rate of a clam from the Galapagos Rise hot spring field using natural radionuclide ratios. *Nature* 280(5721):385–387.

Van Andel, T. H., and R. D. Ballard. The Galapagos Rift at 86 deg. W. II. Volcanism, structure, and evolution of the rift valley. *J. Geophys. Res.* 84(B10):5390–5406.

Wells, G., W. B. Ryan, and T. H. Pearce. Comparative morphology of ancient and modern pillow lavas. *J. Geol.* 87:427–440.

Westwood, J. Surveying from remote controlled submersibles. *Hydrogr. J.* 16:7–13.

Wiebe, P. H., L. P. Madin, L. R. Haury, G. R. Harbison, and L. M. Philbin. Diel vertical migration by *Salpa aspera* and its potential for large-scale particulate organic matter transport to the deep sea. *Mar. Biol.* 53(3):249–255.

Wimbush, H., and B. Lesht. Current-induced sediment movement in the deep Florida Straits: Critical parameters. *J. Geophys. Res.* 84(C5): 2495–2502.

Zeff, M. L., and R. D. Perkins. Microbial alteration of Bahamian deep sea carbonates. *Sedimentology* 26(2):175–202.

1980

Berg, C. J., and R. D. Turner. Description of living specimens of *Calyptogena magnifica* Boss and Turner with notes on their distribution and ecology. *Malacologia* 20(1):183–185.

Boss, K. J., and R. D. Turner. Giant white clam from the Galapagos Rift, *Calyptogena magnifica* species novum. *Malacologia* 20(1):161–194.

Crane, K., and R. D. Ballard. The Galapagos Rift at 86 deg. W: 4. Structure and morphology of hydrothermal fields and their relationship to the volcanic and tectonic processes of the rift valley. *J. Geophys. Res.* 85(B3): 1443–1454.

CYAMEX Scientific Team. Homogenous basalts from the East Pacific Rise at 21 deg. N: Steady state magma reservoirs at moderately fast spreading centers. *Oceanol. Acta* 3(4):487–503.

Filatova, Z. A. Thermophylic communities of deep-sea bottom fauna in rift zones of the Pacific Ocean. *Oceanol. Acad. Sci. USSR* 20(3):339–341.

Finkel, R. C., J. D. Macdougall, and Y. C. Chung. Sulfide precipitates at 21N on the East Pacific Rise: ^{226}Ra, ^{210}Pb and ^{210}Po. *Geophys. Res. Lett.* 7(9):685–688.

Flood, R. D., and Hollister, C. D. Submersible studies of deep-sea furrows and transverse ripples in cohesive sediments. *Mar. Geol.* 36(1/2):M1–M9.

Grassle, J. F. In situ studies of deep-sea communities. In: *Advanced Concepts in Ocean Measurements,* ed. J. P. Diemer, F. J. Vernberg, and D. Z. Mirkes, pp. 321–332. Belle W. Baruch Library of Marine Sciences No. 10. Columbia: University of South Carolina Press.

Green, K. E. Geothermal Processes at the Galapagos Spreading Center. Ph.D. thesis, MIT/WHOI Joint Program in Oceanography and Oceanographic Engineering. WHOI-80-33.

Hekinian, R., M. Fevier, J. L. Bischoff, P. Picot, and W. C. Shanks. Sulphide deposits from the East Pacific Rise near 21 degrees N. *Science* 207:1433–1444.

Hooke, R. LeB., and W. Schlager. Geomorphic evolution of the Tongue of the Ocean and the Providence Channels, Bahamas. *Mar. Geol.* 35(4): 343–366.

Huchon, P., J. Angelier, X. Le Pichon, N. Lyberis, and L. E. Ricou. Manned submersible observations of the tectonic structures in the Hellenic trenches. *Bull. Soc. Geol. Fr. Suppl.* 22(5):162–166.

Jannasch, H. W., and C. O. Wirsen. Studies on the microbial turnover of organic substances in deep sea sediments. *Coll. Int. CNRS* 293:289–290.

Karl, D. M., C. O. Wirsen, and H. W. Jannasch. Deep-sea primary production at the Galapagos hydrothermal vents. *Science* 201(4437):1345–1347.

Killingley, J. S., W. H. Berger, K. C. Macdonald, and W. A. Newman. $^{18}O/^{16}O$ variations in deep-sea carbonate shells from the rise hydrothermal field. *Nature* 287(5779):218–221.

Lonsdale, P., and R. Batiza. Hyaloclastite and lava flows on young seamounts examined with a submersible. *Geol. Soc. Am. Bull.* 91(9):545–554.

Lonsdale, P., and L. A. Lawver. Immature plate boundary zones studied with a submersible in the Gulf of California. *Geol. Soc. Am. Bull.* 91(9):555–569.

Lupton, J. E., G. P. Klinkhammer, W. R. Normark, R. Haymon, K. C. Macdonald, R. F. Weiss, and H. Craig. Helium-3 and manganese at the 21N East Pacific Rise hydrothermal site. *Earth Planet. Sci. Lett.* 50(1):115–127.

MacDonald, K. C., K. Becker, F. N. Spiess, and R. D. Ballard. Hydrothermal heat flux of the "Black Smoker" vents on the East Pacific Rise. *Earth Planet. Sci. Lett.* 48:1–7.

Mottl, M. J. Submarine hydrothermal ore deposits. *Oceanus* 23(2):18–27.

Mullins, H. T., A. C. Neumann, R. J. Weller, and M. R. Boardman. Nodular carbonate sediment on Bahamian slopes: Possible precursors to nodular limestones. *J. Sed. Petrol.* 50(1):117–131.

Mullins, H. T., A. C. Neumann, R. J. Wilber, A. C. Hine, and S. J. Chinburg. Carbonate sediment drifts in the northern Straits of Florida. *Am. Assoc. Pet. Geol. Bull.* 64(10):1701–1717.

Nikalayev, V. P., A. V. Dmitriyev, N. A. Aybulatov, N. N. Grebtsov, Y.Y. Pavlychenko, O. A. Kuprikov, V. V. Bulyga, and V. G. Yakubenko. Combined oceanographic and historical-archeological investigations with the manned submersible *Argus* and diving techniques. *Oceanol. Acad. Sci. USSR* 20(1):124–125.

Ozmidov, R.V., V. S. Belyaev, Y. V. Nozdrin, A. M. Sagalevich, A. M. Podrazhanskij, and V. I. Fedonov. Measurements of fine-structure of hydrophysical fields and turbulence from the *Pisces* submarine. *Okeanologiya* 20(2):235–241.

Ryall, P. J. C. Drilling rock on the deep-ocean floor. *Mar. Technol. Soc. J.* 14(5):23–27.

Sagalevich, A. M., and A. M. Prodrazhanskij. Experience in using the *Pisces* submersibles for oceanographic studies. *Okeanologiya* 20(4):714–721.

Spiess, F. N. Some origins and perspectives in deep ocean instrumentation development. In: *Oceanography: The Past*, ed. M. Sears and D. Merriman, pp. 226–239. New York: Springer-Verlag.

Spiess, F. N., K. C. MacDonald, T. Atwater, R. Ballard, A. Carranza, D. Cordoba, C. Cox, V. M. Diaz Garcia, J. Francheteau, J. Guerrero, J. Hawkins, R. Haymon, R. Hessler, T. Juteau, M. Kastner, R. Larson, B. Luyendyk, J. D. MacDougall, S. Miller, W. Normark, J. Orcutt, and C. Rangin (RISE Project Group). East Pacific Rise: Hot springs and geophysical experiments. *Science* 207:1421–1432.

Thompson, G., W. B. Bryan, and W. G. Melson. Geological and geophysical investigations of the Mid-Cayman Rise spreading center: Geochemical variation and petrogenesis of basalt glasses. *J. Geol.* 88:41–55.

Valentine, P. C., J. R. Uzmann, and R. A. Cooper. Geology and biology of Oceanographer Submarine Canyon. *Marine Geol.* 38(4):283–312.

Waite, S. W., and S. M. O'Grady. Description of a new submersible filter-pump apparatus for sampling plankton. *Hydrobiologia* 74(2):187–191.

1981

Arnold, M., and S. M. F. Sheppard. East Pacific Rise at latitude 21 degrees N: Isotopic composition and origin of the hydrothermal sulphur. *Earth Planet. Sci. Lett.* 56:148–156.

Arp, A. J., and J. J. Childress. Blood function in the hydrothermal vent vestimentiferan tube worm. *Science* 213:342–344.

———. Functional characteristics of the blood of the deep sea hydrothermal vent brachyuran crab. *Science* 214:559–561.

Ballard, R. D., H. Craig, J. Edmond, M. Einaudi, R. Holcomb, H. D. Holland, C. A. Hopson, B. P. Luyendyk, K. Macdonald, J. Morton, J. Orcutt, and N. Sleep (East Pacific Rise Study Group). Crustal processes of the Mid-Ocean Ridge. *Science* 213(4503):31–40.

Ballard, R. D., J. Francheteau, T. Juteau, C. Rangan, and W. Normark. East Pacific Rise at 21°N: The volcanic, tectonic, and hydrothermal processes of the central axis. *Earth Planet. Sci. Lett.* 55(1):1–10.

Boltovskoy, D. Collection of in situ information. In: *Atlas del Zooplancton del Atlantico Sudoccidental y Metodos de Trabajo con el Zooplancton Marino*, ed.

_____ D. Boltovskoy, pp. 7–14. Mar del Plata, Argentina: Instituto Nacional de Investigacion y Desarrollo Pesquera.

Brevart, O., B. Dupre, and C. J. Allegre. Metallogenesis at spreading centers: Lead isotope systematics for sufides, manganese-rich crust, basalts, and sediments from the CYAMEX and *Alvin* areas (East Pacific Rise). *Econ. Geol.* 76(5):1205–1210.

Bryan, W. B., G. Thompson, and J. N. Ludden. Compositional variation in normal MORB from 22–24 deg. N, Mid-Atlantic Ridge and Kane Fracture Zone. *J. Geophys. Res.* 86:11,815–11,836.

Crane, K., and R. D. Ballard. Volcanic and structure of the FAMOUS-Narrowgate Rift: Evidence for cyclic evolution, AMAR-1. *J. Geophys. Res.* 86(B6):5112–5124.

CYMOR Group. Armorican continental margin: Submersible survey and dredging in the Shamrock Canyon, Bay of Biscay. *C. R. Acad. Sci. II* 292(9): 741–748.

East Pacific Rise Study Group. Crustal processes of the Mid-Ocean Ridge. *Science* 213(4503):31–40.

Felbeck, H., G. N. Somero, and J. J. Childress. Calvin-Benson cycle sulphide oxidation enzymes in animals from sulphide-rich habitats. *Nature* 293(5830):291–293.

Francheteau, J., H. D. Needham, P. Choukroune, T. Juteau, M. Seguret, R. D. Ballard, P. J. Fox, W. R. Normark, A. Carranza, D. Cordoba, J. Guerrero, and C. Rangin (CYAMEX Scientific Team). First manned submersible dives on the East Pacific Rise at 21 deg. N (project RITA): General results. *Mar. Geophys. Res.* 4(4):345–379.

Freeman-Lynde, R. P., M. B. Cita, F. Jadoul, E. L. Miller, and W. B. F. Ryan. Marine geology of the Bahama Escarpment. *Mar. Geol.* 44:119–156.

Haymon, R. M., and M. Kastner. Hot spring deposits on the East Pacific Rise at 21 degrees: Preliminary description of mineralogy and genesis. *Earth Planet. Sci. Lett.* 53:363–381.

Hine, A. C., R. J. Wilber, J. M. Bane, A. C. Neumann, and K. R. Lorenson. Offbank transport of carbonate sands along open, leeward bank margins, northern Bahamas. *Mar. Geol.* 42(1-4):327–348.

Lagabrielle, Y. Observations by submersible of oceanic crust outcropping on Gorringe Bank (SW Portugal): Evidence of underwater processes of disintegration of gabbros. *C. R. Acad. Sci. II* 293(11):827–832.

Lambert, D. N., P. A. Rona, R. H., Bennett, and J. W. Kofoed. Two inclinometers for the direct measurement of seafloor gradient from a submersible. *Geo-Mar. Lett.* 1(1):69–72.

Neumann, A. C. Waulsortian mounds and lithotherms compared. *Am. Assoc. Pet. Geol. Bull.* 65(5):965.

Nikolaev, V. P., V. G. Yakubenko, A. A. Zhil'tsov, O. I. Prokopov, and V. V. Bulyga. Hydrooptical studies with the aid of manned submersible *Argus. Okeanologiya* 21(6):1109–1112.

Pastouret, L., and CYAMEX Group. Submersible structural study of Tamayo Transform Fault: East Pacific Rise, 23 degrees N (Project RITA). *Mar. Geophys. Res.* 4(4):381–401.

Rowe, G. T. The deep-sea ecosystems. In: *Analysis of Marine Ecosystems,* ed. A. R. Longhurst, pp. 235–267. New York: Academic Press.

Ryan, W. B. F., and E. L. Miller. Evidence of a carbonate platform beneath Georges Bank. *Mar. Geol.* 44:213–228.

Sagalevich, A. M. Methods for conducting geological studies with the *Pisces XI* submersible in the Red Sea Expedition. *Oceanol. Acad. Sci. USSR* 21(6):781–783.

Silver, M. W., and A. L. Alldredge. Bathypelagic marine snow: Deep-sea algal and detrital community. *J. Mar. Res.* 39(3):501–530.

Smith, K. L., and M. B. Laver. Respiration of the bathypelagic fish *Cyclothone acclinidens. Mar. Biol.* 61(4):261–266.

Stanley, D. J., H. Sheng, D. N. Lambert, P. A. Rona, D. W. McGrail, and J. S. Jenkyns. Current-influenced depositional provinces, continental margin off Cape Hatteras, identified by petrologic method. *Mar. Geol.* 40(3-4):215–235.

Stroup, J. B., and P. J. Fox. Geologic investigations in the Cayman Trough: Evidence for thin oceanic crust along the Mid-Cayman Rise. *J. Geol.* 89(4):395–420.

Tharp, M. Exploring the deep ocean bottom: Chasms beneath the sea (technical note). *Mar. Geod.* 5(1):67–71.

Turner, R. D. "Drevesnye Ostrovki" i Termal'nye Isotochniki kak Tsentry Vozniknoveniya Glubokovoknykh Soobshchestv s Vysokoj Stepen'yu Raznoobraziya. *Biol. Morya* (Vladivostok) 1:3–10.

Van Andel, T. H. *Science at Sea: Tales of an Old Ocean.* San Francisco: W. H. Freeman. (Originally published in 1977 as part of "The Portable Stanford," a series of books published by the Stanford Alumni Association, Stanford, California, under the title *Tales of an Old Ocean.*)

1982

Ballard, R. D., and J. Francheteau. The relationship between active sulfide depositions and the axial processes of the Mid-Ocean Ridge. *Mar. Technol. Soc. J.* 16(3):8–22.

Ballard, R. D., T. H. van Andel, and R. T. Holcomb. The Galapagos Rift at 86°W: 5. Variations in volcanism, structure, and hydrothermal activity along a 30-km segment of the rift valley. *J. Geophys. Res.* 87:1149–1161.

Bellaiche, G., and M. Hoffert. Sedimentary deposits of the inner floor of the rift and transform fault 'A' (FAMOUS area): In-situ observations by submersibles and sample analyses. *Bull. Inst. Geol. Bassin Aquitaine* 31/32: 225–237.

Childress, J. J., and T. J. Mickel. Oxygen and sulfide consumption rates of the vent clam *Calyptogena pacifica. Mar. Biol. Lett.* 3:73–79.

Christopher, P. R. Techniques in diving and submersibles. *Phil. Trans. R. Soc. London A* 307(1499):311–330.

CYAGOR II Group. Gorringe Bank (Atlantic Ocean, SW Portugal): A section in deep oceanic crust and upper mantle surveyed by submersible. *Ofioliti* 7(2/3):267–278.

Gargett, A. E. Turbulence measurements from a submersible. *Deep-Sea Res.* 29(9A):1141–1158.

Huchon, P., N. Lyberis, J. Angelier, X. Le Pichon, and V. Renard. Tectonics of the Hellenic Trench: A synthesis of Sea-Beam and submersible observations. *Tectonophysics* 86(1/3):69–112.

Lalou, C., and E. Brichet. Ages and implications of East Pacific Rise sulphide deposits at 21 deg. N. *Nature* 300(5888):169–171.

Machado, N., J. N. Ludden, C. Brooks, and G. Thompson. Fine scale isotopic heterogeneity in the Sub-Atlantic Mantle. *Nature* 295:226–228.

Malahoff, A., R. W. Embley, and D. J. J. Fornari. Geomorphology of Norfolk and Washington canyons and the surrounding continental slope and upper rise as observed from DSRV *Alvin*. In: *The Ocean Floor: Bruce Heezen Commemorative Volume,* ed. R. A. Scrutton and M. Talwani, pp. 97–111. New York: John Wiley & Sons.

Mickel, T. J., and J. J. Childress. Effects of pressure and temperature on the EKG and heart rate of the hydrothermal vent crab *Bythograea thermydron* (Brachyura). *Biol. Bull.* 162:70–82.

———. Effects of temperature, pressure and oxygen concentration on the oxygen consumption rate of the hydrothermal vent crab *Bythograea thermydron* (Brachyura). *Physiol. Zool.* 55:199–207.

Monin, A. S. Red Sea submersible research expedition. *Deep-Sea Res.* 29(3A):361–373.

Petit, C. Neptune's forge. *Science* 4(1):60–64.

Rimsky-Korsakov, N. A., and A. A. Schreider. Experience of geological mapping at the summit of the Afanasy Nikitin Seamount from the *Zvouk-4* underwater vehicle. *Okeanologiya* 22(4):660–664.

Rowe, G. T., P. T. Polloni, and R. L. Haedrich. The deep-sea macrobenthos on the continental margin of the Northwest Atlantic Ocean. *Deep-Sea Res.* 29(2A):257–278.

Sborshchikov, I. M., A. M. Sagalevich, A. M. Podrazhanskiy, V. S. Kuzin, A. A. Gorlov, and A. S. Rulev. Geologic and geomorphological observations with the *Pisces* manned submersible apparatus. *Oceanol. Acad. Sci. USSR* 22(3):374–377.

Smith, K. L., Jr. Metabolism of two dominant epibenthic echinoderms measured at bathyal depths in the Santa Catalina Basin. *Mar. Biol.* 72(3):249–256.

———. Zooplankton of a bathyal benthic boundary layer: In-situ rates of oxygen consumption and ammonium excretion. *Limnol. Oceanogr.* 27(3):461–471.

Stakes, D. S., and J. R. O'Neill. Mineralogy and stable isotope geochemistry of hydrothermally altered oceanic rocks. *Earth Planet. Sci. Lett.* 57:285–304.

Stubblefield, W. L., B. A. McGregor, E. B. Forde, D. N. Lambert, and G. F. Merrill. Reconnaissance in DSRV *Alvin* of a "fluvial-like" meander system in Wilmington Canyon and slump features in South Wilmington Canyon. *Geology* 10(1):31–36.

Vanney, J. R., and G. Bellaiche. Mediterranean canyons: Geodynamic processes observed from a submersible. *Oceanus* 8(8):729–751.

Vanney, J. R., and M. Gennesseaux. Geodynamic processes observed from submersibles. *Oceanus* 8(8):611–751.

Wimbush, M., L. Nemeth, and B. Birdsall. Current-induced sediment movement in the deep Florida Straits: Observations. In: *The Dynamic Environment of the Ocean Floor,* ed. K. A. Fanning and F. T. Manheim, pp. 77–94. Lexington, Mass.: D. C. Heath.

1983

Arp, A. J., and J. J. Childress. Sulfide binding by the blood of the hydrothermal vent tube worm *Ristia pachyptila. Science* 219:295–297.

Baguet, F., J. Piccard, B. Christophe, and G. Marechal. Bioluminescence and luminescent fish in the Strait of Messina from the mesoscaph *Forel. Mar. Biol.* 18:298–301.

Ballard, R. D., and J. Francheteau. Geologic processes of the mid-ocean ridge and their relation to sulfide deposition. In: *Hydrothermal Processes at Seafloor Spreading Centers,* ed. P. A. Rona et al., pp. 17–25. New York: Plenum Press.

Bertrand, A. R. V., and P. Willm. Underwater operations: Now and in the future. *Rev. Inst. Fr. Pet.* 38(5):555–574.

Biju-Duval, B., and Y. Morel. Examples of condensed sedimentation on escarpments in the Ionian Sea (eastern Mediterranean): Observations from the submersible *Cyana. Rev. Inst. Fr. Pet.* 38(4):427–438.

Byers, C. D., D. W. Muenow, and M. O. Garcia. Volatiles in basalts and andesites from the Galapagos spreading center, 85 degrees to 86 degrees W. *Geochim. Cosmochim. Acta* 47(9):1551–1558.

Cann, J. R. Magnetic reversals from a submersible (report). *Nature* 305 (5930):100.

Cohen, D. H., and R. Haedrich. The fish fauna of the Galapagos thermal vent region. *Deep-Sea Res.* 30(4A):371–379.

Corliss, J. B. The thermal springs of the Galapagos Rift: Their implications for biology and the chemistry of sea water. In: *Patterns of Evolution in Galapagos Organisms,* ed. R. I. Bowman, M. E. Berson, and A. E. Leviton, pp. 25–31. San Francisco: Pacific Division, American Association for the Advancement of Science.

Desbruyeres, D., F. Gaill, L. Laubier, D. Prieur, and G. H. Rau. Unusual nutrition of the "Pompeii worm" *Alvinella pompejana* (polychaetous annelid) from a hydrothermal vent environment: Scanning electron microscope, transmission electron microscope ^{13}C and ^{15}N evidence. *Mar. Biol.* 75(2-3):201–205.

Dymond, J., K. Cobler, L. Gordon, P. Biscayne, and G. Mathieu. ^{226}Ra and ^{222}Rn contents of Galapagos Rift hydrothermal waters. The importance of low-temperature interactions with crustal rocks. *Earth Planet. Sci. Lett.* 64(3):417–429.

Edmond, J. M., and K. Von Damm. Hot springs on the ocean floor. *Sci. Am.* 248(4):78–93.

Farrow, G. E., J. P. M. Syvitski, and V. Tunnicliffe. Suspended particulate loading on the macrobenthos in a highly turbid fjord: Knight Inlet, British Columbia. *Can. J. Fish. Aquat. Sci.* 40(Suppl. 1):273–288.

Felbeck, H., J. J. Childress, and G. N. Somero. Biochemical interaction between molluscs and their algal and bacterial symbionts. In: *The Mollusca,* Vol. 2: *Environmental Biochemistry and Physiology,* ed. P. Hochachka, pp. 331–358. New York: Academic Press.

Francheteau, J., and R. D. Ballard. The East Pacific Rise near 21 deg. N and 20 deg. S: Inferences for along-strike variability of axial processes of the Mid-Ocean Ridge. *Earth Planet. Sci. Lett.* 64:93–116.

Fricke, H. W., and G. Landmann. On the origin of Red Sea submarine canyons: Observations by submersible. *Naturwissenschaften* 70(4):195–197.

Grassle, J. F. Introduction to the biology of hydrothermal vents. In: *Hydrothermal Processes at Sea Floor Spreading Centers,* ed. P. A. Rona et al., pp. 671–682. New York: Plenum Press.

Hawkes, G. S. The future of atmospheric diving systems and associated manipulator technology, with special reference to a new microsubmersible, *Deep Rover. Mar. Technol. Soc. J.* 17(3):51–60.

Hekinian, R. Intense hydrothermal activity at the axis of the East Pacific Rise near 13 degrees N: Submersible witnesses the growth of a sulfide chimney. *Mar. Geophys. Res.* 6(1):1–14.

Hessler, R. R., and W. M. Smithey, Jr. The distribution and community structure of megafauna at the Galapagos Rift hydrothermal vents. In: *Hydrothermal Processes at Seafloor Spreading Centers,* ed. P. A. Rona et al., pp. 735–770. New York: Plenum Press.

Jannasch, H. W. Microbial processes at deep sea hydrothermal vents. In: *Hydrothermal Processes at Seafloor Spreading Centers,* ed. P. A. Rona et al., pp. 677–709. New York: Plenum Press.

Jones, W. J., J. A. Leigh, F. Mayer, C. R. Woese, and R. S. Wolfe. *Methanococcus jannaschii,* new species. An extremely thermophilic methanogen from a submarine hydrothermal vent. *Arch. Microbiol.* 136(4):254–261.

Juteau, T. Structure and petrology of the Red Sea axial rift at 18 degrees North: Results of the Soviet diving program with a submersible (1980). *Bull. Cent. Rech. Explor. Prod. Elf-Aquitaine* 7(1):217–231.

Karson, J. A., and H. J. B. Dick. Tectonics of ridge-transform intersections at the Kane Fracture Zone. *Mar. Geophys. Res.* 6(1):51–98.

Lalou, C., E. Brichet, C. Jehanno, and H. Perez-Leclaire. Hydrothermal manganese oxide deposits from Galapagos mounds, DSDP Leg 70, Hole 509B and *Alvin* dives 729 and 721. *Earth Planet. Sci. Lett.* 63(1):63–75.

Lilley, M. D., J. A. Baross, and L. I. Gordon. Reduced gases and bacteria in hydrothermal fluids: The Galapagos Spreading Center and 21 deg. N East Pacific Rise. In: *Hydrothermal Processes at Seafloor Spreading Centers,* ed. P. A. Rona et al., pp. 411–449. New York: Plenum Press.

Macdonald, K. C. A geophysical comparison between fast and slow spreading centers: Constraints on magma chamber formation and hydrothermal activity. In: *Hydrothermal Processes at Seafloor Spreading Centers,* ed. P. A. Rona et al., pp. 27–51. New York: Plenum Press.

Mackie, G. O., and Mills, C. E. Use of the *Pisces IV* submersible for zooplankton studies in coastal waters of British Columbia. *Can. J. Fish. Aquat. Sci.* 40(6):763–776.

Neis, K. N., and A. M. Sagalevich. Deep-sea octopuses as detected by a manned vehicle. *Priroda* 11:23–25.

Rimsky-Korsakov, N. A., A. S. Rulev, and V. A. Sychev. Geological and geomorphological studies with submersibles. *Okeanologiya* 23(3):513–516.

Robb, J. M., J. R. Kirby, J. C. Hampson, Jr., P. R. Gibson, and B. Hecker. Furrowed outcrops of Eocene chalk on the lower continental slope, offshore New Jersey. *Geology* 11(3):182–186.

Rowe, G. T., and M. Sibuet. Recent advances in instrumentation in deep-sea biological research. In: *The Sea,* Vol. 8: *Deep Sea Biology,* ed. G. T. Rowe, pp. 81–95. New York: John Wiley & Sons.

Ryan, W. B. F., and J. A. Farre. Potential of radioactive and other waste disposals on the continental margin by natural dispersal processes. In:

Wastes in the Ocean, Vol. 3: *Radioactive Wastes and the Ocean,* ed.P. K. Park et al., pp. 215–236. New York: John Wiley & Sons.

Schell, W. R., and A. E. Nevissi. Radionuclides at the Hudson Canyon disposal site. In: *Wastes in the Ocean,* Vol. 3: *Radioactive Wastes and the Ocean,* ed P. K. Park et al., pp. 183–214. New York: John Wiley & Sons.

Simoneit, B. R. T. Effects of hydrothermal activity on sedimentary organic matter: Guaymas Basin, Gulf of California—Petroleum genesis and protokerogen degradation. In: *Hydrothermal Processes at Seafloor Spreading Centers,* ed. P. A. Rona et al., pp. 451–471. New York: Plenum Press.

Smith, C. R. Enrichment, Disturbance and Deep-Sea Community Structure: The Significance of Large Organic Falls to Bathyal Benthos in the Santa Catalina Basin. Ph.D. thesis, University of California, San Diego. University Microfilms International, Ann Arbor, Mich., No. 1984-3592.

Smith, C. R., and S. C. Hamilton. Epibenthic megafauna of a bathyal basin off southern California: Patterns of abundance, biomass, and dispersion. *Deep-Sea Res.* 30(9A):907–928.

Smith, K. L., Jr. Metabolism of two dominant epibenthic echinoderms measured at bathyal depths in the Santa Catalina Basin, Pacific Ocean. *Mar. Biol.* 72(3):249–256.

Smith, K. L., and N. O. Brown. Oxygen consumption of pelagic juveniles and demersal adults of the deep-sea fish *Sebastolobus altivelis,* measured at depth. *Mar. Biol.* 76(3):325–332.

Van Wagoner, N. A. Magnetic properties of three segments of the Mid-Atlantic Ridge at 37 degrees N: FAMOUS, Narrowgate, and AMAR: AMAR 2. *J. Geophys. Res.* 88(B6):5065–5082.

Von Damm, K. L. Chemistry of Submarine Hydrothermal Solutions at 21 deg. North, East Pacific Rise and Guaymas Basin, Gulf of California. Ph.D. thesis, Woods Hole Oceanographic Institution Technical Report No. WHOI-84-3.

Welhan, J. A., and H. Craig. Methane, hydrogen and helium in hydrothermal fluids at 21 deg. N on the East Pacific Rise. In: *Hydrothermal Processes at Seafloor Spreading Centers,* ed. P. A. Rona et al., pp. 391–409. New York: Plenum Press.

1984

Arp, A. J., J. J. Childress, and C. R. Fisher, Jr. Metabolic and blood gas transport characteristics of the hydrothermal vent bivalve *Calyptogena magnifica. Physiol. Zool.* 57:648–662.

Ballard, R. D. The exploits of *Alvin* and ANGUS: Exploring the East Pacific Rise. *Oceanus* 27(3):7–14.

Ballard, R. D., R. Hekinian, and J. Francheteau. Geological setting of hydrothermal activity at 12°50′N on the East Pacific Rise: A submersible study. *Earth Planet. Sci. Lett.* 69:176–186.

Batiza, R., D. J. Fornari, D. A. Vanko, and P. Lonsdale. Craters, calderas, and hyaloclastites on young Pacific seamounts. *J. Geophys. Res.* 89(B10): 8371–8390.

Biggs, D. C., D. E. Smith, R. R. Bidigare, and M. A. Johnson. In situ estimation of the population density of gelatinous planktivores in Gulf of Mexico surface waters. *Mem. Univ. Newfoundland Occ. Pap. Biol.* 9:17–34.

Bougault, H., V. Renard, and R. Hekinian. Oceanic hydrothermal activity: Geocyarise cruise, January-March. Paper presented at the Second International Seminar on Offshore Mineral Resources, Offshore Prospecting and Mining Problems: Current Status and Future Developments, March 19–23, 1984, Centre Océanologique de Bretagne, Brest, France. Pp. 497–505.

Broadus, J. M., and R. E. Bowen. Polymetallic sulfides and policy spheres. *Oceanus* 27(3):26–31.

Childress, J. J., A. J. Arp, and C. R. Fisher, Jr. Metabolic and blood characteristics of the hydrothermal vent tube worm *Riftia pachypitila. Mar. Biol.* 83:109–124.

Choukroune, P., J. Francheteau, and R. Hekinian. Tectonics of the East Pacific Rise near 12 degrees 50′N: A submersible study. *Earth Planet. Sci. Lett.* 68(1):115–127.

CYAMAZ Group. Preliminary results of diving campaign CYAMAZ on Mazagan escarpment (El Jadida–west Morocco). *Bull. Soc. Geol. Fr.* 26(6):1069–1075.

Edmond, J. M. The geochemistry of ridge crest hot springs. *Oceanus* 27(3):15–19.

Fedorov, V. V., and V. E. Ivanov. Geomorphology of the Nazca Ridge seamounts. *Prom. Okeanogr. Isseldov. Produkt. Morej Okeanov.:* 178–189.

Fisher, C. R., Jr., and J. J. Childress. Substrate oxidation by trophosome tissue from *Riftia pachypitila* Jones (Phylum Pogonophora). *Mar. Biol. Lett.* 5:171–184.

Flemming, N. C., ed. Divers, Submersibles and Marine Science: Selected Papers Presented to the Joint Oceanographic Assembly of the Scientific Committee on Oceanic Research Held at Halifax, Nova Scotia, Canada, August 1982. *Mem. Univ. Newfoundland Occ. Pap. Biol.* 9.

Francis, T. J. G. Scientific research on the ocean floor. *Underwater Technol.* 11(1):30.

Gallo, D. G., W. S. F. Kidd, P. J. Fox, J. A. Karson, K. Macdonald, K. Crane, P. Choukroune, M. Seguret, R. Moody, and K. Kastens. Tectonics at the intersection of the East Pacific Rise with Tamayo Transform Fault. *Mar. Geophys. Res.* 6(2):159–185.

Gente, P., J. M. Auzende, H. Bougault, and M. Voisset. Morpho-tectonic variations on the East Pacific Rise axis between 11 degrees N and 13 degrees N: Preliminary results of Geocyarise 2 cruise. *C. R. Acad. Sci. II* 299(19): 1337–1342.

Grassle, J. F. Animals in the soft sediment near the hydrothermal vents. *Oceanus* 27(3):63–66.

Gross, G. M. Introduction (special issue on deep-sea hot springs and cold seeps). *Oceanus* 27(3):2–6.

Hekinian, R. Submersible study of the East Pacific Rise (report). *Nature* 311(5987):606.

Hickman, C. S. A new archaeogastropod (Rhipidoglossa, Trochacea) from hydrothermal vents on the East Pacific Rise. *Zool. Scr.* 13(1):19–25.

Jannasch, H. W. Chemosynthesis: The nutritional basis for life at deep-sea vents. *Oceanus* 27(3):73–78.

Jones, M. L. The giant tube worms. *Oceanus* 27(3):47–52.

Karl, D. M., D. Burns, K. Orrett, and H. W. Jannasch. Themophilic microbial activity in samples from deep sea hydrothermal vents. *Mar. Biol. Lett.* 5:227–231.

Karson, J. A., P. J. Fox, H. Sloan, K. T. Crane, W. S. F. Kidd, E. Bonatti, J. B. Stroup, D. J. Fornari, D. Elthon, P. Hamlyn, J. F. Casey, D. G. Gallo, D. Needham, and R. Sartori. The geology of the Oceanographer Transform: The ridge-transform intersection. *Mar. Geophys. Res.* 6(2):109–141.

Kerr, A. J., and D. F. Dinn. The use of robots in hydrography. *Int. Hydrogr. Rev.* 62(1):41–52.

Lonsdale, P. Hot vents and hydrocarbon seeps in the Sea of Cortez. *Oceanus* 27(3):21–24.

Luyendyk, B. P. On-bottom gravity profile across the East Pacific Rise crest at 21° north. *Geophysics* 49(12):2166–2177.

Main, J., and G. I. Sangster. Observations on the reactions of fish to fishing gear using a towed wet submersible with underwater television. *Progr. Underwater Sci.* 9:99–114.

Messing, C. G. Brooding and paedomorphosis in the deep water feather star *Comatilia iridometriformis* (Echinodermata: Crinoidea). *Mar. Biol.* 80(1): 83–91.

Pastouret, L., et al. (Groupe CYMOR II). Morphology of the Bay of Biscay northern margin: Observations from the submersible *Cyana*. *Bull. Soc. Geol. Fr.* 26(1):81–92.

Peters, R. A., and F. H. Brockett. Application of ocean mining technology to the recovery of mill tailings in deep water. *Mar. Min.* 5(1):75–86.

Pike, D. A gentle touch for the ocean floor (*Deep Rover*). *New Sci.* 101 (1396):16–17.

Ray, G. C. Submersible systems for ecological research in polar regions. *Mar. Technol. Soc. J.* 18(1):54–60.

Rona, P. A., G. Thompson, M. J. Mottl, J. A. Karson, W. J. Jenkins, D. Graham, M. Mallette, K. Von Damm, and J. M. Edmond. Hydrothermal activity at the TAG hydrothermal field, Mid-Atlantic Ridge crest at 26 deg. N. *J. Geophys. Res.* 89:11,365–11,379.

Ryan, P. R. Odyssey to 11 degrees North. *Oceanus* 27(3):34–43.

Stetten, G. D. *Alvin*'s memory. *Oceanus* 27(3):44–46.

Turner, K. D., and K. A. Lutz. Growth and distribution of mollusks at deep-sea vents and seeps. *Oceanus* 27(3):54–62.

Volkov, A. I., B.F. Kurianov, and A. M. Sagalevich. Measurement of the low-frequency noise characteristics of the ocean using a manned submersible. *Dokl. Akad. Nauk SSSR* 276(4):950–953.

Walker, J. A., P. J. C. Ryall, M. Zentilli, I. L. Gibson, and J. Dostal. The origin of compositional variation in basalts recovered by submersible drill from Mount Glooscap, Mid-Atlantic Ridge at 36°25′N. *Can. J. Earth Sci.* 21(8):934–948.

Youngbluth, M. J. Manned submersibles and sophisticated instrumentation: Tools for oceanographic research. In: *Proceedings of SUBTECH 1983 Symposium*, pp. 335–344. London: Society of Underwater Technology.

———. Water column ecology: In situ observations of marine zooplankton from a manned submersible. *Mem. Univ. Newfoundland Occ. Pap. Biol.* 9:45–57.

1985

Amitani, Y., T. Tsuchiya, and A. Shimura. Optional scientific research tools for *Shinkai 2000*. *Tech. Rep. Jpn. Mar. Sci. Technol. Ctr.*, Special Issue, August, pp. 65–66.

Arnold, A. J., F. d'Escrivan, and W. C. Parker. Predation and avoidance responses in the foraminifera of the Galapagos hydrothermal mounds. *J. Foraminiferal Res.* 15(1):38–42.

Auzende, J. M. Le submersible SP 3000 *Cyana* et l'exploration géologique. *Bull. Inst. Oceanogr. Monaco* N.S. 4:153–162.

Barrie, J. V., and W. T. Collins. Submersible investigations of the Grand Banks. *C-CORE News* 10(1):2–3.

Bellaiche, G. Les plongées en submersibles dans les canyons sous-marins de la Méditerranée nord-occidentale: Une illustration photographique. *Rapp. P. V. Reun. Comm. Int. Explor. Sci. Mer Mediterr. Monaco* 29(2):59–60.

———. Les plongées en submersibles dans les canyons sous-marins de la Méditerranée occidentale. *Bull. Inst. Oceanogr. Monaco* N.S. 4:159–162.

Boillot, G. Exploration by submersible of the northwestern Iberian margin. *Bull. Soc. Geol. Fr.* 8(1):89–102.

Chase, R. L., et al. (Canadian-American Seamount Expedition). Hydrothermal vents on an axis seamount of the Juan de Fuca Ridge. *Nature* 313(5999):212–214.

Collins, W. T., and J. V. Barrie. Preliminary submersible observations of an Iceberg Pockmark on the Grand Banks of Newfoundland. C-CORE Publication No. 85-12. Cambridge: Scott Polar Research Institute.

Ebara, S., and O. Asaoka. Direct or optical observations for vertical distribution of suspended material in sea water by the use of the research submersible *Shinkai 2000*. *Tech. Rep. Jpn. Mar. Sci. Technol. Ctr.*, Special Issue, August, pp. 7–16.

Ellis, D. V., and C. Heim. Submersible surveys of benthos near a turbidity cloud. *Mar. Pollut. Bull.* 16(5):197–203.

Fox, P. J., R. H. Moody, J. A. Karson, E. Bonatti, W. S. F. Kidd, K. Crane, D. G. Gallo, J. B. Stroup, D. J. Fornari, D. Elthon, P. Hamlyn, J. F. Casey, D. Needham, and R. Sartori. The geology of the Oceanographer Transform:The transform domain. *Mar. Geophys. Res.* 7(3):329–358.

Francis, T. J. G. Resistivity measurements of an ocean floor sulphide mineral deposit from the submersible *Cyana*. *Mar. Geophys. Res.* 7(3):419–438.

Freeman-Lynde, R. P., and W. B. F. Ryan. Erosional modification of Bahama Escarpment. *Geol. Soc. Am. Bull.* 96(4):481–494.

Fricke, H., and D. Meischner. Depth limits of Bermudan scleractinian corals: A submersible survey. *Mar. Biol.* 88(2):175–187.

Giermann, G. Les submersibles de recherche en géologie marine. *Bull. Inst. Oceanogr. Monaco* N.S. 4:75–83.

Hashimoto, J., and H. Hotta. An attempt of density estimation of megalo-epibenthos by the deep towed TV system and the deep sea research submersible *Shinkai 2000*. *Tech. Rep. Jpn. Mar. Sci. Technol. Ctr.*, Special Issue, August, pp. 23–35.

Hekinian, R., J. M. Auzende, J. Francheteau, P. Gente, W. B. F. Ryan, and E. S. Kappel. Offset spreading centers near 12 degrees 53'N on the East Pacific Rise: Submersible observations and composition of the volcanics. *Mar. Geophys. Res.* 7(3):359–377.

Hekinian, R., and Y. Fouquet. Volcanism and metallogenesis of axial and off-axial structures on the East Pacific Rise near 13 degrees N. *Econ. Geol.* 80:221–249.

Hotta, H., H. Momma, T. Tanaka, K. Ohtsuka, J. Hashimoto, and K. Midorikawa. Investigations on the small scale topography in the western part of the Sagamiwan Bay off the Izu Peninsula. *Tech. Rep. Jpn. Mar. Sci. Technol. Ctr.*, Special Issue, August, pp. 73–82.

Ito, K. Observations on the behaviour and ecology of the pink crab by the cruise of research submersible *Shinkai 2000*. *Tech. Rep. Jpn. Mar. Sci. Technol. Ctr.*, Special Issue, August, pp. 1–6.

Jaffrezo, M. Biostratigraphy of the Jurassic limestones of the CYAMAZ diving campaign. In: *Cyamaz Cruise 1982: Submersible Cyana Studies of the Mazagan Escarpment, Moroccan Continental Margin, September 15–October 15, 1982,* No. 5, ed. J. M. Auzende and U. von Rad, pp. 89–92. Montrouge, France: Gaulthier-Villars.

Jones, M. L., and C. F. Bright. Dive data of certain submersibles at hydrothermal and other sites. *Bull. Biol. Soc. Wash.* 6:539–545.

Kikuchi, S. Topographic and geological survey for Yamada Canyon off Sanriku with *Shinkai 2000. Tech. Rep. Jpn. Mar. Sci. Technol. Ctr.,* Special Issue, August, pp. 89–100.

Kuznetsov, A. Some results of geological investigations of Reykjanes Ridge with submersibles. *Okeanologiya* 25(1):100–107.

Kuznetsov, A. P., Y. A. Bogdanov, A. M. Sagalevich, I. M. Sborshchikov, A. M. Al'Mukhamedov, M. I. Kuz'min, A. M. Podrazhanskiy, A. A. Gorlov. Some results of geological investigation of the Reykjanes Ridge from submersibles. *Oceanology* 25(1):77–82.

Laubier, L., D. Reyss, and M. Sibuet. L'utilisation des sous-marins habités pour l'étude de l'écosystème abyssal. *Bull. Inst. Oceanogr. Monaco* N.S. 4:107–129.

Le Pichon, X. First results of the test dives of the French submersible *Nautile* in the Puerto Rico Trench (Greater Antilles). *C. R. Acad. Sci. II* 301(10): 743–749.

Macdonald, K. C., and B. P. Luyendyk. Investigation of faulting and abyssal hill formation on the flanks of the East Pacific Rise (21 degrees N) using *Alvin. Mar. Geophys. Res.* 7(4):515–535.

Mackie, G. O. Midwater macroplankton of British Columbia studied by submersible *Pisces IV. J. Plankton Res.* 7(6):753–777.

Monin, A. S., and A. P. Lisitsyn. *Underwater Geological Studies from Manned Submersibles.* Moscow: Nauka.

Murav'ev, V. B., and I. P. Kanaeva. A study of behaviour of fishable zooplankters from a manned submersible. *Rybn. Khoz.* (Moscow) 4:34–35.

Nikolayev, V. P., V. V. Bulyga, V. A. Popov, J. Foyo, and R. Claro. Geological-geomorphological and biological research in the Caribbean Sea with submersibles (third cruise of the R/V *Rift,* August 14–December 13, 1983). *Oceanology* 25(5):682–684.

Osborn, T. R., and R. G. Lueck. Turbulence measurements with a submarine. *J. Phys. Oceanogr.* 15(11):1502–1520.

Otsuka, K., and N. Niitsuma. Sedimentary geology and tectonic features of Suruga Trough, off Matsuzaki. The results of Dive 86 by the submersible *Shinkai 2000. Tech. Rep. Jpn. Mar. Sci. Technol. Ctr.,* Special Issue, August, pp. 45–57.

Renard, V., R. Hekinian, J. Francheteau, R. D. Ballard, and H. Backer. Submersible observations at the axis of the ultra-fast-spreading East Pacific Rise (17 degrees 30′ to 21 degrees 30′S). *Earth Planet. Sci. Lett.* 75(4):339–353.

Richards, L. J., and A. J. Cass. Transect counts of rockfish in the Strait of Georgia from the submersible *Pisces IV,* October and November 1984. *Can. Data Rpt. Fish. Aquatic Sci.* 511.

Sagalevich, M. A., M. I. Kuzmin, and A. A. Aksenvo. A method of underwater geological investigation in the rift zone of the Reykjanes Ridge. *Oceanology* 25(1):120–123.

Smith, C. R. Food for the deep sea: Utilization, dispersal, and flux of nekton fall at the Santa Catalina Basin floor. *Deep-Sea Res.* 32(4A):417–442.

Suchanek, T. H., S. L. Williams, J. C. Ogden, D. K. Hubbard, and I. P. Gill. Utilization of shallow-water seagrass detritus by Caribbean deep-sea macrofauna: $\delta\,^{13}C$ evidence. *Deep-Sea Res.* 82(2A):201–214.

Vanney, J. R., F. Rojouan, D. Temine, J. A. Malod, G. Boillot, R. Capdevila, M. Cousin, F. Gonzalez-Loreido, C. Lepvrier, et al. Geomorphological observations during dives at the north-west of the Iberic Peninsula. *Bull. Soc. Geol. Fr. 8 Ser.* 1(2):153–159.

Vecchione, M., and G. R. Gaston. In situ observations on the small-scale distribution of juvenile squids (Cephalopoda: Loliginidae) on the Northwest Florida shelf. Biology and distribution of early juvenile cephalopods. *Vie Milieu* 35(3–4):231–235.

Yanagi, T., and T. Hiraiwa. Investigation of sea bottom gravity. *Tech. Rep. Jpn. Mar. Sci. Technol. Ctr.:* 101–106.

1986

Amitani, Y., T. Tsuchiya, and T. Nakanishi. Study of the underwater telephone for the 6500m deep-submergence research vehicle. *Tech. Rep. Jpn. Mar. Sci. Technol. Ctr.* 17:45–51.

Ballard, R. D. A long last look at *Titanic. Natl. Geogr. Mag.* 170(6):698–705.

Barrie, J. V., W. T. Collins, J. I. Clark, C. F. M. Lewis, and D. R. Parrott. Submersible observations and origin of an iceberg pit on the Grand Banks of Newfoundland. *Geol. Surv. Can. Pap.* 86-1A:251–258.

Boillot, G. Basaltic and ultramafic seafloors along a passive continental margin. Preliminary results of the Galinaute cruise (diving to the west of Spain by the submersible *Nautile*). *C. R. Acad. Sci. II* 303(19):1719–1724.

Brownlee, S. Explorers of dark frontiers. *Discover* 7(2):60–67.

Bruce, A. J. *Periclimenes milleri* new species, a bathyal echinoid-associated pontoniine shrimp from the Bahamas. *Bull. Mar. Sci.* 39(3):637–645.

Carranza-Edwards, A., et al. Sulfuros metalicos submarinos al sur de la peninsula de Baja California. *An. Inst. Cienc. Mar. Limnol. Univ. Nac. Auton. Mex.* 13(1):287–296.

Chestnov, S. V. Utilisation of the towed submersible *Thetis* in the Northern basin. In: *Underwater Fisheries Research,* ed.G. P. Nizovtsev et al., pp. 136–142. Murmansk: PINRO.

Choukroune, P., B. Auvray, J. Francheteau, J. C. Lepine, F. Arthaud, J. P. Brun, J. M. Auzende, B. Sichler, and Y. Khobar. Tectonics of the westernmost Gulf of Aden and the Gulf of Tadjoura from submersible observations. *Nature* 319(6052):396–399.

Fabris, J. G., K. A. Smith, J. E. Atack, G. Hefter, and A. L. Kilpatrick. Submersible integrating water sampler for heavy metals. *Water Res.* 20(11):1393–1396.

Galkin, S. V., L. I. Moskalev, and A. M. Podruzhanskiy. Investigation of bottom fauna with the *Pisces* manned submersible (9th cruise of the R/V *Mstislav Keldysh*). *Oceanology* 26(2):262–263.

Gennesseaux, M. *Cyana* dives: Results on tilted continental blocks and Vavilov volcano (Tyrrhenian Sea). *C. R. Acad. Sci. II* 302(12): 785–792.

Gente, P., J. M. Auzende, V. Renard, Y. Fouquet, and D. Bideau. Detailed geological mapping by submersible of the East Pacific Rise axial graben near 13N. *Earth Planet. Sci. Lett.* 78(2/3):224–236.

Gowing, M. M., and K. F. Wishner. Trophic relationships of deep-sea calanoid copepods from the benthic boundary layer of the Santa Catalina Basin, California. *Deep-Sea Res.* 33(7A):939–961.

Grant, A. C., E. M. Levy, K. Lee, and J. D. Moffat. *Pisces IV* research submersible finds oil on Baffin Shelf. *Curr. Res. Geol. Surv. Can. A*: 65–69.

Haywood, R. M. Acquisition of a micro scale photographic survey using an autonomous submersible. In: *Oceans '86,* Vol. 5, pp. 1423–1426. New York: IEEE/MTS.

Hirsch, B., and Y. Cardozo. Submarines for everyone. *Oceans* 19(2):47–51.

Kulm, L. D., E. Suess, J. C. Moore, B. Carson, B. T. Lewis, S. D. Ritger, D. C. Kadko, T. M. Thornburg, R. W. Embley, et al. Oregon subduction zone: Venting, fauna, and carbonates. *Science* 231(4738):561–566.

Lisitsyn, A. P., Y. A. Bogdanov, L. P. Zonenshayn, M. I. Kuz'min, and A. M. Sagalevich. Structure of the Tadjoura rift, Gulf of Aden, as shown by surveys with manned submersibles. *Dokl. USSR Acad. Sci. Earth Sci. Sect.* 279:103–106.

McConachy, T. F., R. D. Ballard, M. J. Mottl, and R. P. Von Herzen. Geologic form and setting of a hydrothermal vent field at lat 10 degrees 56′N, East Pacific Rise: A detailed study using *Angus* and *Alvin. Geology* 14(4): 295–298.

Metivier, B., T. Okutani, and S. Ohta. *Calpytogena* (*Ectagena*) *phaseoliformis* n. sp., an unusual vesicomyid bivalve collected by the submersible *Nautile* from abyssal depths of the Japan and Kurile Trenches. *Venus Jpn. J. Malacol. Kairuigaku Zasshi* 45(3):161–168.

Moiseyev, S. I. Underwater observations of squid in the North Atlantic from the submersible *Sever-2*. In: *Underwater Fisheries Research,* ed. G. P. Nizovtsev et al., pp. 71–78. Murmansk: PINRO.

Momma, H., and H. Hotta. Comparisons of resolution between a deep tow, a manned submersible and a Sea Beam for topographic survey. *Tech. Rep. Jpn. Mar. Sci. Technol. Ctr.,* Special Issue, November, pp. 63–74.

Morton, J. L., and Ballard, R. D. East Pacific Rise at 19°S: Evidence for a recent ridge jump. *J. Geol.* 14:111–114.

Nakanishi, T., S. Takagawa, T. Tsuchiya, and Y. Amitani. Japanese 6,500m deep manned research submersible project. In: *Oceans '86,* Vol. 5, pp. 1438–1442 New York: IEEE/MTS.

Normark, W. R., J. L. Morton, J. L. Bischoff, R. Brett, R. T. Holcomb, E. S. Kappel, R. A. Koski, S. L. Ross, W. C. Shanks III, J. F. Slack, K. L. Von Damm, and R. A. Zierenberg. Submarine fissure eruptions and hydrothermal vents on the southern Juan de Fuca Ridge: Preliminary observations from the submersible *Alvin. Geology* 14(10):823–827.

Okada, H., and K. Otsuka. Sedimentary environment of the Sagami Trough: A deep-sea submersible survey by *Shinkai 2000* in the Miura Canyon. *Tech. Rep. Jpn. Mar. Sci. Technol. Ctr.,* Special Issue, November, pp. 11–20.

Pritzlaff, J. A. Special issue on manned and unmanned underwater vehicles. *IEEE J. Oceanic Engin.* OE-11(3):347–348.

Smith, M. F., M. R. Perfit, I. R. Jonasson, and R. W. Embley. Geochemical constraints on the development of the Galapagos massive sulfide deposit. *Eos* 91:1185.

Stanley, D. J., S. J. Culver, and W. L. Stubblefield. Petrologic and foraminiferal evidence for active downslope transport in Wilmington Canyon. *Mar. Geol.* 69(3-4):207–218.

Sullivan, W. Deep seeing. *Oceans* 19(1):19–23.

Tamaki, K., and T. Sato. Geological observation of the Yamato Bank in the Japan Sea by Deep Sea Submersible *Shinkai 2000:* Preliminary report of dive 188. *Tech. Rep. Jpn. Mar. Sci. Technol. Ctr.,* Special Issue, November, pp. 123–132.

Tanaka, T., H. Asari, T. Asari, and H. Hotta. Preliminary report on sediments and their chemical compositions collected by the submersible *Shinkai-2000* and deep-towed sampling system. *Tech. Rep. Jpn. Mar. Sci. Technol. Ctr.* 16:31–45.

Ten Hove, H. A., and H. Zibrowius. *Laminatubus alvini* gen. et sp. n. and *Protis hydrothermica* sp. n. (Polychaeta, Serpulidae) from the bathyal hydrothermal vent communities in the eastern Pacific. *Zool. Scr.* 15(1):21–31.

U.S.G.S. Juan De Fuca Study Group. Submarine fissure eruptions and hydrothermal vents on the southern Juan de Fuca Ridge: Preliminary observations from the submersible *Alvin. Geology* 14(10):823–827.

Vanney, J. R. Plongées en submersible et géomorphologie sous-marine. *Inf. Geogr.* 50:195–201.

1987

Askew, T. M. *Johnson Sea-Link* submersible's role in the "Challenger" recovery. In: *Oceans '87: Proceedings,* Vol. 3, pp. 1225–1229. New York: IEEE.

Auffret, G. A., and CYAPORC Groupe. Geology of Porcupine and Goban scarps (N.E. Atlantic): Results of the CYAPORC submersible cruise. *C. R. Acad. Sci. II* 304(16):1003–1008.

Bergstrom, B. I., J. Larsson, and J. O. Pettersson. Use of a remotely operated vehicle (ROV) to study marine phenomena: I. Pandalid shrimp densities. *Mar. Ecol. Progr. Ser.* 37(1):97–101.

Biggs, D. C., P. Laval, J. C. Braconnot, C. Carré, J. Goy, M. Masson, and P. Morand. In situ observations of Mediterranean zooplankton by SCUBA and bathyscaphe in the Ligurian Sea in April 1986. In: *Diving for Science 1986: Proceedings of the American Academy of Underwater Science 6th Annual Scientific Diving Symposium, Oct. 31–Nov. 3, 1986, Tallahasse, FL,* ed. C. T. Mitchell, pp. 153–161. Costa Mesa, Calif.: American Academy of Underwater Science.

Cadet, J. P. Gravity induced erosion and seamount subduction: The result of submersible deep sea dives in the Japan Trench (Kaiko program, Leg III). *C. R. Acad. Sci. II* 305(16):1327–1335.

Carr, H. A., and R. A. Cooper. Manned submersible and ROV assessment of ghost gillnets in the Gulf of Maine. In: *Oceans '87,* Vol. 2, pp. 622–624. New York: IEEE/MTS.

Chambers, F. Canadian manned and unmanned submersibles for science. In: *Oceans '87,* Vol. 3, pp. 1223–1224. New York: IEEE.

Collins,W. T., D. R. Parrott, J. V. Barrie, and B. Imber. Tools and techniques for manned submersible studies of sediment transport and ice scour on the eastern Canadian continental shelf. In: Proceedings of the Nineteenth Annual Offshore Technology Conference, Houston, Texas, pp. 289–294.

Commeau, R. F., C. K. Paull, J. A. Commeau, and L. J. Poppe. Chemistry and mineralogy of pyrite-enriched sediments at a passive margin sulfide brine seep: Abyssal Gulf of Mexico. *Earth Planet. Sci. Lett.* 82(1):62–74.

Emig, C. C. Offshore brachiopods investigated by submersible. *J. Exp. Mar. Biol. Ecol.* 108(3):261–274.

Fricke, H. W., E. Vareschi, and D. Schlichter. Photoecology of the coral *Leptoseris fragilis* in the Red Sea twilight zone. *Oecologia* 73(3):371–381.

Guerrero-Garcia, J. A., D. A. Cordoba-Mendez, and A. Carranza-Edwards. Observación geológica directa del fondo oceánico en la Costa Pacifica Oriental. *Bol. Mineral.* 3(1):31–47.

Hampson, J. C., Jr. Mapping nuclear craters on Enewetak Atoll, Marshall Islands. In: Proceedings of the International Symposium on Marine Positioning, "Positioning the Future" (Insmap 86), pp. 249–258.

Hanson, L. C., and S. A. Earle. Submersibles for scientists. *Oceanus* 30(3): 31–38.

Harbison, G. R., and J. Janssen. Encounters with a swordfish (*Xiphias gladius*) and sharptail mola (*Masturus lanceolatus*) at depths greater than 600 meters. *Copeia* 1987(2):511–513.

Josenhans, H. W., J. V. Barrie, and L. A. Kiely. Mass wasting along the Labrador Shelf margin: Submersible observations. *Geo-Mar. Lett.* 7(4):199–205.

Karson, J. A., G. Thompson, S. E. Humphris, J. M. Edmond, W. B. Bryan, J. R. Brown, A. T. Winters, R. A. Pockalny, and J. F. Casey. Along-axis variations in seafloor spreading in the MARK area. *Nature* 328(6132):681–685.

Kimura, M., et al. Submersible *Shinkai 2000* study on the central rift in the middle Okinawa Trough. *Tech. Rep. Jpn. Mar. Sci. Technol. Ctr.*, Special Issue, August, pp. 165–196.

Le Pichon, X. Manned submersible observations of tectonic features and fluid venting along the Nankai subduction complex (eastern Nankai accretionary prism, southern Japan): Results of the *Kaiko* cruise (Leg I). *C. R. Acad. Sci. II* 305(16):1285–1293.

Matsuzawa, S., and J. Hashimoto. A survey of deep-sea megalo-epibenthos in the Mogami Trough using the deep-sea research submersible *Shinkai 2000* and the deep-towed color TV system. *Tech. Rep. Jpn. Mar. Sci. Technol. Ctr.*, Special Issue, August, pp. 251–260.

Normark, W. R., J. L. Morton, and S. L. Ross. Submersible observations along the southern Juan de Fuca Ridge: 1984 *Alvin* program. *J. Geophys. Res.* 92(B11):11,283–11,290.

Otsuka, K. The results of a diving survey by the submersible *Shinkai 2000* at the upper slope of the northernmost end of Suruga Trough, off the mouth of the Fuji River. *Tech. Rep. Jpn. Mar. Sci. Technol. Ctr.*, Special Issue, August, pp. 1–14.

Parrott, D. R., R. G. Campanella, and B. Imber. Seacone: A cone penetrometer for use with the *Pisces* submersible. In: *Oceans '87*, Vol. 3, pp. 1290–1294. New York: IEEE.

Pautot, G. Submersible observations in subduction trenches off SE Japan: French-Japanese 1985 Kaiko cruise, Leg II. *C. R. Acad. Sci. II* 305(16): 1321–1326.

Pugh, P. R., and G. R. Harbison. Three new species of prayine siphonophore (Calycophorae, Prayidae) collected by a submersible, with notes on related species. *Bull. Mar. Sci.* 41(1):68–91.

Rechnitzer, A. B. Beebe Project: Looking for sixgill sharks . . . and more. *Sea Technol.* 28(12):10–17.

Richardson, J. G. Submarine exploration and the tale of "Saga." *Impact Sci. Soc.* 147:277–286.

Rosman, I., G. S. Boland, L. Martin, and C. Chandler. Underwater sightings of sea turtles in the northern Gulf of Mexico. *OCS Rpt. U.S. Miner. Manage. Serv.*

Sebens, K. P. Applications of unmanned submersibles in benthic marine ecological research. In: *Undersea Teleoperators and Intelligent Autonomous Vehicles,* ed. N. Doelling and E. T. Harding, pp. 83–100. Cambridge, Mass.: MIT Sea Grant Program.

Siapno, W. D. 1968 Vinogradov expedition. *Mar. Mineral.* 6(3):223–229.

Sibuet, M. *Structure of Benthic Populations in Relation to Trophic Conditions in the Abyssal Environment of the Atlantic Ocean.* Paris: Université Paris 6.

Smith, K. L., Jr. Food energy supply and demand: A discrepancy between particulate organic carbon flux and sediment community oxygen consumption in the deep ocean. *Limnol. Oceanogr.* 32(1):201–220.

Swinbanks, D. Joint survey by Japan and France in South Pacific. *Nature* 328(6125):4.

Takagawa, S. Deep submersible project (6,500 m). *Oceanus* 30(1):29–32.

Taras, B. D., and S. R. Hart. Geochemical evolution of the New England Seamount chain: Isotopic and trace-element constraints. *Chem. Geol.* 64(1-2):35–54.

Vanney, J. R., and M. Gennesseaux. Versants escarpés dans une mer jeune: Reconnaissance en submersible en mer Tyrrhenienne. *Acta Geogr. 3 Ser.* 71:28–39.

Wakutsubo, T., and E. Koganezaki. Distribution and ecology of deep-sea organisms in the Japan Sea—especially the western region of the Tugaru Strait. In: Proceedings of the 3rd Symposium on Deep Sea Research Using the Submersible *Shinkai 2000* System, pp. 261–272.

Zonenshajn, L. P., I. O. Murdmaa, B. V. Baranov, A. P. Kuznetsov, V. S. Kuzin, M. I. Kuz'min, G. P. Avdejko, P. A. Stunzhas, V. N. Lukashin, et al. Gas emission source in the Sea of Okhotsk. *Okeanologiya* 27(5):795–800.

1988

Bazylinski, D. A., J. W. Farrington, and H. W. Jannasch. Hydrocarbons in surface sediments from a Guaymas Basin hydrothermal vent site. Woods Hole Oceanographic Institution Technical Report.

Boyer, L. F. Video-sediment-profile camera imagery in marine and freshwater benthic environments. In: *Oceans '88,* Vol. 2, pp. 443–447. New York: IEEE.

Bukin, S. D., and K. A. Zgurovskij. Distribution pattern, biology and behaviour of humpback shrimp *Pandalus hypsinotus* from the northwestern Sea of Japan. *Morsk. Prom. Bespozvonochnye,* pp. 108–119.

Clark, E. Down the Cayman Wall. *Natl. Geogr. Mag.* 174(5):712–731.

Cooper, R. A., and I. G. Babb. Manned submersibles support a wide range of underwater research in New England and the Great Lakes. In: *Oceans '88,* Vol. 1, pp. 112–118. New York: IEEE.

Cooper, R. A., H. A. Carr, and A. H. Hulbert. Manned submersible and ROV assessment of ghost gillnets on Jeffreys and Stellwagen Banks, Gulf of Maine. *NOAA Natl. Undersea Res. Prog. Res. Rep.* 88-4:429–442.

Culver, S. J., C. A. Brunner, and C. A. Nittrouer. Observations of a fast burst of the Deep Western Boundary Undercurrent and sediment transport in South Wilmington Canyon from DSRV *Alvin. Geo-Mar. Lett.* 8(3):159–165.

Embley, R. W., S. R. Hammond, and K. Murphy. The caldera of Axial Volcano: Remote sensing and submersible studies of a hydrothermally active submarine volcano. *NOAA Natl. Undersea Res. Prog. Res. Rep.* 88-4:61–70.

Embley, R. W., I. R. Jonasson, M. R. Perfit, J. M. Francheteau, M. A. Tivey, A. Malahoff, M. F. Smith, and T. J. G. Francis. Submersible investigation of an extinct hydrothermal system on the Galapagos Ridge: Sulfide mounds, stockwork zone, and differentiated lavas. *Can. Mineral.* 26(3):517–539.

Emig, C. C., and P. M. Arnaud. Observations by submersible of population densities of *Gryphus vitreus* (Brachiopoda) along the continental slope in Provence (northwestern Mediterranean). *C. R. Acad. Sci. III* 306(16): 501–505.

Fornari, D. J., M. R. Perfit, J. F. Allan, and R. Batiza. Small-scale heterogeneities in depleted mantle sources: Near-ridge seamount lava geochemistry and implications for mid-ocean-ridge magmatic processes. *Nature* 331(6156):511–513.

Fornari, D. J., M. R. Perfit, J. F. Allan, R. Batiza, R. Haymon, A. Barone, W. B. F. Ryan, T. Smith, T. Simkin, and M. A. Luckman. Geochemical and structural studies of the Lamont Seamounts: Seamounts as indicators of mantle processes. *Earth Planet. Sci. Lett.* 89(1):63–83.

Galerne, A. Development of deep water technology as it relates to future salvage. In: *Oceans '88,* Vol. 4., pp. 1573–1575. New York: IEEE.

Galkin, S. V. Use of a towed submersible to identify hydrothermals from biological anomalies and for investigation of bottom fauna. *Oceanology* 28(5):661–663.

Gooding, R. M., J. J. Polovina, and M. D. Dailey. Observations of deepwater shrimp, *Heterocarpus ensifer,* from a submersible off the Island of Hawaii. *Mar. Fish. Rev.* 50(1):32–38.

Ivanov, S. V., B. F. Kel'Balikhanov, V. V. Chernyy, and A. F. Genkina. A submersible sea water refractometer with a three-dimensional waveguide. *Oceanology* 28(5):664–667.

Jordon, M. B. A new submersible recording scalar light sensor array. *Deep-Sea Res.* 35A(8):1411–1423.

Kitazato, H. Geology of the axial part of the Suruga Trough. Diving survey reports of the submersible *Shinkai 2000,* Dive No. 309. *Tech. Rep. Jpn. Mar. Sci. Technol. Ctr.,* Special Issue, September, pp. 89–100.

Kobayashi, K. Igneous and sedimentary rocks on the NW flank of the Yamato Tai in the Sea of Japan observed and collected by the submersible *Shinkai 2000*. *Tech. Rep. Jpn. Mar. Sci. Technol. Ctr.*, Special Issue, September, pp. 1–8.

Krezoski, J. R. In-situ tracer studies of surficial sediment transport in the Great Lakes using a manned submersible. In: *Oceans '88*, Vol. 2, p. 442. New York: IEEE.

Kyo, M., S. Takagawa, J. Hashimoto, and S. Matsuzawa. Development of a pressure-retaining sampler for deep sea microorganism. *Tech. Rep. Jpn. Mar. Sci. Technol. Ctr.* 20:19–30.

Langton, R. W., and J. R. Uzmann. A survey of the macrobenthos in the Gulf of Maine using manned submersibles. *NOAA Natl. Undersea Res. Prog. Res. Rep.* 88-3:131–138.

Larson, R. J., G. R. Harbison, P. R. Pugh, J. A. Janssen, R. H. Gibbs, J. E. Craddock, C. E. Mills, R. L. Miller, and R. W. Gilmer. Midwater community studies off New England using the *Johnson Sea-Link* submersibles. *NOAA Natl. Undersea Res. Prog. Res. Rep.* 88-4:265–281.

Laval, P., and C. Carré. Comparison between observations from the submersible *Cyana* and midwater trawl sampling during the MIGRAGEL I cruise in the Ligurian Sea (North-western Mediterranean Sea). *Bull. Soc. R. Liege* 57(4/5):249–257.

Lisitsyn, A. P., and Y. U. A. Bogdanov. *The Geology of the Tadjura Rift: Observations from Submersibles.* Moscow: Nauka.

Mevel, C. Metamorphism in oceanic layer 3, Gorringe Bank, Eastern Atlantic. *Cont. Mineral. Pet.* 100(4):496–509.

Midorikawa, K., H. Momma, K. Mistuzawa, and H. Hotta. Measurement of the deep-sea current near the bottom in the Suruga Trough. *Tech. Rep. Jpn. Mar. Sci. Technol. Ctr.*, Special Issue, September, pp. 101–109.

Mills, C. E., and J. Goy. In situ observations of the behavior of mesopelagic *Solmissus narcomedusae* (Cnidaria, Hydrozoa). *Bull. Mar. Sci.* 43:914–937.

Nayak, B.U. Role of ocean technology in marine archaeological explorations: Perspectives and prospects. In: *Marine Archaeology of Indian Ocean Countries,* ed. S. R. Rao, pp. 157–163. Goa: National Institute of Oceanography.

Okutani, T., and S. Ohta. A new gastropod mollusk associated with hydrothermal vents in the Mariana Back-Arc Basin, western Pacific. *Venus Jpn. J. Malacol. Kairuigaku Zasshi* 47(1):1–9.

Pugh, P. R., and M. J. Youngbluth. A new species of Halistemma (Siphonophora:Physonectae:Agalmidae) collected by submersible. *J. Mar. Biol. Assoc. U.K.* 68:1–14.

———. Two new species of prayine siphonophore (Calycophorae, Prayidae) collected by the submersibles *Johnson Sea-Link I* and *II*. *J. Plankton Res.* 10(4):637–657.

Roonwal, G. S., and A. Mitra. Hydrothermal sulphide mineralization on seafloor spreading centres. *Ind. J. Mar. Sci.* 17(4):249–257.

Segawa, J., and H. Fujimoto. Observation of an ocean bottom station installed in the Sagami Bay and replacement of the acoustic transponder attached to it. *Tech. Rep. Jpn. Mar. Sci. Technol. Ctr.,* Special Issue, September, pp. 251–257.

Shimamura, H., and T. Kanazawa. Ocean bottom tiltmeter with acoustic data retrieval system implanted by a submersible. *Mar. Geophys. Res.* 9(3): 237–254.

Tsutsui, T., M. Ando, and S. Kaneshima. The ocean-bottom seismograph installed on hard rock by the *Shinkai 2000. Tech. Rep. Jpn. Mar. Sci. Technol. Ctr.,* Special Issue, September, pp. 259–266.

Tyler, P. A., and R. S. Lampitt. Submersible observations of echinoderms at bathyal depths in the N.E. Atlantic. In: *Echinoderm Biology,* ed. R. D. Burke et al., pp. 431–434. Rotterdam: A. A. Balkema.

Uchupi, E., M. Muck, and R. D. Ballard. Geology of the *Titanic* site and vicinity. *Deep-Sea Res.* 35(7):1093–1110.

Van Dover, C. L. Dive 2000 (of WHOI's *Alvin*). *Sea Frontiers* 34(6):326–331.

Yamazaki, H., and S. Kato. Submarine topography, geology and tectonic movement in the northern part of the Suruga Trough. *Tech. Rep. Jpn. Mar. Sci. Technol. Ctr.,* Special Issue, September, pp. 67–81.

Yuasa, M., T. Urabe, and F. Murakami. A submersible study of hydrothermal fields at the Kaikata Seamount, Izu-Ogasawara Arc. *Tech. Rep. Jpn. Mar. Sci. Technol. Ctr.,* Special Issue, September, pp. 129–139.

1989

Auzende, J. M. Preliminary results of the STARMER 1 cruise of the submersible *Nautile* in the North Fiji Basin. *C. R. Acad. Sci. II* 309 (18):1787–1795.

Batiza, R., T. L. Smith, and Y. Niu. Geological and petrologic evolution of seamounts near the EPR (East Pacific Rise) based on submersible and camera study. *Mar. Geophys. Res.* 11(3):169–236.

Belkin, S., and H. W. Jannasch. Microbial mats at deep-sea hydrothermal vents: New observations. In: *Microbial Mats: Physiological Ecology of Benthic Microbial Communities,* ed. Y. Cohen and E. Rosenberg, pp. 16–21. Washington, D.C.: American Society for Microbiology.

Brown, J. R., and J. A. Karson. Variations in axial processes on the Mid-Atlantic Ridge: The median valley of the MARK area. *Mar. Geophys. Res.* 10(1-2):109–138.

Child, C. A. Pycnogonida of the western Pacific Islands. 6. *Sericosura cochleifovea,* a new hydrothermal vent species from the Marianas Back-Arc Basin. *Proc. Biol. Soc. Wash.* 102(3):732–737.

Childress, J. J., D. L. Gluck, R. S. Carney, and M. M. Gowing. Benthopelagic biomass distribution and oxygen consumption in a deep-sea benthic boundary layer dominated by gelatinous organisms. *Limnol. Oceanogr.* 34(5):913–930.

Eittreim, S. L., R. W. Embley, W. R. Normark, H. G. Greene, C. M. McHugh, and W. B. F. Ryan. Observations in Monterey Canyon and Fan Valley using the submersible *Alvin* and a photographic sled. *U.S. Geol. Surv. Open-File Rep.* 89-291.

Gershanovich, D. E., G. A. Golovan, and V. B. Murav'ev. *Underwater Observations in Biooceanological and Fisheries Research.* Moscow: VNIRO.

Grandperrin, R., and B. Richer-de-Forges. Observations made aboard submersible *Cyana* in the epibathyal zone of New Caledonia. *Rapp. Miss. Sci. Mer Biol. Mar. Cent. Noumea Orstom* 3.

Gutsal, D. K. Underwater observations on distribution and behaviour of cuttlefish *Sepia pharaonis* in the western Arabian Sea. *Biol. Morya* (Vladivostok) 1:48–55.

Harris, P. T., and P. J. Davies. Submerged reefs and terraces on the shelf edge of the Great Barrier Reef, Australia: Morphology, occurrence and implications for reef evolution. *Coral Reefs* 8(2):87–98.

Horkowitz, J., D. Stakes, and R. Enrlich. Unmixing mid-ocean ridge basalts with extended Q model. *Tectonophysics* 165(1-4):1–19.

Huber, R., M. Kurr, H. W. Jannasch, and K. O. Stetter. A novel group of abyssal methanogenic archaebacteria (Methanopyrus) growing at 110 degrees C. *Nature* 342(6251):833–834.

Il'Yushenok, A. V., S. N. Drakov, S. P. Katsevich, M. Y. Kostko, V. N. Knyukshto, and B. P. Primshits. The BF-2 submersible fluorimeter. *Oceanology* 29(2):250–253.

Jannasch, H. W. Lessons from the *Alvin* lunch. *Oceanus* 31(4):28–33.

Karson, J. A., and J. R. Brown. Geological setting of the Snake Pit hydrothermal site: An active vent field on the Mid-Atlantic Ridge. *Mar. Geophys. Res.* 10(1-2):91–107.

Kato, S., K.-I. Nakamura, Y. Iwabuchi, K. Kawai, and H. Seta. Topographic and geologic characteristics of the middle part of the Okinawa Trough revealed by Seabeam survey and the geology around the Aguni Knoll studied with a submersible in 1986. *Tech. Rep. Jpn. Mar. Sci. Technol. Ctr.*, Special Issue, September, pp. 145–162.

Kimura, M., T. Tanaka, M. Kyo, M. Ando, T. Oomori, E. Izawa, and I. Yoshikawa. Study of topography, hydrothermal deposits and animal colonies in the middle Okinawa Trough hydrothermal areas using the submersible *Shinkai 2000* system. *Tech. Rep. Jpn. Mar. Sci. Technol. Ctr.*, Special Issue, September, pp. 223–244.

Kondyurin, A. V., V. V. Sochelnikov, M. D. Khutorskoy, and R. N. Chzhu. First geothermal investigations from the *Argus* manned submersible. *Oceanology* 29(3):392–394.

Lapshin, A. I., G. I. Nesvetova, S. B. Stepin, and A. A. Shavykin. Utilization of a submersible chemiluminograph for studies of Barents Sea waters. *Rapp. P. V. Reun. ICES* 188:78.

Laval, P. The use of the *Cyana* submersible for plankton studies. *Oceanus* 15(1):25–30.

Laval, P., J. C. Braconnot, C. Carré, J. Goy, P. Morand, and C. E. Mills. Small-scale distribution of macroplankton and micronekton in the Ligurian Sea (Mediterranean Sea) as observed from the manned submersible *Cyana*. *J. Plankton Res.* 11(4):665–685.

Le Pichon, X., and K. Kobayashi. *The Japanese Trenches Kaiko Program* Nautile *Submersible Cruise, June 1st–August 11th 1985.* IFREMER, Campagnes Océanographiques Françaises No. 10. Plouzane, France: IFREMER.

Legrand, J., and L. Floury. Re-entry of deep sea boreholes using *Nautile* and Nadia: An application of heavy operations with a manned submersible. In: Proceedings of the EEZ Resources: Technology Assessment Conference, January, pp. 7-40–7-50.

McCamis, M. J. Captain Hook's hunt for the H-bomb. *Oceanus* 31(4):22–27.

Masson, D. G., M. R. Dobson, J. M. Auzende, M. Cousin, A. Coutelle, J. Rolet, and P. Vaillant. Geology of Porcupine Bank and Goban Spur, northeastern Atlantic: Preliminary results of the *Cyaporc* submersible cruise. *Mar. Geol.* 87(2/4):105–119.

Miyashita, S., T. Tanaka, H. Momma, H. Tokuyama, W. Soh, S.-I. Kuramoto, and J. Ishii. Observations using the submersible *Shinkai 2000* at the northern part of the Okushiri Ridge, a section of the oceanic crust of the Japan Sea. *Tech. Rep. Jpn. Mar. Sci. Technol. Ctr.*, Special Issue, September, pp. 85–100.

Moiseev, S. I., and B.V. Kolodnitskii. Behavioral and distribution patterns of *Sthenoteuthis oualiniensis* in the equatorial zone of the Indian Ocean. In: *Underwater Observations in Biooceanological and Fisheries Research,* ed. D. E. Gershanovich, G. A. Golovan, and V. B. Murav'ev, pp. 61–65. Moscow: VNIRO.

Monniot, C., and F. Monniot. Ascidians collected around the Galapagos Islands using the *Johnson Sea-Link* research submersible. *Proc. Biol. Soc. Wash.* 102(1):14–32.

Nelson, D. C., C. O. Wirsen, and H. W. Jannasch. Characterization of large, autotrophic *Beggiatoa* spp. abundant at hydrothermal vents of the Guaymas Basin. *Appl. Environ. Microbiol.* 55(11):2909–2917.

Pearcy, W. G., D. L. Stein, M. A. Hixon, E. K. Pikitch, W. H. Barss, and R. M. Starr. Submersible observations of deep-reef fishes of Heceta Bank, Oregon. *Fish. Bull.* 87(4):955–965.

Sachs, P. L., T. R. Hammar, and M. P. Bacon. A Large-Volume, Deep-Sea Submersible Pumping System. Woods Hole Oceanographic Institution Technical Report No. WHOI-89-55.

Sagalevitch, A. M. Methods of ocean research with manned submersibles. In: *Oceans '89,* Vol. 3, pp. 728–733. New York: IEEE.

Shibanov, V. N., A. N. Kalugin, and A. S. Yarovoj. Methods of assessing stocks of roundnose grenadier on the North Atlantic Ridge. In: *Mesopelagic and Bathypelagic Living Resources of the High Seas in the North Atlantic,* pp. 156–165. Murmansk: PINRO.

Shinn, E. A., and R. I. Wicklund. Artificial reef observations from a manned submersible off southeast Florida. *Bull. Mar. Sci.* 44(2):1041–1050.

Smith, T. L., and R. Batiza. New field and laboratory evidence for the origin of hyaloclastite flows on seamount summits. *Bull. Volcanol.* 51(2): 96–114.

Takagawa, S., K. Takahashi, T. Sano, M. Kyo, Y. Mori, and T. Nakanishi. 6,500m Deep manned research submersible *Shinkai 6500* system. In: *Oceans '89,* Vol. 3, pp. 741–746. New York: IEEE.

Tokuyama, H., S. Kuramoto, W. Soh, S. Miyashita, A. Takeuchi, H. Monma, and T. Tanaka. Deep-sea submersible survey of the exposed profile of the Japan Basin crust: Observation on the fault scarps bounded by the western margin of the Okushiri Ridge off West Hokkaido. *Tech. Rep. Jpn. Mar. Sci. Technol. Ctr.,* Special Issue, September, pp. 101–110.

Uchupi, E., and R. D. Ballard. Evidence of hydrothermal activity on Marsili Seamount, Tyrrhenian Basin. *Deep-Sea Res.* 36(9):1443–1448.

Uchupi, E., R. D. Ballard, and W. N. Lange. New evidence about *Titanic*'s final moments: Resting in pieces. *Oceanus* 31(4):53–60.

Widder, E. A., S. A. Bernstein, D. F. Bracher, J. F. Case, K. R. Reisenbichler, J. J. Torres, and B. A. Robinson. Bioluminescence in the Monterey submarine canyon: Image analysis of video recordings from a midwater submersible. *Mar. Biol.* 100(4):541–551.

Youngbluth, M. J. Species diversity, vertical distribution, relative abundance and oxygen consumption of midwater gelatinous zooplankton: Investigations with manned submersibles. *Oceanus* 15(1):9–15.

1990

Anderson, A. E., H. Felbeck, and J. J. Childress. Aerobic metabolism is maintained in animal tissues during rapid sulphide oxidation activity in the symbiont-containing clam *Solemya reidi. J. Exp. Zool.* 256:130–134.

Auzende, J. M. The MAR-Vema Fracture Zone intersection surveyed by deep submersible *Nautile. Terra Nova* 2(1):68–73.

Bagley, P. M., I. G. Priede, and J. D. Armstrong. An autonomous deep ocean vehicle for acoustic tracking of bottom living fishes. In: *Electronics Division Colloquium on Monitoring the Sea, 18 December 1990,* pp. 2.1–2.3. London: Institution of Electrical Engineers.

Biryukov, S. G., V. N. Maryatkin, A. S. Matveyev, V. A. Popov, and Y. A. Rudyakov. Use of the *Zvuk* towed submersible to investigate benthic plankton. *Oceanology* 30(1):114–117.

Burggraf, S., H. W. Jannasch, B. Nicolaus, and K. O. Stetter. *Archaeoglobus profundus* sp. nov. represents a new species within the sulfate reducing archaebacteria. *Syst. Appl. Microbiol.* 13:24–28.

Childress, J. J., D. L. Cowles, J. A. Favuzzi, and T. J. Mickel. The metabolic rates of deep-sea benthic decapod crustaceans decline with increasing depth primarily due to the decline in temperature. *Deep-Sea Res.* 37:929–949.

Clark, E., and E. Kristof. Deep sea elasmobranchs observed from submersibles off Bermuda, Grand Cayman, and Freeport, Bahamas. *NOAA Tech. Rep.* 90:269–284.

Cohen, D. M., R. M. Rosenblatt, and H. G. Moser. Biology and description of a bythitid fish from deep-sea thermal vents in the eastern tropical Pacific. *Deep-Sea Res.* 37A:267–283.

Colbourne, E. B., and A. E. Hay. An acoustic remote sensing and submersible study of an Arctic submarine spring plume. *J. Geophys. Res.* 95(C8): 13,219–13,234, and plates, 13,581–13,582.

Corell, R. W. Commentary: A critical need (deepest ocean research). *Mar. Technol. Soc. J.* 24(2):42.

DeBevoise, A., J. J. Childress, and N. Withers. Carotenoids indicate differences in diet of hydrothermal vent crab, *Bythograea thermydron. Mar. Biol.* 105:109–115.

Edmond, J. M. Deep-sea science needs in marine geochemistry. *Mar. Technol. Soc. J.* 24(2):32–33.

Embley, R. W., S. L. Eittreim, C. H. McHugh, W. R. Normark, G. H. Rau, B. Hecker, A. E. DeBevoise, H. G. Greene, W. B. F. Ryan, C. Harrold, and C. Baxter. Geological setting of chemosynthetic communities in the Monterey Fan Valley system. *Deep-Sea Res.* 37(11A):1651–1667.

Embley, R. W., K. M. Murphy, and C. G. Fox. High resolution studies of the summit of Axial Volcano. *J. Geophys. Res.* 95:12,785–12,812.

Felbeck, H. Symbiosis of bacteria with invertebrates in the deep sea. In: *Endocytobiology IV,* ed. P. Nardon, V. Gianinazzi-Pearson, A. M. Grenier, L. Margulis, and D. C. Smith, pp. 327–334. Paris: Institut National de la Recherche Agronomique.

Fiala-Medioni, A., and H. Felbeck. Autotrophic processes in invertebrate nutrition: Bacterial symbiosis in bivalve molluscs. In: *Animal Nutrition and Transport Processes*, Comparative Physiology, Vol. 5, ed. J. Mellinger, pp. 49–69. Basel: Karger Verlag.

Fiala-Medioni, A., H. Felbeck, J. J. Childress, C. R. Fisher, and R. D. Vetter. Lysosomic resorption of bacterial symbionts in deep-sea bivalves. In: *Endocytobiology IV*, ed. P. Nardon, V. Gianinazzi-Pearson, A. M. Grenier, L. Margulis, and D. C. Smith, pp. 335–338. Paris: Institut National de la Recherche Agronomique.

Fisher, C. R. Chemoautotrophic and methanotrophic symbioses in marine invertebrates. *Rev. Aquatic Sci.* 2:399–436.

Fisher, C. R., M. C. Kennicutt II, and J. M. Brooks. Stable carbon isotopic evidence for carbon limitation in hydrothermal vent vestimentiferans. *Science* 247:1094–1096.

Fisher, D. E., and M. R. Perfit. Evidence from rare gases for magma-chamber degassing of highly evolved mid-ocean-ridge basalt. *Nature* 343(6257): 450–452.

Fricke, H. Coelacanths: Exploring the future of an ancient fish by submersible. In: *Spirit of Enterprise: The 1990 Rolex Awards*, pp. 452–454. Bern: Buri.

Fryer, P. Deep submersibles and potential marine geological research. *Mar. Technol. Soc. J.* 24(2):22–31.

Fryer, P., K. L. Saboda, L. E. Johnson, M. E. Mackay, G. F. Moore, and P. Stoffers. Conical Seamount: SeaMARC II, *Alvin* submersible, and seismic-reflection studies. *Proc. Ocean Drilling Prog. Init. Rep.* 125: 69–80.

Galkin, S. V., and L. I. Mostalev. Study of the abyssal fauna of the North Atlantic Ocean from deep-water manned submersibles. *Oceanology* 30(4):502–507.

Hochstaedter, A. G., J. B. Gill, M. Kusakabe, S. Newman, M. Pringle, B. Taylor, and P. Fryer. Volcanism in the Sumisu Rift I: The effect of distinctive volatiles on fractionation. *Earth Planet. Sci. Lett.* 100:179–194.

Hochstaedter, A. G., J. B. Gill, J. D. Morris, and C. H. Langmuir. Volcanism in the Sumisu Rift II: Subduction related and unrelated sources. *Earth Planet. Sci. Lett.* 100:195–209.

Holmes, M. L., and R. A. Zierenberg. Submersible observations in Escanaba Trough, southern Gorda Ridge. In: *Gorda Ridge: A Seafloor Spreading Center in the United States' Exclusive Economic Zone*, ed. G. R. McMurray, pp. 93–115. New York: Springer-Verlag.

Hurst, S. D., J. A. Karson, and E. M. Moores. Nodal basins at slow spreading ridge-transform intersections: A comparison to the central portion of the Troodos ophiolite. In: *Oceanic Crustal Analogues: Proceedings of the Symposium "TROODOS 1987,"* ed. J. Malpas, E. M. Moores, A. Panayiotou, and C. Xenophontos, pp. 125–130. Nicosia, Cyprus: Geological Survey Department.

Jannasch, H. W. Isolation of extremely thermophilic, fermentative archaebacteria from deep-sea geothermal sediments. In: *Bioprocessing and Biotreatment of Coal,* ed. D. L. Wise, pp. 417–428. New York: Marcel Dekker.

———. Microbiology of deep sea hydrothermal vents. *Austral. Microbiol.* 11: 370–372.

Joergensen, B. B., L. X. Zawacki, and H. W. Jannasch. Thermophilic bacterial sulfate reduction in deep-sea sediments at the Guaymas Basin hydrothermal vent site (Gulf of California). *Deep-Sea Res.* 37(4A):695–710.

Johnson, H. P., and R.W. Embley. Axial Seamount: An active ridge-axis volcano on the central Juan de Fuca Ridge. *J. Geophys. Res.* 95:12,689–12,696.

Jollivet, D., J.-C. Faugeres, R. Griboulard, D. Desbruyeres, and G. Blanc. Composition and spatial organization of a cold seep community on the Barbados Accretionary Prism: Tectonic, geochemical and sedimentary context. *Prog. Oceanog.* 24:25–45.

Jones, A. T., and K. J. Sulak. First Central Pacific Plate and Hawaiian record of the deep-sea tripod fish *Bathypterois grallator* (Pisces: Chlorophthalmidae). *Pac. Sci.* 44(3):254–257.

Karson, J. A. Seafloor spreading on the Mid-Atlantic Ridge: Implications for the structure of ophiolites and oceanic lithosphere produced in slow-spreading environments. In: *Oceanic Crustal Analogues: Proceedings of the Symposium "TROODOS 1987,"* ed. J. Malpas, E. M. Moores, A. Panayiotou, and C. Xenophontos, pp. 547–555. Nicosia, Cyprus: Geological Survey Department.

———. Tectonic disruption of volcanic units of the oceanic crust. In: *Proceedings of a Workshop on the Physical Properties of Volcanic Seafloor,* ed. G. M. Purdy and G. J. Fryer, pp. 138–142. Woods Hole, Mass.: Woods Hole Oceanographic Institution.

Karson, J. A., and P. A. Rona. Block-tilting, transfer faults, and structural control of magmatic and hydrothermal processes in the TAG area, Mid-Atlantic Ridge 26°N, *Geol. Soc. Am. Bull.* 102:1635–1645.

Kaspar, M., and C. Anklin. Miniature manned submersibles for scientific research with comparison to ROVs. In: *Oceanology International '90, Brighton, March 1990,* Vol. 1: *Civil Applications.* Kingston-upon-Thames, U.K.: Spearhead Exhibitions.

Keller, N. B. Fauna of seamounts. *Priroda* 3:51–53.

Laval, P., and T. Baussant. Effect of the lights from an approaching submersible on the 15 kHz deep scattering layer in the Ligurian Sea (Mediterranean). *C. R. Acad. Sci. III* 311(5):181–186.

Le Pichon, X., J.-P. Foucher, J. Boulegue, P. Henry, S. Lallemant, M. Benedetti, F. Avedik, and A. Mariotti. Mud volcano field seaward of the Barbados accretionary complex: A submersible survey. *J. Geophys. Res.* 95(B6): 8931–8943.

Lein, A. Y., N. I. Konova, and A. P. Lisityn. New data on the nature of naphthoids in the hydrothermal system of the Guaymas Rift (Gulf of California). *Dokl. USSR Acad. Sci. Earth Sci. Sect.* 305(1-6):178–181.

Lisitsyn, A. P., A. M. Sagalevich, G. A. Cherkashev, and N. L. Shashkov. Investigation of a hydrothermal vent in the Atlantic Ocean from the *Mir* submersibles. *Dokl. USSR Acad. Sci. Earth Science Sect.* 311:248–252.

Little, S. A., K. D. Stolzenbach, and G. M. Purdy, The sound field near hydrothermal vents on Axial Seamount, Juan de Fuca Ridge. *J. Geophys. Res.* 95(B8):12,927–12,945.

Macdonald, K. C., and P. J. Fox. The mid-ocean ridge. *Sci. Am.* 262:72–79.

Madin, L. P. Overview: Being there—the role of in situ science in oceanography. *Mar. Technol. Soc. J.* 24(2):19–22.

Moore, J. C., D. Orange, and L. D. Kulm. Interrelationship of fluid venting and structural evolution: *Alvin* observations from the frontal accretionary prism. *J. Geophys. Res.* 95(6B):8795–8808.

Moore, J. G., W. R. Normark, and B. J. Szabo. Reef growth and volcanism on the submarine southwest rift zone of Mauna Loa, Hawaii. *Bull. Volcanol.* 52:375–380.

Morton, J. L., R. A. Koski, W. R. Normark, and S. L. Ross. Distribution and composition of massive sulfide deposits at Escanaba Trough, southern Gorda Ridge. In: *Gorda Ridge: A Seafloor Spreading Center in the United States Exclusive Economic Zone,* ed. G. R. McMurray, pp. 77–92. New York: Springer-Verlag.

Mullineaux, L. S., and C. A. Butman. Recruitment of benthic invertebrates in boundary-layer flows: A deep water experiment on Cross Seamount. *Limnol. Oceanogr.* 35:409–423.

Natland, J., P. Lonsdale, J. A. Karson, and D. Sims. High-level gabbros of the East Pacific Rise sampled by submersible at fault exposures in Hess Deep, Eastern Equatorial Pacific. *Eos* 71:1647.

Neuner, A., H. W. Jannasch, S. Belkin, and K. O. Stetter. *Thermococcus litoralis* sp. nov.: A novel species of extremely thermophilic marine archaebacteria. *Arch. Microbiol.* 153:205–207.

Newman, J. B. Fiber-optic data network for the *Argo/Jason* vehicle system. *IEEE J. Oceanic Engin.* 15(2):66–71.

Niitsuma, N., K. I. Otsuka, K. I. Kano, H. Wada, R.-I. Sato, T. Shibutani, S. Takeuchi, T. Yoshida, and K. Oourasaka. Submersible observations of plate-relations in the subduction zone of the Suruga Trough. *Tech. Rep. Jpn. Mar. Sci. Technol. Ctr.,* Special Issue, September, pp. 261–276.

Ohta, S. Deep-sea submersible survey of the hydrothermal vent community on the northeastern slope of the Iheya Ridge, the Okinawa Trough. *Tech. Rep. Jpn. Mar. Sci. Technol. Ctr.,* Special Issue, September, pp. 145–156.

Okutani, T. Two new species of *Provanna* (Gastropoda: Cerithiacea) from "snail pit" in the hydrothermal vent site at the Mariana Back-Arc Basin. *Venus Jpn. J. Malacol. Kairuigaku Zasshi* 49(1):19–24.

Page, H. M., C. R. Fisher, and J. J. Childress. The role of suspension-feeding in the nutritional biology of a deep-sea mussel with methanotrophic symbionts. *Mar. Biol.* 104:251–257.

Paull, C. K., R. Freeman-Lynde, T. J. Bralower, J. M. Garsemal, A. C. Neumann, B. D'Argenio, and E. Marsella. Geology of the strata exposed on the Florida Escarpment. *Mar. Geol.* 91(3):177–194.

Rau, G. H., C. M. McHugh, C. Harrold, C. Baxter, B. Hecker, and R. W. Embley. *Calyptogena phaseoliformis* (bivalve mollusc) from the Ascension Fan-valley near Monterey, California. *Deep-Sea Res.* 37:1669–1676.

Roberts, H. H., P. Aharon, R. Carney, J. Larkin, and R. Sassen. Seafloor responses to hydrocarbon seeps, Louisiana continental slope. *Geo-Mar. Lett.* 10(4):232–243 (special issue).

Robison, B. H., and K. Wishner. Biological research needs for submersible access to the greatest ocean depths. *Mar. Technol. Soc. J.* 24(2):34–37.

Roux, M. Observation of steep rocky bathyal slopes (biology and geology) using diving saucer. In: Proceedings of Undersea Technology Research and Development, ISM 90, Toulon, France, 3–5 December, 12: 334–342.

Sanders, N. K., and J. J. Childress. The use of single column ion chromatography to measure the concentrations of the major ions in invertebrate body fluids. *Comp. Biochem. Physiol.* 98A:97–100.

Sanford, M. W., S. A. Kuehl, and C. A. Nittrouer. Modern sedimentary processes in the Wilmington Canyon area, U.S. East Coast. *Mar. Geol.* 92(3-4):205–226.

Savoye, B. Submarine instabilities and submersibles. In: Proceedings of Undersea Technology Research and Development, ISM 90, Toulon, France, 3–5 December, 12:225–244.

Sborshchikov, I. M., A. M. Gorodnitskiy, V. V. Matveyenkov, M. G. Ushakova, and A. A. Shreyder. Ophiolites on the floor of the Tyrrhenian Sea. *Dokl. Earth Sci. Sect.* 304(1-6):252–254.

Shor, A. N., D. J. W. Piper, C. J. E. Hughes, and L. A. Mayer. Giant flute-like scour and other erosional features formed by the 1929 Grand Banks turbidity current. *Sedimentology* 37(4):631–645.

Sibuet, M., L. Floury, A.-M. Alayse-Danet, A. Echardour, T. Lemoign, and R. Perron. In situ experimentation at the water/sediment interface in the deep sea. 1. Submersible experimental instrumentation developed for sampling and incubation. *Prog. Oceanogr.* 24(1/4):161–167.

Smith, J. R., B. Taylor, A. Malahoff, and L. Petersen. Submarine volcanism in the Sumisu Rift, Izu-Bonin arc: Submersible and deep-tow camera results. *Earth Planet. Sci. Lett.* 100(1-3):148–160.

Stein, J. L., M. Haygood, and H. Felbeck. Diversity of ribulose bisphosphate carboxylase genes in sulfur-oxidizing symbioses. In: *Endocytobiology IV*, ed. P. Nardon, V. Gianinazzi-Pearson, A. M. Grenier, L. Margulis, and D. C. Smith, pp. 343–348. Paris: Institut National de la Recherche Agronomique.

Straube, W. L., J. W. Deming, C. C. Somerville, R. R. Colwell, and J. A. Baross. Particulate DNA in smoker fluids: Evidence for existence of microbial populations in hot hydrothermal systems. *Appl. Environ. Microbiol.* 56(5):1440–1447.

Taylor, B., G. Brown, P. Fryer, J. B. Gill, A. G. Hochstaedter, H. Hotta, C. H. Langmuir, M. Leinen, A. Nishimura, and T. Urabe. *Alvin*-Seabeam studies of the Sumisu Rift, Izu-Bonin Arc. *Earth Planet. Sci. Lett.* 100(1-3):127–147.

Taylor, C. D., and K. W. Doherty. Submersible Incubation Device (SID), autonomous instrumentation for the in situ measurement of primary production and other microbial rate processes. *Deep-Sea Res.* 37A(2):343–358.

Tokunaga, S. The Japanese tourist submarine *Moglyn*. In: *Tourist Oceanology International '90: Proceedings*, Vol. 1. Kingston-Upon-Thames, U.K.: Spearhead Exhibitions.

Tunnicliffe, V., J. F. Garrett, and H. P. Johnson. Physical and biological factors affecting the behaviour and mortality of hydrothermal vent tubeworms. *Deep-Sea Res.* 37(1):103–125.

Urabe, T., and M. Kusakabe. Barite-silica chimneys from the Sumisu Rift, Izu-Bonin arc: Possible analog to hematitic chert associated with Kuroko deposits. *Earth Planet. Sci. Lett.* 100:283–290.

Von Damm, K. L. Seafloor hydrothermal activity: Black smoker chemistry and chimneys. *Annu. Rev. Earth Planet. Sci.* 18:173–204.

Wishner, K., L. Levin, M. Gowing, and L. Mullineaux. Involvement of the oxygen minimum in benthic zonation of a deep seamount. *Nature* 346:57–59.

Yamamoto, H. Geological observations of the Oki Ridge, southern Japan Sea, by the deep sea submersible *Shinkai 2000*. *Tech. Rep. Jpn. Mar. Sci. Technol. Ctr.*, Special Issue, September, pp. 309–315.

Zierenberg, R. A., and P. Schiffman. Microbial control of silver mineralization at a sea-floor hydrothermal site on the northern Gorda Ridge. *Nature* 348:155–157.

1991

Alayse-Danet, A. M., et al. Utilization of submersibles in deep-sea biology. *Actes Coll. IFREMER* 12:345–352.

Anderson, C. A. Underwater exploration: *Jason* lost at sea. *Nature* 354(6351): 260.

Atema, J., P. A. Moore, L. P. Madin, and G. A. Gerhardt. Subnose-1: Electrochemical tracking of odor plumes at 900 m beneath the ocean surface. *Mar. Ecol. Progr. Ser.* 74(2/3):303–306.

Auster, P. J., R. J. Malatesta, S. C. LaRosa, R. A. Cooper, and L. L. Stewart. Microhabitat utilization by the megafaunal assemblage at a low relief outer continental shelf site: Middle Atlantic Bight, USA. *J. Northwestern Atlantic Fish. Sci.* 11:59–69.

Bivaud, J. P. Évolution des concepts d'exploration minière en mer. *Actes Coll. IFREMER* 12:111–120.

Butler, J. L., W. W. Wakefield, P. B. Adams, B. H. Robison, and C. H. Baxter. Application of line transect methods to surveying demersal communities with ROVs and manned submersibles. In: *Oceans '91*, Vol. 2, pp. 689–696. New York: IEEE.

Casanova, J. P. Chaetognaths from the *Alvin* dives on the Seamount Volcano 7 (east tropical Pacific). *J. Plankton Res.* 13(3):539–548.

Chadwick, W. W., Jr., R. W. Embley, and C. G. Fox. Evidence for volcanic eruption on the southern Juan de Fuca Ridge between 1981 and 1987. *Nature* 350:416–418.

Chave, E. H., and A. T. Jones. Deep-water megafauna of the Kohala and Haleakala slopes, Alenuihaha Channel, Hawaii. *Deep-Sea Res.* 38A(7): 781–803.

Childress, J. J., C. R. Fisher, J. A. Favuzzi, R. Kochevar, N. K. Sanders, and A. M. Alayse. Sulfide-driven autotrophic balance in the bacterial symbiont–containing hydrothermal vent tubeworm, *Riftia pachyptila* Jones. *Biol. Bull.* 180:135–153.

Childress, J. J., C. R. Fisher, J. A. Favuzzi, and N. K. Sanders. Sulfide and carbon dioxide uptake by the hydrothermal vent clam, *Calyptogena magnifica*, and its chemoautotrophic symbionts. *Physiol. Zool.* 64:1444–1470.

Edwards, D. B., and D. C. Nelson. DNA-DNA solution hybridization studies of the bacterial symbionts of hydrothermal vent tube worms (*Riftia pachyptila* and *Tevnia jerichonana*). *Appl. Environ. Microbiol.* 57:1082–1088.

Embley, R. W., W. W. Chadwick, M. Perfit, and E. T. Baker. Geology of the northern Cleft Segment, Juan de Fuca Ridge: Recent lava flows, seafloor spreading and the formation of megaplumes. *Geology* 19:771–775.

Emig, C. C., and M. A. Garcia-Carrascosa. Distribution of *Gryphus vitreus* (Born, 1778) (Brachiopoda) on transect P2 (continental margin, French Mediterranean coast) investigated by submersible. *Sci. Mar.* 55(2):385–388.

Fedorov, V. V., and A. K. Karamyshev. Trophic structure of benthos on the Walvis Ridge and conditions of its formation. In: *Biotopicheskie Osnovy Raspredeleniya Promyslovykh i Kormovykh-Morskikh Zhivotnykh*, ed. A. A. Nejman, pp. 6–18.

Felbeck, H., and D. L. Distel. Prokaryotic symbionts in marine invertebrates. In: *The Prokaryotes,* 2d ed., Vol. IV, ed. A. Balows, H. G. Trüper, M. Dworkin, W. Harder, and K. H. Schleifer, pp. 3891–3906. New York: Springer-Verlag.

Gage, J. D., and P. A. Tyler. *Deep-Sea Biology: A Natural History of Organisms at the Deep-Sea Floor.* New York: Cambridge University Press.

Gente, P., C. Mevel, J. M. Auzende, J. A. Karson, and Y. Fouquet. An example of a recent accretion on the Mid-Atlantic Ridge: The Snake Pit neovolcanic ridge (MARK area, 23 degrees 22′N). *Tectonophysics* 190(1):1–29.

Goff, J. A., and M. C. Kleinrock. Quantitative comparisons of bathymetric survey systems. *Geophys. Res. Lett.* 18:1243–1256.

Gortan, C. Exploration sousmarine: Engins d'hier, d'aujourd'hui et de demain. *Sci. Vie* HS(176):4–12.

Grassle, J. F. Deep-sea benthic biodiversity. *BioScience* 41(7):464–469.

Gutsal, D. K. New information on the vertical distribution of the squid *Sthenoteuthis oualaniensis. Biol. Morya* (Vladivostok) 4:94–98.

Haymon, R. M., D. J. Fornari, M. H. Edwards, S. Carbotte, D. Wright, and K. C. Macdonald. Hydrothermal vent distribution along the East Pacific Rise Crest (9 09′–54′N) and its relationship to magmatic and tectonic processes on fast-spreading Mid-Ocean Ridge. *Earth Planet. Sci. Lett.* 104:513–534.

Herman, A. W., D. D. Sameoto, S. Chen, M. R. Mitchell, B. Petrie, and N. Cochrane. Sources of zooplankton on the Nova Scotia shelf and their aggregations within deep-shelf basins. *Cont. Shelf Res.* 11(3):211–238.

Hoffert, M., R. Le Suave, G. Pautot, A. Schaaf, P. Cochonat, Y. Morel, and P. Larque. Characterization by direct observations with the aid of the submersible *Nautile* of actual erosive structures in Equatorial Northern Pacific's manganese nodule area: "Grey Patches." In: Proceedings of the Third French Symposium on Sedimentology, Brest, 18–20 November, pp. 161–162.

Itzkowitz, M., M. Haley, C. Otis, and D. Evers. A reconnaissance of the deeper Jamaican coral reef fish communities. *Northeast Gulf Sci.* 12(1):25–34.

Karson, J. A. Geologic processes along slow-spreading ridge axes and their potential as seismic sources. In: *Characterization of Mid-Ocean Ridge Earthquake Activity Using Acoustic Data from U.S. Navy Permanent Hydrophone Arrays,* conv. G. M. Purdy, pp. 26–29. Ridge Inter-Disciplinary Global Experiments. Woods Hole, Mass.: Woods Hole Oceanographic Institution.

———. Accommodation zones and transfer faults: integral components of Mid-Atlantic Ridge extensional systems. In: *Ophiolites: Genesis and Evolution of Oceanic Lithosphere,* edited by T. Peters, A. Nicolas, and R. G. Coleman, pp. 21–37. Boston: Kluwer.

Kim, D. S., and S. Ohta. Submersible observations and comparison of the biological communities of the two hydrothermal vents on the Iheya Ridge of the Mid-Okinawa Trough. *Tech. Rep. Jpn. Mar. Sci. Technol. Ctr.,* Special Issue, September, pp. 221–233.

Kurr, M., R. Huber, H. Konig, H. W. Jannasch, H. Fricke, A. Trincone, J. K. Kristiansson, and K. O. Stetter. *Methanopyrus kandleri,* gen. and sp. nov. represents a novel group of hyperthermophilic methanogens growing at 110°C. *Arch. Microbiol.* 156:239–247.

Larson, R. J., C. E. Mills, and G. R. Harbison. Western Atlantic midwater hydrozoan and scyphozoan medusae: In-situ studies using manned submersibles. *Hydrobiologia* 216/217:311–317.

Lutz, R. A. The biology of deep-sea vents and seeps. *Oceanus* 34(4):75–83.

Macintyre, I. G. An Early Holocene reef in the western Atlantic: Submersible investigations. *Coral Reefs* 10(3):167–174.

McLean, J. H. Four new pseudococculinid limpets collected by the deep-submersible *Alvin* in the eastern Pacific. *Veliger* 34(1):38–47.

Mamaloukas-Frangoulis, V., J. M. Auzende, D. Bideau, E. Bonatti, M. Cannat, J. Honnorez, Y. Lagabrielle, J. Malavieille, C. Mevel, and H. D. Needham. In-situ study of the eastern ridge-transform intersection of the Vema Fracture Zone. *Tectonophysics* 190(1):55–71.

Manson, M. L., and H. C. Halls. An investigation of superior shoal, central Lake Superior, with a manned submersible. *Can. J. Earth Sci.* 28(1): 145–150.

Masuzawa, T., H. Kitagawa, and N. Handa. Pore water sampling with an in situ pore water squeezer from sediments within a deep sea giant clam colony off Hatsushima Island, Sagami Bay, Japan: Part 2. Dive 521 of the submersible *Shinkai 2000. Tech. Rep. Jpn. Mar. Sci. Technol. Ctr.,* Special Issue, September, pp. 7–15.

Matlock, G. C., W. R. Nelson, R. S. Jones, A. W. Green, T. J. Cody, E. Gutherz, and J. Doerzbacher. Comparison of two techniques for estimating tilefish, yellowedge grouper, and other deepwater fish populations. *Fish. Bull.* 89(1):91–99.

Mevel, C., M. Cannat, P. Gente, E. Marion, J. M. Auzende, and J. A. Karson. Emplacement of deep crustal and mantle rocks on the west median valley wall of the MARK area (MAR, 23 degrees N). *Tectonophysics* 190(1):31–53.

Moiseev, S. I. Observation of the vertical distribution and behavior of nektonic squids using manned submersibles. *Bull. Mar. Sci.* 49(1/2):446–456.

Momma, H., K. Mitsuzawa, and H. Hotta. Was long term observatory in the Suruga Trough lost by strong bottom currents? *Tech. Rep. Jpn. Mar. Sci. Technol. Ctr.,* Special Issue, September, pp. 51–61.

Paffenhoefer, G. A., T. B. Stewart, M. J. Youngbluth, and T. G. Bailey. High-resolution vertical profiles of pelagic tunicates. *J. Plankton Res.* 13(5):971–981.

Page, H. M., A. Fiala-Medioni, C. R. Fisher, and J. J. Childress. Experimental evidence for filter-feeding by the hydrothermal vent mussel, *Bathymodiolus thermophilus. Deep-Sea Res.* 38:1455–1461.

Pavlov, A. I. Distribution pattern and behaviour of the deepwater redfish *Sebastes mentella* Travin from the Reykjanes Ridge as observed from the *Sever-2* manned submersible. In: *Biological Resources of the World Ocean Thalassobathyal: Collected Papers,* ed. A. S. Grechina, T. B. Agaphonova, A. N. Kotljar, A. T. Mandych, and T. G. Tariverdieva, pp. 49–61. Moscow: VNIRO.

Phipps Morgan, J., and M. C. Kleinrock. Transform zone migration: Implications of bookshelf faulting for oceanic and Icelandic propagating rifts. *Tectonics* 10:920–935.

Pley, Y., J. Schipka, A. Gambacorta, H. W. Jannasch, H. Fricke, R. Rachel, and K. O. Stetter. *Pyrodictium abyssi* sp. nov. represents a novel heterotrophic marine archaeal hyperthermophile growing at 110°C. *Syst. Appl. Microbiol.* 14:245–253.

Renosto, F., R. L. Martin, J. L. Borrell, D. C. Nelson, and I. H. Segel. ATP sulfurylase from trophosome tissue of *Riftia pachyptila* (hydrothermal vent tube worm). *Arch. Biochem. Biophys.* 290:66–78.

Rice, A. L., and J. E. Miller. Chirostylid and galatheid crustacean associates of coelenterates and echinoderms collected from the *Johnson Sea-Link* submersible, including a new species of *Gastroptychus. Proc. Biol. Soc. Wash.* 104(2):299–308.

Ryan, W. B. F., W. F. Haxby, L. Pratson, and C. McHugh. Intercomparison of co-registered Seabeam bathymetry, Hydrosweep bathymetry, SEAMARC I imagery and submersible observations on the continental slope of the eastern U.S. In: *Oceans '91,* Vol. 2, pp. 1159–1164. New York: IEEE.

Sagalevich, A. M., L. I. Moskalev, and E. A. Shushkina. Biological research with the deep-water *Mir* submersible (22nd cruise of the R/V *Akademik Matislav Keldlysh*). *Oceanology* 31(5):647–651.

Shushkina, E. A., Y. G. Chindonova, M. Y. Vinogradov, and A. M. Sagalevich. Investigation of ocean zooplankton in the Kuril-Kamchatka region from the *Mir* deep-water submersible. *Oceanology* 31(4):442–446.

Staehle, C. M., J. E. Noakes, and J. Spaulding. Survey and analyses of deep water mineral deposits using nuclear methods. In: *Oceans '91,* Vol. 3, pp. 1316–1318. New York: IEEE.

Taylor, B., A. Klaus, G. R. Brown, G. F. Moore, Y. Okamura, and F. Murakami. Structural development of Sumisu Rift, Izu-Bonin Arc. *J. Geophys. Res.* 96:16,113–16,129.

Tsuji, Y., and S. Sukizaki. Measurement of marine snow abundance using the submersible. *La Mer* 29(4):159–164.

Vanney, J. R. Le modèle des pentes sous-marines observées par submersible lors de la campagne CALSUB. *Doc. Trav. Inst. Geol. Albert-de-Lapparent* 15:38–55.

Vecchione, M., and C. F. E. Roper. Cephalopods observed from submersibles in the western North Atlantic. *Bull. Mar. Sci.* 49(1/2):433–445.

Von Damm, K. L. A comparison of the Guaymas Basin hydrothermal solutions to other sedimented systems and experimental results. *Am. Assoc. Pet. Geol. Mem.* 47:743–751.

Wada, K., and S. Takagawa. Development of a rock drill system for a submersible. *Rep. Jpn. Mar. Sci. Technol. Ctr.* 26:55–63.

Weigel, A. M. DSV *Sea Cliff* recovery of the UAL 811 cargo door. In: *Oceans '91*, Vol. 1, pp. 175–178. New York: IEEE.

Wheatcroft, R. A. Conservative tracer study of horizontal sediment mixing rates in a bathyal basin, California borderland. *J. Mar. Res.* 49(3):565–588.

Zaferman, M. L. On the behaviour of the roundnose grenadier *Coryphaenoides rupestris* (from underwater observation data). *Vopr. Ikhtiol.* 31(6):1028–1033.

Zaferman, M. L., V. N. Shibanov, A. N. Kalugin, and A. S. Yarovoj. Studies on behaviour and distribution of rock grenadier from the Mid-Atlantic Ridge using the *Sever-2* manned submersible. In: *Biological Resources of the World Ocean Thalassobathyal: Collected Papers*, ed. A. S. Grechina, T. B. Agaphonova, A. N. Kotljar, A. T. Mandych, and T. G. Tariverdieva, pp. 62–69. Moscow: VNIRO.

INDEX

Page numbers in *italics* indicate illustrations.